Solvent Effects

on

Reaction Rates and Mechanisms

Solvent Effects on Reaction Rates and Mechanisms

by

EDWARD S. AMIS

Department of Chemistry
University of Arkansas
Fayetteville, Arkansas

1966

ACADEMIC PRESS   NEW YORK and LONDON

541.39
A 517

ACADEMIC PRESS INC.
111 Fifth Avenue, New York, New York 10003

United Kingdom Edition published by
ACADEMIC PRESS INC. (LONDON) LTD.
Berkeley Square House, London W.1

LIBRARY OF CONGRESS CATALOG CARD NUMBER: 64-24652

PRINTED IN THE UNITED STATES OF AMERICA

This book is dedicated to my wife, Velma,
my son, Edward Stephen, Jr., and my daughter, Velma Dianne

PREFACE

"Solvent Effects on Reaction Rates and Mechanisms" is a title that will conjure up visions of different things to different investigators in the field of reaction kinetics. The physical chemist will envision the effect on rates of reactions of dielectric constant, viscosity, internal cohesion, and external pressure as these are influenced by the solvent. The physical-organic chemist will perhaps call to mind acidity, basicity, hydrogen bonding, structure effects, electro-negativity, and solvating ability as related to the solvent. The strictly organic chemist may simply think in terms of a medium in which reactants can be made to form products merely because of solubility relations, and his choice of solvent may depend on the ease of obtaining in a reasonable length of time a relatively pure product by extraction or other procedures.

And, in fact, the topic includes all these and much more. Some of the phenomena are merely recorded as experimental observations. Some factors are subject to theoretical explanation, but even when theoretically explained or mathematically formulated they may not be sufficiently dominant to justify the application of the theory. Other effects may not be subject to theoretical explanations but may be included in correlations that are widely applicable. The explanation of some effects may be purely conjectural, but at least give some satisfaction to the seekers of the answers to the question, why? The material presented in the following pages will run the gamut of all of the above possibilities.

Theoretical attempts at explaining solvent effects on reactions between various charge types of reactants are presented in the first five chapters of the book. In Chapter VI correlative relations are related to solvent effects. In Chapters VII and VIII various solvent effects, classified as being observed experimentally or explained hypothetically or theoretically are presented.

In spite of all the efforts to explain solvent effects on reaction rates and mechanisms, much about these effects still remain obscure. The microscopic dielectric constant in the immediate vicinity of a reactant particle, the existence, extent, and nature of selective solvation, the effect of solvation on the electron distribution in and reactivity of reactant particles, and other factors in solvent effects, remain vague.

It is hoped that modern tools and methods, for example, nuclear magnetic resonance and electron spin resonance, may help to clarify some of these uncertainties. At any rate the inspired, devoted, persistent efforts of dedicated investigators will continue to unravel the riddle of solvent effects, until some day, it is hoped, all will stand revealed.

Thanks are due to Dr. Nirmal K. Shastri for proofreading the manuscript, and to Miss Gayle Garrison for assisting with the preparation of the subject index.

EDWARD S. AMIS

Fayetteville, Arkansas

CONTENTS

CHAPTER II

SOLVENT EFFECTS ON ION-DIPOLAR MOLECULE REACTIONS

CHAPTER III

SOLVENT EFFECTS ON DIPOLAR MOLECULE-DIPOLAR MOLECULE REACTIONS

CHAPTER IV

THE INFLUENCE OF THE SOLVENT ON ELECTRON EXCHANGE REACTIONS

CHAPTER V

SOLVENT EFFECTS COMMON TO REACTIONS INVOLVING VARIOUS CHARGE TYPE REACTANTS

CHAPTER VI

CORRELATIONS INVOLVING SOLVENT EFFECTS

CHAPTER VII

VARIOUS TYPES OF SOLVENT EFFECTS ON DIFFERENT REACTIONS

CHAPTER VIII

FURTHER CONSIDERATION OF SOLVENT EFFECTS

Solvent Effects

on

Reaction Rates and Mechanisms

CHAPTER I

SOLVENT EFFECTS ON ION-ION REACTIONS

THEORETICAL

Introduction

There are theoretical expressions extant for the effect of the solvent on reaction rates. Some of these theoretical expressions deal with the influence of the dielectric constant of the solvent, others with the influence of viscosity, and others with the influences of yet other properties of the solvent on reaction rates. In addition there are many empirical expressions relating the properties of the solvent with the rates of chemical reactions taking place in the solvent.

In this chapter we shall be interested in theoretical expressions relating physical properties of the solvent and reaction rate, and especially with theoretical expressions relating physical properties of the solvent with the rate of reaction between ionic reactants.

The Influence of Dielectric Constant on the Rates-Scatchard Equation

Scatchard (1), using the expression for the potential ψ in the vicinity of an ion from the Debye-Hückel theory, derived an expression for the effect of dielectric constant of the solvent on ion–ion reactions. The expression for ψ is

$$\psi = \frac{Z_i \varepsilon}{Dr} \frac{\exp\left[\varkappa(a_i - r)\right]}{1 + \varkappa a_i}. \tag{1.1}$$

In this equation Z_i is the valence of the ith type of ion, ε is the electronic charge, D the dielectric constant of the medium, exp the character for exponential, a_i the distance of closest approach to the ith type of ion, r the distance from the ion at which the potential is ψ, and $\varkappa$ is the Debye kappa, namely,

$$\varkappa = \sqrt{\frac{4\pi\varepsilon^2}{DkT} \Sigma\, n_i Z_i^2}, \qquad (1.2)$$

where k is the Boltzmann gas constant, T is the absolute temperature, n_i is the number of the ith type of ion per cubic milliliter, and the other items are as defined above.

The activity coefficient f_i of the ith type of ion can be found from the equation

$$\ln f_i = \frac{1}{kT} \int_0^{Z_i\varepsilon} \psi\, d(Z_i\varepsilon) = \frac{1}{kT} \int_0^{Z_i\varepsilon} \frac{Z_i\varepsilon}{Dr} \frac{\exp\,[\varkappa(a_i-r)]}{1+\varkappa a_i}\, d(Z_i\varepsilon)$$

$$= \frac{Z_i^2\varepsilon^2 \exp\,[\varkappa(a_i-r)]}{2DkT(1+\varkappa a_i)r}. \qquad (1.3)$$

Now, for the reaction

$$A + B \; \rightleftarrows \; X \; \longrightarrow \; \text{Products}, \qquad (1.4)$$

we have, from Eq. (1.3),

$$\ln \frac{f_A f_B}{f_X} = \frac{[Z_A^2 + Z_B^2 - (Z_A + Z_B)^2]\, \varepsilon^2 \exp\,[\varkappa(a_i-r)]}{2DkT(1+\varkappa a_i)r}$$

$$= - \frac{Z_A Z_B\, \varepsilon^2 \exp\,[\varkappa(a_i-r)]}{DkT\,(1+\varkappa a_i)r} \qquad (1.5)$$

or

$$\frac{f_A f_B}{f_X} = \exp\left(-\frac{Z_A Z_B \varepsilon^2}{DkT}\right) \frac{\exp\,(-\varkappa r)}{r} \frac{\exp\,(\varkappa a_i)}{1+\varkappa a_i}, \qquad (1.6)$$

which for $\varkappa = 0$ becomes

$$\frac{f_A f_B}{f_X} = \exp\left(-\frac{Z_A Z_B \varepsilon^2}{DkTr}\right). \qquad (1.7)$$

Using the procedure of Brønsted (2), applied to the reaction at zero ionic strength given in Eq. (1.4),

$$\frac{C_X^0 f_X}{C_A^0 f_A C_B^0 f_B} = K \qquad (1.8)$$

and

$$\ln \frac{f_A f_B}{f_X} = \ln \frac{1}{K} \frac{C_X^0}{C_A^0 C_B^0} = -\frac{Z_A Z_B \varepsilon^2}{DkTr}. \tag{1.9}$$

Taking the difference between the logarithmic term when the dielectric constant of the solvent is D and when the dielectric constant of the solvent is some standard reference value D_0 where the activity coefficients of solutes, reactants, and complex become unity, we have

$$\ln \frac{f_A f_B}{f_X} - \ln \left(\frac{f_A f_B}{f_X}\right)_0 = \ln \frac{f_A f_B}{f_X} = \ln \frac{C_X^0}{C_A^0 C_B^0} - \ln \left(\frac{C_X^0}{C_A^0 C_B^0}\right)_0$$

$$= -\frac{Z_A Z_B \varepsilon^2}{DkTr} - \left(-\frac{Z_A Z_B \varepsilon^2}{D_0 kTr}\right) = \frac{Z_A Z_B \varepsilon^2}{kTr} \left(\frac{1}{D_0} - \frac{1}{D}\right). \tag{1.10}$$

Now using the Brønsted (2) approach, the velocity constant $k_{\varkappa=0}$ for a reaction at absolute temperature T and zero ionic strength is given by the product of the specific velocity constant $k'_{\varkappa=0}$, freed from electrostatic $_{D=D_0}$ effects arising from ion atmospheres and from electrical forces between charged reactants, and of the activity coefficients term $f_A f_B / f_X$, obtainable from Eq. (1.10) which corrects for the change of specific velocity due to a transfer of the reactants from the media of standard reference dielectric constant D_0 to a media of dielectric constant D. That is,

$$\ln k'_{\varkappa=0} = \ln k'_{\varkappa=0} + \ln \frac{f_A f_B}{f_X} \tag{1.11}$$
$$\phantom{\ln k'_{\varkappa=0} = \ln k'}{}_{D=D_0}$$

$$= \ln k'_{\varkappa=0} + \frac{Z_A Z_B \varepsilon^2}{kTr} \left(\frac{1}{D_0} - \frac{1}{D}\right). \tag{1.12}$$
$${}_{D=D_0}$$

This equation would give the change in the specific velocity constant at zero ionic strength for a reaction with changing dielectric constant of the solvent at constant temperature, if the change in rate with changing dielectric constant, and therefore changing composition of the solvent, is governed primarily by electrostatic considerations. The choice of the standard reference state of dielectric constant is not uniquely prescribed. Laidler and Eyring (3) choose the reference state as the gas phase where the dielectric constant is unity. Using this assumption, Eq. (1.12) becomes

$$\ln k'_{\varkappa=0} = \ln k'_{\varkappa=0} + \frac{Z_A Z_B \varepsilon^2}{kTr} \left(1 - \frac{1}{D}\right). \tag{1.13}$$
$$\phantom{\ln k'_{\varkappa=0} = \ln k'}{}_{D=1}$$

If the standard reference state of dielectric constant is taken as infinity (4, 5), Eq. (1.12) becomes

$$\ln k'_{\substack{x=0}} = \ln k'_{\substack{x=0 \\ D=\infty}} - \frac{Z_A Z_B \varepsilon^2}{kTrD}.$$
(1.14)

Scatchard used the Christiansen (6) approach for the calculation of the complex C_X^0 given in Eq. (1.8). This method calculates the concentration of the complex directly from the equations given by Debye and Hückel.

The r used in the above equations is the radius of the complex and can be written $r = r_A + r_B$. It is the distance to which two ionic reactants must approach in order to react.

The Laidler-Eyring Equation

Laidler and Eyring (3) write the constant K^* for the equilibrium between the reactants and the activated complex as

$$K^* = \frac{a_X a_N a_O \ldots}{a_A a_B a_C \ldots} = \frac{C_X C_N C_O \ldots}{C_A C_B C_C \ldots} \frac{f_X' f_N' f_O' \ldots}{f_A' f_B' f_C' \ldots}$$
(1.15)

for the reaction

$$A + B + C \longrightarrow X + N + O \longrightarrow K + L + M, \quad (1.16)$$

where X is the complex, N and O are intermediates formed simultaneously with the complex, and K, L, M are final products.

The complex differs from an ordinary molecule in that it is unstable and has a fourth degree of translational freedom replacing the degree of vibrational freedom of the bond which is broken when the molecule of complex decomposes. With respect to this extra degree of translational freedom, which corresponds to the decomposition of the complex along the abscissa of the energy coordinate of decomposition plot, the complex is considered to be a free particle in a one-dimensional space of length S. Within this space the partition function F_X of the complex is $(2\pi m_X kT)^{1/2} S/h$.

Writing K^* in terms of the partition functions of the molecules gives

$$K^* = \frac{F_X F_M F_N \ldots}{F_A F_B F_C \ldots}.$$
(1.17)

Separating out from the partition function F_X of the activated complex that part appertaining to the degree of translational freedom corresponding to decomposition gives

$$F_X = F_X' \frac{(2\pi m_X kT)^{1/2}S}{h} , \tag{1.18}$$

and the equilibrium constant can be written

$$K^* = \frac{F_X' F_N F_O...}{F_A F_B F_C...} \frac{(2\pi m_X kT)^{1/2}S}{h} . \tag{1.19}$$

Therefore,

$$K^* = K_0^* \frac{(2\pi m_X kT)^{1/2}S}{h} , \tag{1.20}$$

where

$$K_0^* = \frac{F_X' F_N F_O...}{F_A F_B F_C...} . \tag{1.21}$$

From Eqs. (1.15) and (1.20) it can be shown that

$$C_X = K_0^* \frac{(2\pi m_X kT)^{1/2}}{h} S \frac{C_A C_B C_C... f_A' f_B' f_C'...}{C_N C_O... f_X' f_N' f_O'...} . \tag{1.22}$$

The rate of the reaction is given by the concentration of complex per unit length of one-dimensional space C_X/S multiplied by the translational velocity in the direction of decomposition, which is given by the kinetic theory expression $(kT/2\pi m_X)^{1/2}$ and by a transmission coefficient $\varkappa$ which is equal to the probability of decomposition in the forward direction. Hence,

$$\begin{aligned}
\text{Rate} &= K_0^* \frac{(2\pi m_X kT)^{1/2}}{h} \frac{C_A C_B C_C...}{C_N C_O...} \frac{f_A' f_B' f_C'}{f_X' f_A' f_B'} \left(\frac{kT}{2\pi m_X}\right)^{1/2} \varkappa \\
&= \varkappa \frac{kT}{h} K_0^* \frac{C_A C_B C_C...}{C_N C_O...} \frac{f_A' f_B' f_C'...}{f_X' f_N' f_O'...} .
\end{aligned} \tag{1.23}$$

The specific velocity constant k' is that rate at which the concentrations of A, B, C, ... and N, O, ... are all unity. Thus

$$k' = \varkappa \frac{kT}{h} K_0^* \frac{f_A' f_B' f_C'...}{f_X' f_N' f_O'...} . \tag{1.24}$$

In dilute gas the activity coefficients are unity and

$$k'_{(gas)} = \varkappa \frac{kT}{h} K_0^* . \tag{1.25}$$

Now consider the reaction

$$A^{Z_A} + B^{Z_B} \longrightarrow X^{(Z_A+Z_B)} \longrightarrow K + L. \qquad (1.26)$$

The equation for the logarithm of the specific velocity constant is as before

$$\ln k' = \log \left(\varkappa \frac{kT}{h} K_0{}^* \right) + \ln \frac{f_A' f_B'}{f_X'}. \qquad (1.27)$$

If the activity coefficient f' is split into a coefficient β, the activity coefficient of the infinitely dilute solution with reference to the infinitely dilute gas, and a coefficient f, the activity coefficient of the solution being studied with reference to the infinitely dilute solution, then

$$f' = \beta f \qquad (1.28)$$

and

$$\ln f' = \ln \beta + \ln f. \qquad (1.29)$$

The free energy of transfer of an ion from a vacuum to a medium of dielectric constant D is made up of four parts as follows: (1) $-Z^2\varepsilon^2/2r$ free energy change due to discharging the ion in a vacuum, (2) Φ' due to transfer of the uncharged particle to infinitely dilute solution where it encounters van der Waals and possibly other forces, (3) $Z^2\varepsilon^2/2rD$ due to charging the ion in a medium of dielectric constant D, (4) Φ'' due to reorientation of the dipoles around the charged ion. The total free energy change ΔF of the transfer is, therefore,

$$\Delta F = -\frac{Z^2\varepsilon^2}{2r} + \frac{Z^2\varepsilon^2}{2rD} + \Phi' + \Phi'' \qquad (1.30)$$

and the activity coefficient of the ion is

$$\ln \beta = \frac{Z^2\varepsilon^2}{2kTr} \left(\frac{1}{D} - 1 \right) + \frac{\Phi'}{kT} + \frac{\Phi''}{kT} \qquad (1.31)$$

and

$$\ln \frac{\beta_A \beta_B}{\beta_X} = \frac{\varepsilon^2}{2kT} \left(\frac{1}{D} - 1 \right) \left[\frac{Z_A{}^2}{r_A} + \frac{Z_B{}^2}{r_B} - \frac{(Z_A + Z_B)^2}{r_X} \right]$$
$$+ \frac{\Phi_A' + \Phi_A'' + \Phi_B' + \Phi_B'' - \Phi_X' - \Phi_X''}{kT}. \qquad (1.32)$$

In deriving the solvent effect we can dismiss the contribution of the activity coefficients f which refer the solution of ionic strength μ to infinitely dilute solution, since in the solvent effect we will assume infinitely dilute solution can be used.

From Eq. (1.24),

$$k' = \varkappa \, \frac{kT}{h} \, K_0^* \, \frac{\beta_A \beta_B}{\beta_X} \, , \qquad (1.33)$$

and from Eqs. (1.33) and (1.32),

$$\ln k' = \ln \left(\varkappa \, \frac{kT}{h} \, K_0^* \right) + \ln \frac{\beta_A \beta_B}{\beta_X}$$

$$= \ln \left(\varkappa \, \frac{kT}{h} \, K_0^* \right) + \frac{\varepsilon^2}{2kT} \left(\frac{1}{D} - 1 \right) \left[\frac{Z_A^2}{r_A} + \frac{Z_B^2}{r_B} - \frac{(Z_A + Z_B)^2}{r_X} \right]$$

$$+ \frac{\Phi_A' + \Phi_A'' + \Phi_B' + \Phi_B'' - \Phi_X' - \Phi_X''}{kT} . \qquad (1.34)$$

Applying Eq. (1.14), Amis and LaMer (4) plotted $\log k'_{\varkappa=0}$ versus $1/D$ for the reaction between negative divalent tetrabromophenolsulfonphthalein ions and negative univalent hydroxide ions in water–ethyl alcohol and water–methyl alcohol at 25°C. The lines, as shown in Fig. 1, were straight and of negative slopes, as required by Eq. (1.14), down to a dielectric constant of around 65. The values of r calculated from these slopes were 1.22 Å and 1.49 Å in water–ethanol and water–methanol, respectively. Amis and

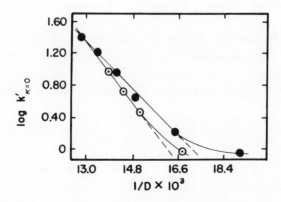

FIG. 1. Log $k'_{k=0}$ versus $1/D$ at 25°C for the reaction $(B\Phi B)^{2-} + OH^-$. ●, H_2O—CH_3OH; ⊙, H_2O—C_2H_5OH.

Price (7) plotted $\ln k_{\varkappa=0}$ versus $1/D$ for the ammonium ion–cyanate ion reaction in water–methanol and water–glycol at 30°C. The lines were, for the most part, straight with positive slope, as expected from Eq. (1.14), down to a dielectric constant of about 50. In the water–methanol case there was a slight curvature in the region of pure water. The r-values obtained from the slopes were 2.2 Å and 2.5 Å for water–methanol and water–glycol, respectively. LaMer (8) applied the Scatchard equation to the reaction between the negative univalent bromoacetate ion and the negative bivalent thiosulfate ion. The r-value found at 25°C was 5.1 Å.

Laidler and Eyring applied Eq. (1.34) to the bromoacetate–thiosulfate reaction. The straight line through the points in the higher-dielectric-constant region of the $\ln k'_{\varkappa=0}$ versus $1/D$ plot corresponded to $r_{\text{bromoacetate}} = 3.3$ Å, $r_{\text{thiosulfate}} = 1.7$ Å, and $r_{\text{activated complex}} = 5.0$ Å at 25°C.

The deviation of the $\ln k'_{\varkappa=0}$ versus $1/D$ plots from linearity in low-dielectric-constant regions of the solvent are such as to cause the rates to be more like those at high dielectric constants than is predicted by theory. These deviations, according to Laidler and Eyring (3), are due to preferential adsorption of water on the ions. Thus when solvents of lower dielectric constants are added the effects are not as great as if the solvent molecules were randomly oriented. These authors explain on a similar basis the data on the hydroxyl ion hydrolysis of the triphenylsulfonium cation (9).

Evidence of selective solvation is available from other sources (10, 11).

Amis (11a) used a Coulombic energy approach to obtain an equation identical with the Scatchard equation for the dependence of $\ln k'$ on $1/D$.

The Influence of Dielectric Constant on Energies of Activation

Moelwyn-Hughes (12) makes the total energy of activation, E, equal to the sum of two components: E_n, of which we know nothing, and E_e, an electrostatic contribution. In dilute solutions E_e may be evaluated in terms of the Debye-Hückel expression. The equation for E is then

$$E = E_n + E_e = E_n + \frac{Z_A Z_B \varepsilon^2}{Dr} (1 - \varkappa r). \tag{1.35}$$

The second-order rate constant k' can be written in terms of Z_0 the standard collision frequency and E. Thus

$$k' = Z_0 \exp(-E_n/kT) \exp\left(-\frac{Z_A Z_B \varepsilon^2}{DrkT}\right)(1 - \varkappa r). \tag{1.36}$$

For the ionic strength influence on reaction rates we can include all the terms not involving $\varkappa$ in a constant k_0' and obtain

$$k' = k_0' \exp\left(\frac{Z_A Z_B \varepsilon^2 \varkappa}{DkT}\right), \tag{1.37}$$

which is Brønsted's equation for the salt effect.

If the logarithmic form of Eq. (1.36) is differentiated with respect to temperature at constant pressure and ionic strength, assuming Z_0 constant, and allowing for the fact that $\varkappa$ is a function of temperature and dielectric constant, Eq. (1.38) for the apparent energy of activation E_A can be obtained by multiplying through by kT^2:

$$E_A = E_n + \frac{Z_A Z_B \varepsilon^2}{Dr}\left(1 + \frac{\partial \ln D}{\partial \ln T}\right)(1 - \tfrac{3}{2}\varkappa r), \tag{1.38}$$

and if

$$L = -\left(\frac{\partial \ln D}{\partial T}\right)_P, \tag{1.39}$$

Eq. 1.38 can be written

$$E_A = E_n + \frac{Z_A Z_B \varepsilon^2}{Dr}(1 - LT)(1 - \tfrac{3}{2}\varkappa r). \tag{1.40}$$

Elimination of E_n using Eqs. (1.35) and (1.40) yields, for the second-order rate constant,

$$k' = Z_0 \exp\left(-E_A/kT\right)\exp\left(-\frac{Z_A Z_B \varepsilon^2 L}{Dkr}\right)\exp\left(-\frac{Z_A Z_B \varepsilon^2 \varkappa}{2DkT}\right)(1 - 3LT). \tag{1.41}$$

Equation (1.38) for zero ionic strength becomes

$$E_A_{\varkappa=0} = E_n + \frac{Z_A Z_B \varepsilon^2}{Dr}(1 - LT). \tag{1.42}$$

Differentiation of the logarithmic form of Eq. (1.36) with respect to temperature at constant pressure, ionic strength, and dielectric constant yields, when multiplied through by kT^2,

$$E_D = E_n + \frac{Z_A Z_B \varepsilon^2}{Dr}(1 - \tfrac{3}{2}\varkappa r), \tag{1.43}$$

where E_D is the energy of activation for the reaction when the rate is measured in constant dielectric constant media. At infinite dilution $\varkappa = 0$, and Eq. (1.43) becomes

$$E_D = E_n + \frac{Z_A Z_B \varepsilon^2}{Dr}. \tag{1.44}$$

True Energy of Activation

Svirbely and Warner (*13*) first obtained what they termed true energy of activation, E_D, in isodielectric solvents at infinite dilution of solute species by setting

$$\ln k'_{\varkappa=0} = f(D, T). \tag{1.45}$$

Differentiating with respect to T and multiplying the resulting expression through by $2.303RT^2$ gave

$$E_A_{\varkappa=0} = E_D_{\varkappa=0} + 2.303RT^2 \left(\frac{\partial \ln k'_{\varkappa=0}}{\partial D}\right) \frac{dD}{dT}. \tag{1.46}$$

These authors calculated the difference $E_A_{\varkappa=0} - E_D_{\varkappa=0}$ for the ammonium ion–cyanate ion reaction at infinite dilution in methyl alcohol–water solvents, and found excellent agreement between observed and calculated values.

The Electrostatic Contribution to Energy of Activation

LaMer (*14*) gives the electrostatic contribution of the energy of activation $\varDelta E_D$ as

$$\varDelta E_D = \frac{\partial(\varDelta F_{D,T})}{\partial(1/T)} \tag{1.47}$$

and since

$$\varDelta F_D = \frac{Z_A Z_B N \varepsilon^2}{D} \frac{1}{r}, \tag{1.48}$$

where again $r = r_A + r_B$, then

$$\varDelta E_D = \varDelta F_D \left(1 + \frac{\partial \ln D}{\partial \ln T}\right). \tag{1.49}$$

Difference in Energy of Activation at Constant Dielectric Constant and Apparent Energy of Activation at any Dielectric Constant

By subtracting (Eq. 1.40) from Eq. (1.43) there results

$$E_D - E_A = \frac{Z_A Z_B \varepsilon^2 LT}{Dr} (1 - \tfrac{3}{2} \varkappa r), \qquad (1.50)$$

which can be used to calculate the difference between the energy of activation for a reaction at constant dielectric constant and the apparent energy of activation of the reaction at any dielectric constant when the ionic strength is finite. At infinite dilution the difference in the two energies of activation would be

$$E_D_{\varkappa=0} - E_A_{\varkappa=0} = \frac{Z_A Z_B \varepsilon^2 LT}{Dr}. \qquad (1.51)$$

Amis and Holmes (15) obtained an equation comparable to Eq. (1.50) by differentiating the Brønsted-Christiansen-Scatchard equation with respect to T (T and D variable), differentiating the equation with respect to T (D constant), taking the difference, and multiplying by RT^2. The equation, using ionic strength μ, becomes

$$E_A - E_D = \frac{Z_A Z_B \varepsilon^2 NT}{D^2 J} \left(\frac{1}{r} - \frac{3\varepsilon}{10} \sqrt{\frac{2\pi N\mu}{10DkT}} \right) \frac{dD}{dT}. \qquad (1.52)$$

In the above equation, L of Eq. (1.50) has been replaced by $(-\partial \ln D/\partial T)_P$, and division by J (the mechanical equivalent of heat in ergs per calorie) has been performed so that the equation gives the difference in energy of activation in calories per mole. Amis and Potts (16) further modified the equation to contain the Coulombic energy of activation and obtained

$$E_A - E_D = \frac{Z_A Z_B \varepsilon^2 N}{D_c J} \left[\left(\frac{1}{r} - \frac{3}{10} \sqrt{\frac{2\pi N\mu}{10D_c kT}} \right) \frac{T}{D_c} \frac{dD}{dT} - \frac{\Delta D}{D_D r} \right], \qquad (1.53)$$

where $\Delta D = D_c - D_D$, and D_c and D_D are the dielectric constants of the isocomposition and isodielectric runs, respectively. Both Eq. (1.52) and Eq. (1.53) have been applied to experimental data (15, 16).

Coulombic Energy of Activation

The Coulombic energy of activation mentioned above was calculated using the equation proposed by Amis (17) which is

$$E_c = -329.7 \frac{Z_A Z_B}{D_1 D_2 r} \Delta D .$$ (1.54)

This energy is the energy of activation contribution when the temperature coefficient of the reaction is measured in two different isodielectric media of dielectric constants D_1 and D_2, respectively, and for which ΔD is the difference between the dielectric constants.

The Ionic Atmosphere Contribution to the Energy of Activation

The above equations indicate that the energies of activation of reactions are influenced by the dielectric constant of the solvent. Even the contribution of the ionic atmosphere to the energy of activation is influenced by the change of dielectric constant and volume of the solvent with temperature, as can be seen from the equation derived by LaMer and Kamner (18) for the contribution of the ionic atmosphere to the energy of activation. From the equations

$$E_{In} = \frac{\partial (F_{In}/T)}{\partial (1/T)}$$ (1.55)

and

$$F_{In} = -\frac{N Z_A Z_B \varepsilon^2}{2D} ,$$ (1.56)

they obtained

$$E_{In} = -\frac{3 N Z_A Z_B \varepsilon^2}{4D} \varkappa \left[1 + \frac{\partial \ln D}{\partial \ln T} + \frac{\partial \ln V}{\partial \ln T} \right].$$ (1.57)

Moelwyn-Hughes (19) more explicitly by the collision theory has derived a similar equation. Amis and LaMer (4) plot the total energy of activation minus a constant quantity versus the square root of the ionic strength for the reaction of the negative bivalent tetrabromophenolsulfonphthalein reacting with negative univalent hydroxide ion. The slopes of the curves in regions of low ionic strengths correspond to the limiting slopes predicted by the Debye-Hückel limiting law. These data are reproduced in Fig. 2.

The Application of Dielectric Constant Influence on Rates in Calculating Nonelectrostatic Effects

In Eq. (1.34) of Laidler and Eyring, terms of the nature Φ/kT are included to account for the nonelectrostatic effects, which can be treated as repulsions that begin suddenly at the distance of closest approach and are modified to allow for van der Waals attractions. These attractions may be of considerable importance. Amis and Jaffé (20) developed a procedure for calculating

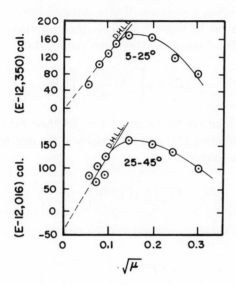

FIG. 2. Energy of activation of the reaction $(B\Phi B)^{2-} + OH^-$ as a function of the ionic strength. Broken lines correspond to the Debye-Hückel limiting law for activation energy for reaction between bivalent and univalent ionic reactants as a function of ionic strength.

the term. These authors assume a Brønsted complex is formed between reactants undergoing reaction. They further assume that spherical symmetry exists around each particle and that the mutual potential of the two particles at a distance r can be represented by the equation

$$\psi(r) = \Phi_1(r) + \Phi_2(r), \tag{1.58}$$

$\Phi_2(r)$ arises from electrostatic forces in the case that the reactants A and B are charged. Thus if A and B are ions of valencies Z_A and Z_B, respectively, then

$$\Phi_2(r) = \frac{Z_A Z_B \varepsilon^2 \exp\left[\varkappa(a_i - r)\right]}{Dr(1 + \varkappa a_i)}, \tag{1.59}$$

$\Phi_1(r)$ was assumed by Amis and Jaffé to be positive for distances r at which formation of the intermediate complex X ensues and to be independent of dielectric constant of the solvent. This latter assumption is reasonable since, at distances necessary for complex formation, the dielectric constant of the solvent would have no significance.

Let C_A and C_B be the concentrations of A and B, respectively. The probability of finding a molecule B at a distance between r_0 and $r_0 + dr_0$ from molecule A can be found from these concentrations and the potential given in Eq. (1.58). This probability is proportional to

$$C_A C_B \exp\left[-\frac{\psi(r_0)}{kT}\right] r_0^2 \, dr_0. \tag{1.60}$$

If formation of X ensues for values of r within these limits, then using the method of Christiansen (21) and of Scatchard, the rate of reaction can be expressed by the equation

$$\text{rate} = KT^{1/2} C_A C_B \exp\left[-\frac{\psi(r_0)}{kT}\right], \tag{1.61}$$

and the specific velocity constant can be written

$$k' = KT^{1/2} \exp\left[-\frac{\psi(r_0)}{kT}\right]. \tag{1.62}$$

If the rate constant for some standard reference temperature T_0 is k_0' and can be considered as including the $T^{1/2}$ term, the final formula becomes

$$\ln k' = \ln k_0' + \frac{\Phi_1(r_0)}{kT_0} - \frac{\Phi_1(r_0)}{kT} - \frac{Z_A Z_B \varepsilon^2 \exp\left[\varkappa(a_i - r_0)\right]}{DkTr_0(1 + \varkappa a_i)}. \tag{1.63}$$

For our present purpose it is not necessary to extrapolate k_0' to standard reference states of concentration or of dielectric constant.

The differentiation of Eq. (1.60) separately with respect to $1/T$ and with respect to $1/D$ yields

$$a_T = \left[\frac{d \ln k'}{d(1/T)}\right]_D = -\frac{1}{k}\left[\Phi_1 r_0 + \frac{Z_A Z_B \varepsilon^2 \exp\left[\varkappa(a_i - r_0)\right]}{Dr_0(1 + \varkappa a_i)}\right] \tag{1.64}$$

and

$$a_D = \left[\frac{d \ln k'}{d(1/D)} \right]_T = - \frac{Z_A Z_B \varepsilon^2 \exp \left[\varkappa (a_i - r_0) \right]}{kTr_0 (1 + \varkappa a)}. \tag{1.65}$$

If the nonelectrostatic potential is sufficiently large compared to the electrostatic potential, a_T will be negative whatever the charges Z_A and Z_B are.

From Eqs. (1.64) and (1.65) the following inequalities, relating measurable quantities only, can be found:

$$a_T/a_D > T/D \tag{1.66}$$

when A and B are of like charge sign, and

$$a_T/a_D < T/D \tag{1.67}$$

when A and B are of unlike charge sign.

From Eqs. (1.64) and (1.65) and from measured values of a_T and a_D, both the electrostatic potential $Z_A Z_B \varepsilon^2/Dr_0$ and the nonelectrostatic potential $\Phi_1(r_0)$ for the critical distance can be calculated.

Tests for the inequality (1.66) were made on the persulfate ion–iodide ion reaction in water–ethanol and on tetrabromophenolsulfonphthalein ion–hydroxide ion reaction in water–methanol and in water–ethanol. Tests for the inequality (1.67) were made on the ammonium ion–cyanate ion reaction in water–methanol and water–glycol. In all cases theory was found to check observed results.

Electrostatic potentials $\Phi_2(r_0)$ and nonelectrostatic potentials $\Phi_1(r_0)$ were calculated using Eqs. (1.64) and (1.65) for the same reactants in the same solvents as used for testing the inequalities discussed above. The ratio $\Phi_1(r_0)/\Phi_2(r_0)$ for the ammonium ion–cyanate ion reaction was found to be 7.0 and 10.9, respectively, for the water–methyl alcohol and water–glycol, respectively. Thus there is a noticeable specific solvent effect evident in the reaction, the nonelectrostatic effect being more dominant in the water–glycol as compared to the water–methanol. It has been pointed out that specific solvation by water in water–organic solvent mixtures reduced the electrostatic effect arising from changing dielectric constant to a value less than expected from theory. In the present case, if a reduction of electrostatic effect causes the higher nonelectrostatic effect–electrostatic effect ratio, then there must be greater specific solvation of the ions by water in the water–glycol than in the water–methanol solvent. On the other hand the increased ratio of $\Phi_1(r_0)/\Phi_2(r_0)$ in water–glycol compared to that in water–methanol might be due to a relative enhancement of nonelectrostatic effect in the former solvent. This would imply relatively greater repulsive forces as compared to van der Waals attractions at distance r_0 in water–glycol as compared to water–methanol. The larger amount of water of solvation in this case must occur in the water–methanol rather than in the water–glycol. The large amount of water on the ions increases the attractive and decreases the repulsive forces among the ions due to the hydrogen bonding tendencies in their solvent sheaths. Thus in the water–methanol case there is relatively less repulsive forces as compared to van der Waals attractions, and the nonelectrostatic effect is reduced relative to the electrostatic effect. The opposite is true in water–glycol with its lesser solvation of the ions by water.

In the case of reactions between bivalent and univalent ions, the electrostatic influences are greater than in the case of univalent ion–univalent ion reaction, and hence the ratio of nonelectrostatic potential to electrostatic potential is less for the persulfate–iodide reaction than for the ammonium ion–cyanate ion reaction. In water–ethyl alcohol the ratio is 3.8 for the former reaction, and in the case of the bivalent tetrabromophenolsulfonphthalein–univalent hydroxide ion reaction the ratio is 2.9 and 3.6 in water–ethanol and water–methanol, respectively. The ratio is 7.0 and 10.9 for the ammonium ion-cyanate ion reaction in water–methanol and water–glycol, respectively.

If a_D is not known $\Phi_1(r_0)$ can be calculated from a_T alone using Eq. (1.64) provided r_0 is known from some other source or provided the second term on the right-hand side of Eq. (1.64) is neglected because of its relatively small magnitude. In this case the values of $\Phi_1(r_0)$ should be the same for all types of reactants, charged or uncharged, in solution or in the gaseous state. Such calculations have been made and confirmed the above statement. As can be expected from Eq. (1.64), $\Phi_1(r_0)$ calculated from a_T alone should be greater than $\Phi_1(r_0)$ calculated from both a_T and a_D when the ionic reactants are of like charge sign, and $\Phi_1(r_0)$ calculated from a_T alone should be less than $\Phi_1(r_0)$ calculated from both a_T and a_D when the ionic reactants are of unlike charge sign. Calculations confirmed these expectations. Calculated from a_T alone, all charge-type reactants give $\Phi_1(r_0)$ values which do not vary by a power of ten for all reactions to which the theory was applied.

Components of the Energy of Activation

LaMer (*14*) shows that, from the Brønsted-Christiansen-Scatchard equation, we may write the total free energy of activation F as the sum of the free energy F_0 freed from all charge effects, the free energy F_D arising from electrostatic forces between reactant particles, and the free energy F_μ arising from ion atmosphere effects.

Thus

$$F = F_0 + F_D + F_\mu , \qquad (1.68)$$

and therefore the energy of activation E can be written

$$E = E_0 + E_D + E_\mu . \qquad (1.69)$$

For $\mu = 0$, $E_\mu = 0$, and for isodielectric solvents,

$$E_D = F_D = - RT \ln k_D' \qquad (1.70)$$

from Eq. (1.49). Therefore for isodielectric solvents and for $\mu = 0$

$$E_0 = E + RT \ln k_D' . \qquad (1.71)$$

Therefore from experimental values of total energy of activation and k_D', E_0 may be calculated. Such a calculation for the tetrabromophenol-phthalein–hydroxide ionic reaction was made by Amis (5) who found E_D to be 6330 cal compared with a total energy of activation of 21,100 cal.

Thus we see that from the standpoint of theory and of experiment the dielectric constant of the solvent influences markedly the energy of activation and hence the rates of chemical reactions. Variation from theory which becomes more pronounced as the dielectric constant becomes continually lower can be attributed in part to a selective solvation of the ionic reactants by one component, in general the more polar component, of the solvent. Other solvent properties, such as viscosity, cohesion, hydrogen bonding tendencies, and solvolysis propensities, which will be discussed later, no doubt enter the picture.

The Influence of the Dielectric Constant on Entropy of Activation and the Arrhenius Frequency Factor

The influence of the dielectric constant of the solvent on the contribution of the ionic strength to the entropy of activation was formulated by LaMer (14). Since

$$S = - \frac{\partial F}{\partial T} \qquad (1.72)$$

and since for dilute solutions the contribution to the free energy of activation due to ionic strength effects, F_μ, is

$$F_\mu = -\frac{NZ^2\varepsilon^2}{2D}\,\varkappa \qquad (1.73)$$

from the Debye-Hückel theory, and remembering that $\varkappa$ is proportional to $(DTV)^{-1/2}$, then the contribution to the entropy of activation, S_μ, of the ionic strength is given by

$$S_\mu = \frac{F_\mu}{T}\left[\frac{3}{2}\,\frac{\partial \ln D}{\partial \ln T} + \frac{1}{2}\,\frac{\partial \ln V}{\partial \ln T} + \frac{1}{2}\right]. \qquad (1.74)$$

From the Arrhenius equation

$$k' = Z \exp\left(-E/RT\right) \qquad (1.75)$$

and the absolute rate equation

$$k' = \varkappa\,\frac{kT}{h}\,\exp\left(S/R\right)\exp\left(-E/RT\right), \qquad (1.76)$$

it is seen that the Arrhenius frequency factor Z is related to the total entropy of activation S by the equation

$$\ln Z = \ln \varkappa\,\frac{kT}{h} + \frac{S}{R}. \qquad (1.77)$$

Now $\varkappa\,kT/h$ is independent of concentration and at constant temperature is constant, while S depends on both temperature and concentration. Furthermore S is made up of contributions to the entropy of activation: S_0, arising from reaction between uncharged reactants; S_D, due to interionic forces between the charged reactants; and S_μ, coming from the ion atmospheres of reactants. Thus

$$S = S_0 + S_D + S_\mu. \qquad (1.78)$$

Now S_0 is constant at a given temperature for given reactants, S_D will be constant for a given media at a given temperature since the dielectric constant will not change. Hence, a change in S, and therefore of $\ln Z$, with changing ionic strength will arise only from a change in S_μ for a given reaction, in a given solvent, at a given temperature. But from Eqs. (1.73) and (1.74),

F_μ and also S_μ is linear with $\sqrt{\mu}$ at constant T, V, and D. Hence a plot of $\ln Z$ versus $\sqrt{\mu}$ should be a straight line with a slope predicted by Eq. (1.74). If the values for water for $\partial \ln D/\partial \ln T$ and $\partial \ln V/\partial \ln T$ are substituted into Eq. (1.74) an equation

$$\frac{S_\mu}{2.303 \, R} = Z_A Z_B \varepsilon^2 A \sqrt{\mu} \qquad (1.79)$$

can be written, where A depends on the dielectric constant and the temperature and equals 1.44, 1.53, and 1.64 for water at 15°, 25°, and 35°C, respectively. Plots of log Z versus $\sqrt{\mu}$ for the tetrabromophenolsulfonphthalein ion–hydroxide ion reaction in the temperature ranges 5–25°C and 25–45°C were made for aqueous solutions. In dilute solutions the plots were straight lines with the predicted slopes.

When the volume change with temperature is negligible but when $a\varkappa$ cannot be neglected in the first approximation of Debye and Hückel, the expression for S_μ becomes

$$S_\mu = \frac{3}{2} \frac{F_\mu}{T} \left[\left(\frac{1}{3} + \frac{\partial \ln D}{\partial \ln T} \right) - \frac{1}{3} \frac{a\varkappa}{1 + a\varkappa} \left(1 + \frac{\partial \ln D}{\partial \ln T} \right) \right]. \qquad (1.80)$$

The dielectric constant makes a contribution S_D to the entropy of activation. This contribution arises from the effect of the dielectric constant of the media on the interionic forces between the charged reactants. The contribution F_D to the free energy of activation of the dielectric constant influence is

$$F_D = \frac{N Z_A Z_B \varepsilon^2}{Dr}, \qquad (1.81)$$

where $r = r_A + r_B$. Applying Eq. (1.72), S_D is found to be (14)

$$S_D = \frac{F_D}{T} \left[\frac{\partial \ln D}{\partial \ln T} \right]. \qquad (1.82)$$

When temperature coefficients of reaction rates are measured in isodielectric solvents, $\partial \ln D/\partial \ln T$ is zero, and therefore S_D is zero. If the rate constants are extrapolated to $\mu = 0$, S_μ vanishes and $S = S_0$, the entropy of activation for uncharged reactants possessing the same chemical characteristics except for charge as those of the ions. Likewise the frequency factor calculated under these conditions should, except perhaps from some minor dipolar effects, be that for reactants with like chemical characteristics to those of the ions except for charge. This approach should allow comparison of data for an ionic reaction with that predicted by collision theory for reaction between uncharged particles.

An equation for the difference between the logarithm of the frequency factor at fixed composition, $\ln Z_A$, and the logarithm of the frequency factor at fixed dielectric constant, $\ln Z_D$, was obtained by Amis and Cook (22). They derived the equation by multiplying the Brønsted-Christiansen-Scatchard equation through by T, differentiating the resulting equation by T (considering T and D variables), then differentiating the equation with respect to T holding D constant, and substracting the latter differential equation from the former. The resulting equation is

$$\ln Z_A - \ln Z_D = \frac{Z_A Z_B N \varepsilon^2}{D^2 R} \left[\frac{1}{r} - \frac{3\varepsilon}{10} \sqrt{\frac{2\pi N^2 \mu}{10 D R T}} \right] \frac{dD}{dT}. \quad (1.83)$$

For ordinary solvents dD/dT is negative, and Eq. (1.83) predicts, for reactants of like sign, that the Arrhenius frequency factor in fixed composition media is less than the factor in isodielectric media. For ions of unlike sign the opposite is true. The statements imply that the first term in the brackets on the right-hand side of Eq. (1.83) is larger than second term. This is generally the case. The predictions of this equation were verified using experimental data; quantitative as well as qualitative agreement was found.

Deviations between Theory and Experiment

It might be pointed out that selective solvation and a different microscopic dielectric constant around the reactant particles than the macroscopic dielectric constant of the bulk media would cause deviations between theory and experiment in all the theoretical treatments presented above. The deviations would probably be such as to favor the measured results found in the medium selectively absorbed by the ions. In general this will be the more polar, higher dielectric component of the medium, so that in general the deviations observed at lower dielectric constants will be in the direction of results obtained in this higher dielectric component when used as a single solvent. Thus in water–alcohol or water–dioxane solvents, the kinetic rates in organic-component rich solvents tend to deviate toward the results found in water.

Scatchard (23) indicates that, in considering change in rate with changing dielectric constant it should be remembered that theories of solution give the ratio of activities to mole fraction and not the ratio of activities to relative volume concentrations as activities. The correction of rate constants in terms of concentrations to rate constants in terms of mole fractions is made by multiplying the former by $\Sigma (N/V)^{\nu-1}$, where N is the total number of moles, including solvents, in volume V and ν is the order of the reaction. The sudden change in slope of the $\log k'$ versus $1/D$ for the bromoacetate–thiosulfate reaction given by Laidler and Eyring (3) is eliminated by such calculations. Less marked changes in curvature at lower dielectric constants for some other reactions are removed or greatly decreased by a more exact extrapolation to zero concentration. However, for still other reactions these corrections do not eliminate the change in slope at lower dielectric constants, and Laidler and Eyring's assumption of selective solvation is probably correct according to Scatchard.

Scatchard further points out that the difference between his treatment and that of Laidler and Eyring is the difference in models for the critical complex. Scatchard assumed two spheres, each with the charge of one of the original ions distributed symmetrically about its center. Laidler and Eyring assume one sphere with the net charge symmetrically distributed around its center. These are the two extremes with respect to both shape and charge distribution. Scatchard has discussed intermediate possibilities (24).

The Salt Effect

The solvent influence on the salt effect is a secondary effect arising from the dielectric constant modification of the activity coefficients of the reactant particles. This influence of the dielectric constant will be discussed here relative to ion–ion reactions.

Brønsted (25), Bjerrum (26), and Christiansen (27) have applied the Debye-Hückel (28) theory to the influence of neutral salts upon the velocity of reactions in solution.

We shall here deal with Brønsted's treatment of the neutral salt effects. These effects are of two kinds. In the first case the activities of the reactants, whether ions or polar molecules, may be altered by added electrolyte. This is the primary salt effect. In the second case the effective concentration of a reactant ion coming from a weak electrolyte may be decreased by the decreased ionization of the electrolyte due to added salt. This is the secondary salt effect. This latter effect is well illustrated by the decreased catalytic effect of acetic acid upon the inversion of cane sugar in the presence of alkali acetates. In this case the activity of the hydrogen ion reactant is increased by the added salt, but the effective concentration of ion is so reduced by the common ion effect that the inversion rate constant may be decreased by as much as 40 to 50% at ordinary concentrations of acid and added salt.

The Brønsted Theory

The primary salt effect can be understood on the basis of Brønsted's theory. The theory assumes that there is an intermediate complex formed which may decompose reversibly to give reactants, or which may decompose irreversibly to yield products. The reaction may be written

$$A + B \; \rightleftharpoons \; X \; \longrightarrow \; C + D. \tag{1.84}$$

The first step in the reaction may be considered a thermodynamic equilibrium, and the equilibrium constant from the mass law is

$$K = \frac{a_X}{a_A a_B} = \frac{C_X f_X}{C_A f_A C_B f_B}. \tag{1.85}$$

The reaction rate is proportional to the concentration of X, so that

$$r = k'' C_X = k'' K C_A C_B \frac{f_A f_B}{f_X}. \tag{1.86}$$

Now $k'' K$ is a constant which we may call k_0', so that

$$r = k_0' C_A C_B \frac{f_A f_B}{f_X}. \tag{1.87}$$

Here $r = dC_C/dt = dC_D/dt$, and since k_0' is independent of concentration, the factor $f_A f_B / f_X = F$ is the correction to the classical rate equation due to added salt. Brønsted restricts his theory to dilute solutions, since in concentrated solution nonthermodynamic factors, with which this theory is not concerned, may influence the rate.

Now the observed bimolecular rate constant k' is given by

$$k' = \frac{dC}{dt} \frac{1}{C_A C_B}. \tag{1.88}$$

Therefore, from Eqs. (1.87) and (1.88),

$$k' = k_0' \frac{f_A f_B}{f_X} = k_0' F.$$ (1.89)

Although k_0' is independent of concentration, it must be emphasized that it is nevertheless, composite, and must yet be referred to a standard reference state of dielectric constant and of temperature before it is independent of these factors which are so influential in rate processes.

If we consider the reactants to be ions, the equilibrium equation for the reaction is

$$A^{Z_A} + B^{Z_B} \rightleftharpoons X^{(Z_A + Z_B)},$$ (1.90)

where z_A, z_B, and $(z_A + z_B)$ are the charges on A, B, and X, respectively. Now from the Debye-Hückel theory

$$-\ln f_A = \frac{z_A^2 A \sqrt{\mu}}{1 + \beta a_i \sqrt{\mu}},$$ (1.91)

$$-\ln f_B = \frac{z_B^2 A \sqrt{\mu}}{1 + \beta a_i \sqrt{\mu}},$$ (1.92)

and

$$-\ln f_X = \frac{(z_A + z_B)^2 A \sqrt{\mu}}{1 + \beta a_i \sqrt{\mu}};$$ (1.93)

as a result

$$\ln \frac{f_A f_B}{f_X} = \frac{2 z_A z_B A \sqrt{\mu}}{1 + \beta a_i \sqrt{\mu}}$$ (1.94)

or

$$\frac{f_A f_B}{f_X} = \exp\left(\frac{2 z_A z_B A \sqrt{\mu}}{1 + \beta a_i \sqrt{\mu}}\right),$$ (1.95)

and from Eqs. (1.89) and (1.95) we obtain

$$k' = k_0' \exp\left(\frac{2 z_A z_B A \sqrt{\mu}}{1 + \beta a_i \sqrt{\mu}}\right).$$ (1.96)

In logarithmic form, Eq. (1.96) becomes

$$\ln k' = \ln k_0' + \frac{2 z_A z_B A \sqrt{\mu}}{1 + \beta a_i \sqrt{\mu}}.$$ (1.97)

For very dilute solutions, where μ is small, Eqs. (1.96) and (1.97) may be written as limiting forms, thus

$$k' = k_0' \exp (2z_A z_B A \sqrt{\mu}) \tag{1.98}$$

and

$$\ln k' = \ln k_0' + 2z_A z_B A \sqrt{\mu}. \tag{1.99}$$

In the above equations A and β are the Debye-Hückel constants. Decadic logarithms may be used if, for the constant A, we write the constant A', where $A' = 2A/2.303$. Then Eqs. (1.97) and (1.99) become

$$\log k' = \log k_0' + \frac{z_A z_B A' \sqrt{\mu}}{1 + \beta a_i \sqrt{\mu}} \tag{1.100}$$

and

$$\log k' = \log k_0' + z_A z_B A' \sqrt{\mu}. \tag{1.101}$$

From Eq. (1.101) a plot of $\log k'$ vs $\sqrt{\mu}$ should, for dilute solutions of ions, yield a straight line with intercept $\log k_0'$ and slope $z_A z_B A'$.

While it is generally assumed that, if one of the reactants is molecular rather than ionic, the slope of the $\log k'$ vs any function of μ should be zero, since either z_A or z_B is zero, yet, as will be shown in a later chapter, k' is a function of μ in the case of ion–dipolar molecule reactions.

Application of the Equation to Data

The classical example of the agreement of data with the predictions of Eq. (1.101) is the figure given by LaMer (28) which, except for the data on the inversion of sucrose, is given below in Fig. 3. The data for sucrose are omitted since ion–dipolar molecule reactions will be discussed under the theory dealing with those reactions.

These data show that the slopes of the $\log k'$ vs $\sqrt{\mu}$ curves for dilute solutions of ionic reactants and of other ions are about equal to the product $z_A z_B$ predicted by Eq. (1.101), since for water at 25°C the constant A' is about unity. These data are for solutions of ions of low charge. Some of the data fail to reach the theoretical slopes, and in these cases, due to specific solvent effect or other cause, the Debye-Hückel expression fails to give the correct activity coefficients. The presence of ions of higher valence sometimes causes the slope of the $\log k'$ vs $\sqrt{\mu}$ empirical curve to be several times that predicted by theory.

The slopes of the lines log k' versus $\sqrt{\mu}$ based on the Debye-Hückel theory involve constant A, where

$$A = \frac{\varepsilon^3}{DkT} \sqrt{\frac{2\pi N}{1000 DkT}} \tag{1.102}$$

and are dependent upon the dielectric constant D of the media. The other terms in Eq. (1.102) are the charge on the electron ε, the Boltzmann gas

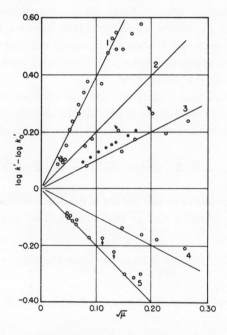

FIG. 3. The influence of ionic strength on the velocity of ionic reactions. Curves: 1. $2[Co(NH_3)_5Br]^{++} + Hg^{++} + 2H_2O \rightarrow 2[Co(NH_3)_5H_2O]^{+++} + Hg\,Br_2$ (Bimolecular). No foreign salt added. 2. Circles. $CH_2BrCOO^- + S_2O_3^= \rightarrow CH_2S_2O_3COO^= + Br^-$ as the sodium salt. No foreign salt added. Dots. $S_2O_8^= + 2I^- \rightarrow I_2 + 2SO_4^=$ as $Na_2S_2O_8$ and KI. 3. Saponification of nitrourethane ion by hydroxyl ion. $[NO_2^= N\text{-}COOC_2H_5]^- + OH^- \rightarrow N_2O + CO_3^= + C_2H_5OH$. 4. $H_2O_2 + 2H^+ + 2Br^- \rightarrow 2H_2O + Br_2$. 5. $[Co(NH_3)_5Br]^{++} + OH^- \rightarrow [Co(NH_3)_5OH]^{++} + Br^-$.

constant k, Avogadro's number N, and the absolute temperature T. When the first approximation, Eq. (1.100), is used, the constant β is also dependent on dielectric constant

$$\beta = \sqrt{\frac{8\pi N\varepsilon^2}{1000 DkT}}. \tag{1.103}$$

Amis (29) has discussed the Debye-Hückel theory in detail. Reference can be made to this work for activity coefficient expressions and equations for A and β.

Bjerrum's Derivation

Although the treatment in this chapter has dealt only with Brønsted's derivation of reaction velocity based on the hypothesis that the velocity depends only on the difference in potentials between initial and critical states of the reactants, yet it should be pointed out that both Bjerrum and Christiansen have made satisfactory derivations of reaction velocity based on somewhat different assumptions.

Bjerrum assumes that a thermally induced spontaneous monomolecular decomposition of a collision complex determines the reaction rate. This Stosscomplex is not a chemical compound but is merely of a physical nature. Thus

$$A + B \; \rightleftharpoons \; X_S \; \longrightarrow \; C + D . \tag{1.104}$$

The rate of the reaction at any instant is proportional to the concentration of the collision complex C_{X_S} at that instant. This gives

$$-\frac{dC_A}{dt} = -\frac{dC_B}{dt} = k'_{X_S} C_{X_S} . \tag{1.105}$$

The collision equilibrium in terms of activities is from the mass law

$$\frac{a_{X_S}}{a_A a_B} = \frac{C_{X_S} f_{X_S}}{C_A f_A C_B f_B} = K_a . \tag{1.106}$$

Solving for C_{X_S} out of Eq. (1.106) and substituting this into Eq. (1.105) yields a formula identical with the Brønsted formula so long as no distinction is made between the critical complex X of Brønsted and the Stosscomplex X_S of Bjerrum.

According to Brønsted it is the rate of formation of the complex which determines the reaction velocity. Brønsted makes this rate of formation of the complex depend on activities. Furthermore, the temperature coefficient of the reaction depends entirely upon the rate of complex formation. From Bjerrum's viewpoint, reaction velocities are proportional to concentrations. The ionic strength influences only the primary equilibrium according to

Bjerrum, and also the reaction of the X_S complex to yield products is strongly dependent upon temperature.

LaMer (28) criticizes the concentration hypothesis of Bjerrum from the standpoint of reversible monomolecular reactions such as

$$A \rightleftharpoons Y. \tag{1.107}$$

Assuming the validity of Bjerrum's hypothesis, the mass law in terms of concentrations must hold, and hence

$$\frac{C_A}{C_Y} = \text{constant}, \tag{1.108}$$

irrespective of change in the properties of the medium. But displacement does occur upon addition of salts, except in the insignificant case where the two activity coefficients are affected equally. LaMer also points out that the theory of Bjerrum is hardly analogous to radioactivity, since such transformations are due to instability of the nucleus for which there is a finite probability of decomposition, quite independent of external conditions.

Christiansen's Theory

Christiansen avoids the troublesome factors of concentrations of a critical complex or of a collision complex by basing his theory upon the number of collisions of reacting molecules. In his treatment the reaction velocity is set equal to

$$V = C_A C_B Z_{AB} \sqrt{T} \exp\left(- \frac{Q_{AB}}{RT}\right) \Omega \frac{\varrho}{\varrho + \Sigma NC}. \tag{1.109}$$

Here $C_A C_B Z_{AB} \sqrt{T}$ is the number of collisions per second per liter, Q_{AB} is the energy of activation, Ω is the steric factor, ϱ is the probability that the complex will react to yield products, and ΣNC is the probability that it will be deactivated again to reactants by molecules of concentration C each with an individual probability N. It must be emphasized that C is not the bulk concentration but the concentration of the various molecules in the vicinity of the reacting molecules.

For very small concentrations the factor $\varrho/(\varrho + \Sigma NC)$ approaches unity, and for high concentrations it approaches the value $\varrho/\Sigma NC$. For limiting cases of low concentrations Eq. (1.109) is therefore simplified.

The concentrations C_A and C_B in the vicinity of colliding reactant ions used in Eq. (1.109) are not the bulk concentrations C_A' and C_B' of the reactants, but are calculated in the case of ionic reactants, from these bulk concentrations by means of the Debye theory,

$$C_A = C_A' \exp\left(\frac{\varepsilon z_A \Phi_B}{kT}\right), \qquad (1.110)$$

where z_A is the valence of ion A and Φ_B is the electrostatic potential at the collision distance a from ion B and is given by the equation

$$\Phi_B = \frac{\varepsilon z_B}{D} \frac{\exp(-\varkappa a)}{a}. \qquad (1.111)$$

In this equation $\varkappa$ is the Debye kappa and D is the dielectric constant of the medium.

According to Christiansen, Eq. (1.110) requires a concentration of ions around an ion of opposite sign, but ions of opposite sign do not collide with any greater average kinetic energy because of electrostatic attraction between them.

By setting Ω and $\varrho/(\varrho + \Sigma\, NC)$ each equal to unity, Christiansen finally obtains

$$\ln k' = \ln k_0'' - \frac{z_A z_B \varepsilon^2}{DkTa} + 2.31\, z_A z_B \sqrt{\mu}, \qquad (1.112)$$

where $\ln k_0''$ is the sum of several terms which, for a given set of conditions, are constants. At infinite dilution

$$\ln k'_{\varkappa=0} = \ln k_0'' - \frac{z_A z_B \varepsilon^2}{DkTa}. \qquad (1.113)$$

If the dielectric constant D becomes infinite or if any other condition prevails which would eliminate the effect of the electric charges, k_0'' would then be the observed rate constant at infinite dilution. LaMer points out that though k_0'' is thus the value of k' extrapolated to infinite dilution and to a condition freed from all effects due to net electric charges, yet k_0'' will still be influenced by effects arising from electric moments.

Thus in Christiansen's theory it is the distribution of concentrations arising from interionic attractions which predominately influences the velocity of ionic reactions. Further the last term in Eq. (1.112) is concentration dependent and corresponds to the Brønsted activity coefficient term, i.e., to $-\ln (f_A f_B/ f_X)$.

Scatchard (*1*) concludes that it is unimportant whether the rate is calculated from the concentration of reacting complexes or from the number of collisions with the necessary orientation and energy multiplied by a factor which allows for the duration of a collision. If the molecules obtain the necessary energy and become sufficiently deformed and correctly orientated before collision, reaction will ensue when the molecules collide. If the molecules collide before any of the other steps have taken place, there will be preliminary complex formation.

REFERENCES

1. G. Scatchard, *Chem. Rev.* **10**, 229 (1932).

2. J. N. Brønsted, *Z. physik. Chem.* **102**, 169 (1922); **115**, 337 (1925).

3. K. J. Laidler and H. Eyring, *Ann. N. Y. Acad. Sci.* **39**, 303 (1940).

4. E. S. Amis and V. K. LaMer, *J. Am. Chem. Soc.* **61**, 905 (1939).

5. E. S. Amis, "Kinetics of Chemical Change in Solution," Chapter IV. Macmillan, New York, 1949.

6. J. A. Christiansen, *Z. physik. Chem.* **113**, 35 (1924).

7. E. S. Amis and J. B. Price, *J. Phys. Chem.* **47**, 338 (1943).

8. V. K. LaMer, *J. Franklin Inst.* **225**, 709 (1938).

9. J. L. Gleave, E. D. Hughes, and C. K. Ingold, *J. Chem. Soc.* **1935**, 236.

10. E. S. Amis, *J. Phys. Chem.* **60**, 428 (1956).

11. N. Goldenberg and E. S. Amis, *Z. physik. Chem.* [N. S.] **31**, 145 (1962).

11a. E. S. Amis, *J. Chem. Educ.* **29**, 337 (1952).

12. E. A. Moelwyn-Hughes, "Physical Chemistry," Chapter XXIV. Pergamon Press, New York, 1957.

13. W. J. Svirbely and J. C. Warner, *J. Am. Chem. Soc.* **57**, 1883 (1935).

14. V. K. LaMer, *J. Franklin Inst.*, **225**, 709 (1938).

15. E. S. Amis and F. C. Holmes, *J. Am. Chem. Soc.* **63**, 2231 (1941).

16. E. S. Amis and J. E. Potts, *J. Am. Chem. Soc.* **63**, 2833 (1941).

17. E. S. Amis, *J. Am. Chem. Soc.* **63**, 1606 (1941).

18. V. K. LaMer and M. E. Kamner, *J. Am. Chem. Soc.* **47**, 2662 (1935).

19. E. A. Moelwyn-Hughes, *Proc. Roy. Soc.* (*London*), **A155**, 308 (1936); **A157**, 667 (1936).

20. E. S. Amis and G. Jaffé, *J. Chem. Phys.* **10**, 646 (1942).

21. J. A. Christiansen, *Z. physik. Chem.* **113**, 35 (1924).

22. E. S. Amis and S. Cook, *J. Am. Chem. Soc.* **63**, 2621 (1941).

23. G. Scatchard, *Ann. N. Y. Acad. Sci.* **39**, 341 (1940).

24. G. Scatchard, *J. Chem. Phys.* **7**, 657 (1939).

25. J. N. Brønsted, *Z. physik. Chem.* **102**, 169 (1922); **115**, 337 (1925).

26. N. Bjerrum, *Z. physik. Chem.* **108**, 82 (1924); **118**, 251 (1925).

27. J. A. Christiansen, *Z. physik. Chem.* **113**, 35 (1924).

28. V. K. LaMer, *Chem. Rev.* **10**, 179 (1932).

29. E. S. Amis, "Kinetics of Chemical Change in Solution." Macmillan, New York, 1949.

CHAPTER II

SOLVENT EFFECTS ON ION-DIPOLAR MOLECULE REACTIONS

THEORETICAL

Introduction

Since there are electrical forces between ions and dipolar molecules, and since such forces are modified by the solvent through its dielectric constant and polar properties, there should be a dielectric-constant-influenced electrostatic term in a rate expression for an ion–dipolar molecule reaction just as there is such a term in the rate expression for ion–ion reactions. The electrostatic interactions between ions and dipolar molecules will not be so pronounced as in the case of electrostatic interactions between ions and ions, and will therefore be more readily obscured by various other solvent and structural effects. The effect, however, is quite pronounced and should be calculable. Theoretical expression for the influence of the solvent on electrostatic forces which modify rate processes between ions and dipolar molecules have been derived. Some of these theories are presented in the following pages.

The Laidler-Eyring Equation

Laidler and Eyring (1) took the type reaction

$$A^{Z_A} + B^0 \longrightarrow M^{*Z_A} \longrightarrow X + Y \qquad (2.1)$$

where Z_A is the charge on A and M*. They write for the activity coefficients of A and M*, including the semi-empirical $b\mu$ terms of Hückel (2), the equation

$$\ln f_A = \frac{Z_A^2 \varepsilon^2}{2r_A kT} \left(\frac{1}{D} - 1 \right) - \frac{Z_A^2 \varepsilon^2}{2DkT} \frac{\varkappa}{1 + a_A \varkappa} + b_A + \frac{\Phi_A}{kT}, \qquad (2.2)$$

31

$$\log f_{\text{M*}} = \frac{Z_{\text{A}}^2 \varepsilon^2}{2 r_{\text{A}} kT} \left(\frac{1}{D} - 1 \right) - \frac{Z_{\text{A}}^2 \varepsilon^2}{2 DkT} \frac{\varkappa}{1 + a_{\text{M*}} \varkappa} + b_{\text{M*}} \mu + \frac{\Phi_{\text{M*}}}{kT}, \quad (2.3)$$

where μ is the ionic strength and the b's are constants. To determine f_{B}, the activity coefficient of the dipolar molecule in a medium of dielectric constant D, having n_i ions of the type i per cubic centimeter, charge Z_i, and radius r_i, we use Laidler and Eyring's modification of Kirkwood's (3) expression for the free energy of transfer F_1 of a strong dipole of moment m_{B} from a vacuum to the medium having dielectric constant D_0 after the dipolar molecules, but not the ions, have been added, and sum with this the approximate expression of Debye and McAulay (4) for the free energy increase, F_2 brought about by addition of the ions.

The free energy of transfer is given by

$$F_1 = kT \ln \beta = - \frac{1}{kT} \frac{m_{\text{B}}^2}{a_{\text{B}}^3} \left[\frac{D_0 - 1}{2 D_0 + 1} \right] + \frac{\Phi_{\text{B}}}{kT} \qquad (2.4)$$

and for the increase of free energy F_2 caused by adding the ions, the equation

$$F_2 = kT \log f = \frac{\bar{\alpha} \varepsilon^2}{2 D} \sum_i \frac{Z_i^2 n_i}{r_i} \qquad (2.5)$$

holds, where $\bar{\alpha}$ is defined by

$$D' = D_0' (1 - \bar{\alpha} \, n_{\text{B}}). \qquad (2.6)$$

In Eq. (2.6) D_0' is the dielectric constant of the pure solvent, and D' that of a solution containing n_{B} molecules of the dipolar molecule per cubic centimeter. The activity coefficient f_{B} of the dipolar molecule is given by the equation

$$\ln f_{\text{B}} = - \frac{1}{kT} \frac{m_{\text{B}}^2}{a_{\text{B}}^3} \left[\frac{D_0 - 1}{2 D_0 + 1} \right] + \frac{\bar{\alpha} \varepsilon^2}{2 DkT} \sum_i \frac{Z_i^2 n_i}{r_i} + \frac{\Phi_{\text{B}}}{kT} \quad (2.7)$$

and the rate expression can then be written

$$\ln k' = \ln \left(K \frac{kT}{h} K_0^* \right) + \frac{Z_{\text{A}}^2 \varepsilon^2}{2 kT} \left(\frac{1}{D} - 1 \right) \left(\frac{1}{r_{\text{A}}} - \frac{1}{r_{\text{M*}}} \right)$$

$$- \frac{Z_{\text{A}}^2 \varepsilon^2}{2 DkT} \left(\frac{\varkappa}{1 + a_{\text{A}} \varkappa} - \frac{\varkappa}{1 + a_{\text{M*}} \varkappa} \right) - \frac{1}{kT} \left[\frac{D_0 - 1}{2 D_0 + 1} \right] \frac{m_{\text{B}}^2}{a_{\text{B}}^3}$$

$$+ \left(b_{\text{B}} - b_{\text{M*}} + \frac{\bar{\alpha} \varepsilon^2}{DrkT} \right) \mu + \frac{\Phi_{\text{A}} + \Phi_{\text{B}} - \Phi_{\text{M*}}}{kT},$$

$$(2.8)$$

where r is the average values of the r_i's. Setting $a_A = a_{M^*}$ since their effects at $\mu \to 0$ becomes vanishingly small, the rate then becomes

$$\ln k' = \ln \left(K \frac{kT}{h} K_0^* \right) + \frac{Z_A^2 \varepsilon^2}{2kT} \left(\frac{1}{D} - 1 \right) \left(\frac{1}{r_A} - \frac{1}{r_{M^*}} \right)$$

$$- \frac{1}{kT} \frac{m_B^2}{a_B^3} \left[\frac{D_0 - 1}{2D_0 + 2} \right] + \left(b_A - b_{M^*} + \frac{\bar{\alpha} \varepsilon^2}{DrkT} \right) \mu \qquad (2.9)$$

$$+ \frac{\Phi_A + \Phi_B - \Phi_{M^*}}{kT},$$

and at zero ionic strength, $\mu = 0$, the equation can be written

$$\ln k' = \ln \left(K \frac{kT}{h} K_0^* \right) + \frac{Z_A^2 \varepsilon^2}{2kT} \left(\frac{1}{D} - 1 \right) \left(\frac{1}{r_A} - \frac{1}{r_{M^*}} \right)$$

$$- \frac{1}{kT} \frac{m_B^2}{a_B^3} \left[\frac{D_0 - 1}{2D_0 + 1} \right] + \frac{\Phi_A + \Phi_B - \Phi_{M^*}}{kT}. \qquad (2.10)$$

To test this equation $\ln k'$ extrapolated to zero ionic strength for a molecule of moment zero reacting with an ion of charge Z_A can be plotted versus $1/D$. The slope of the resulting straight line should be

$$\frac{Z_A^2 \varepsilon^2}{2kT} \left(\frac{1}{r_A} - \frac{1}{r_{M^*}} \right), \qquad (2.11)$$

if theory is obeyed.

Laidler and Eyring (1) point out that data involving changing dielectric constants that can be extrapolated to zero ionic strength are scarce. They reason that since the Hückel and Debye-McAulay terms in Eq. (2.9) tend to cancel, the ionic strength influence on these types of reactions tend to be very small. Hence at plot of $\ln k'$ unextrapolated to zero ionic strength can be plotted versus $1/D$ and tested for obedience to theory. They actually plotted the data for $\ln k'$ versus $1/D$ in the case of the conversion of N-chloroacetanilide and N-chloropropionanilide into the p-chloro compounds in the presence of hydrochloric acid (5), reactions which may involve an initial attack by the chloride ion. Good straight lines with positive slopes were obtained The slopes of such lines should, irrespective of the charge sign of the ion, always be positive according to Eq. (2.9), since the charge on the ion is squared. While the N-chloroacetanalide and N-chloro-propionanalide are polar, yet the Kirkwood term, as compared to the ionic term, is only slightly influenced by dielectric constant, and hence although the moments of the dipolar anilide molecules are not zero, there is a linear dependence of $\ln k'$ on $1/D$. Thus when the dielectric constant changes from 100 to 50, the change in the Kirkwood term due to changing dielectric constant is only from $(100 - 1)/(200 + 2)$ or 0.49 to $(50 - 1)/(100 + 2)$ or 0.48 while the term $Z_A^2\varepsilon^2/2kTD$ doubles in value. At low dielectric constants the influence of the Kirkwood term would become more pronounced. To quantitatively test the obedience of the data to theory the values of r_A and r_{M*} would have to be known.

The Amis and Jaffé Equation

Amis and Jaffé (6) have derived an equation for the rates of reactions between ions and dipolar molecules. In deriving the equation Amis and Jaffé first obtained the potential ψ_0 in the neighborhood of any one dipolar molecule. Neglecting the interaction between dipoles, the differential equation for ψ_0 is the same as the potential around an ion, except for the boundary conditions at the surface of the molecule. For a species of ions let n_i be the number per cubic centimeter and z_i the valence of species i. The differential equation for the potential, as in the potential around an ion, is

$$\Delta^2\psi_0 = \varkappa^2\psi \tag{2.12}$$

when ψ_0 is small. $\varkappa$ is the Debye kappa.

A particular solution of Eq. (2.12), assuming the dipole as a point singularity, is

$$\psi_0 = \exp(-\varkappa r/r). \tag{2.13}$$

If the idealized dipole is represented as a sphere of radius a, if the position of the dipole coincides with the origin, and if its direction is that of the positive z-axis, then, for $r > a$, a solution having the same dependence as that of a dipole potential can be found by differentiating Eq. (2.13) with respect to z. Thus

$$\frac{\partial\psi_0}{\partial z} = \frac{\partial\psi_0}{\partial r}\frac{\partial r}{\partial z}, \tag{2.14}$$

but

$$r^2 = x^2 + y^2 + z^2 \tag{2.15}$$

and

$$\frac{\partial r}{\partial z} = \frac{z}{r} = \cos\theta, \tag{2.16}$$

also

$$\frac{\partial \psi'_0}{\partial z} = - \frac{C_1 \exp{(- \varkappa r)}}{r^2} (1 + \varkappa r) \frac{\partial r}{\partial z}. \tag{2.17}$$

Hence,

$$\psi'_0 = \frac{C_1 \exp{(- \varkappa r)}}{r^2} (1 + \varkappa r) \cos \theta, \; r \geq a. \tag{2.18}$$

Bateman et al. (7) gave the solution [Eq. (2.18)] of the special case considered above. Kirkwood (3) gave the solution for the potential of a particle with the most general distribution of charge.

Following the Debye-Hückel procedure and holding D constant for the interior of the dipole molecule we obtain the potential ψ_i for the interior of the molecule from the equation

$$\varDelta^2 \psi_i = 0, \quad r \leq a. \tag{2.19}$$

Except for a singularity at the origin corresponding to a dipole of a given strength, ψ in this domain must be continuous. A reasonable expression for the ion atmosphere results from this procedure; however, the dipoles have been chosen about five times greater than ordinary magnitudes of dipoles to fit the formula to experimental kinetic data. The interaction of the dipoles with the dielectric solvent or some other influential electrostatic forces are not being accounted for properly.

The external moment with which an immersed dipole acts upon distant charges is different from its moment in vacuo as has been demonstrated by Onsager (8). The action of both the permanent and induced moments are modified by the medium. It is necessary therefore to combine the Onsager model of a dipole immersed in a dielectric liquid with the Debye-Hückel theory of ionic atmospheres.

Our model, like that of Onsager, will be a spherical molecule of permanent moment in vacuo μ_0 and polarizability α, which is related to an internal refractive index n by the equation

$$\alpha = \frac{n^2 - 1}{n^2 + 2} a^3. \tag{2.20}$$

The solution for ψ_0 in Eq. (2.18) must be joined properly at $r = a$ to a solution of Eq. (2.19) having a dipole singularity at $r = 0$. This requires that we find the external characteristic moment μ^* such as it becomes due to the action of the polarization generated by itself.

For a rigid dipole of moment m placed in a cavity of radius a, the exterior potential is given by Eq. (2.18), and the interior potential can be written

$$\psi_i = \left(\frac{m}{r^2} + Br\right)\cos\theta, \qquad r \leq a, \tag{2.21}$$

where B is a constant.

By setting $\psi_i = \psi_0$ and $\partial\psi_i/\partial r = \partial\psi_0'/\partial r$, the usual boundary conditions, constants C_1 and B, can be determined. Thus

$$\frac{C_1 \exp(\varkappa a)}{a^2}(1 + \varkappa a) = \frac{m}{a^2} + Ba \tag{2.22}$$

and

$$\frac{DC_1 \exp(\varkappa a)}{a^3}(2 + 2\varkappa a + \varkappa^2 a^2) = \frac{2m}{a^3} - B, \tag{2.23}$$

and therefore

$$C_1 = \frac{3m\exp(\varkappa a)}{D(2 + 2\varkappa a + \varkappa^2 a^2) + (1 + \varkappa a)} \tag{2.24}$$

and

$$B = -\frac{m}{a^3}\frac{[D(2 + 2\varkappa a + \varkappa^2 a^2) - 2(1 + \varkappa a)]}{[D(2 + 2\varkappa a + \varkappa^2 a^2) + (1 + \varkappa a)]}. \tag{2.25}$$

Onsager gives the condition for internal equilibrium as

$$m = \mu_0 + \alpha F_z, \tag{2.26}$$

where F_z is the force on the dipole arising from the polarization created by itself. Thus the total electric moment of a molecule in an electric field F_z is the sum of the permanent moment μ_0 and the induced moment αF_z. F_z, the local field, is

$$F_z = -\frac{\partial}{\partial z}\left(\psi_i - \frac{m\cos\theta}{r^2}\right) = -B. \tag{2.27}$$

Therefore, from Eqs. (2.20), (2.26), and (2.27),

$$m = \mu_0 - \frac{n^2 - 1}{n^2 + 2}a^3 B. \tag{2.28}$$

B from Eq. (2.25) substituted into Eq. (2.28) yields

$$m = \mu_0\frac{[n^2 + 2][D(2 + 2\varkappa a + \varkappa^2 a^2) + (1 + \varkappa a)]}{3[D(2 + 2\varkappa a + \varkappa^2 a^2) + n^2(1 + \varkappa a)]}. \tag{2.29}$$

m is thus determined as a function of μ_0 and n^2. If m from Eq. (2.29) is substituted into Eq. (2.24) there results

$$C_1 = \frac{\mu^*}{D},$$ (2.30)

where

$$\mu^* = \frac{\mu_0 (n^2 + 2) D \exp (\varkappa a)}{D(2 + 2\varkappa a + \varkappa^2 a^2) + n^2(1 + \varkappa a)},$$ (2.31)

and C_1 from Eq. (2.30) substituted into Eq. (2.18) gives

$$\psi_0 = \frac{\mu^* \exp (- \varkappa r) (1 + \varkappa r)}{Dr^2} \cos \theta,$$ (2.32)

where ψ_0 is the potential of the ionic atmosphere around a dipolar molecule of permanent moment μ_0, the external moment of the molecule in the dielectric solvent being μ^*.

In the absence of an ionic atmosphere, $\varkappa = 0$, Eqs. (2.32) and (2.21) became, as they should, identical with Onsager's solutions.

Equations (2.31) and (2.32) for $n^2 = 1$ represent dipole with no polarizability.

Letting C_X, C_A, and C_B be the concentrations of the intermediate complex X, and reactants A and B, respectively, and using the procedure of Christiansen and Scatchard, the probability of finding an ion B in a specified element of volume, defined by the limits r and $r + dr$, θ and $\theta + d\theta$, and φ and $\varphi + d\varphi$, is by Boltzmann's theorem proportional to

$$C_A C_B \exp \left(- \frac{\psi_0 \varepsilon Z_B}{kT}\right) r^2 \theta \, dr \, d\theta \, d\varphi,$$ (2.33)

where r, θ, and φ are the polar coordinates about the center of the dipole A, and ψ_0 is defined by Eq. (2.32).

To obtain the velocity of the reaction, the rate of formation of X must be determined. To do this we may assume (1) sensitive zones on the surface of the molecule A which, if touched by the ion B, produce complex X, or (2) there is a critical distance of approach from each direction of ion B to molecule A in order for the formation of X to ensue. The first alternative which necessitates that r, θ, and φ must assume specified values between rather narrow limits r_0 and $r_0 + dr_0$, θ_0 and $\theta_0 + d\theta_0$, φ_0 and $\varphi_0 + d\varphi_0$ will be followed first.

If there is rotational symmetry about the z-axis, φ need not be specified, and Eq. (2.33) can be integrated with respect to φ, giving 2π.

Therefore the C_X is given by

$$C_X = 2\pi k' C_A C_B r_0{}^2 \sin \theta_0 \, \varDelta r_0 \, \varDelta \theta_0 \exp\left(- \frac{\psi_0 \varepsilon Z_B}{kT}\right), \tag{2.34}$$

where k' is a proportionality constant, and letting

$$K = 2\pi k' r_0{}^2 \sin \theta_0 \, \varDelta r_0 \, \varDelta \theta_0 \tag{2.35}$$

we have

$$C_X = K C_A C_B \exp\left[- \frac{\psi_0(r_0, \theta_0)\varepsilon Z_B}{kT}\right] \tag{2.36}$$

which can be put in the form

$$\ln \frac{C_X}{C_A C_B} = \ln K - \frac{\psi_0(r_0, \theta_0)\varepsilon Z_B}{kT}. \tag{2.37}$$

But from Eq. (2.32) φ_0 becomes, when $\varkappa = 0$,

$$\psi_0 = \frac{\mu_0{}^*}{D r_0{}^2} \cos \theta \tag{2.38}$$

in which $\mu_0{}^*$ is Onsager's value and from Eq. (2.31) is

$$\mu_0{}^* = \frac{\mu_0(n^2 + 2)D}{2D + n^2}. \tag{2.39}$$

From Eqs. (2.37) and (2.38)

$$\ln \frac{C_X}{C_A C_B} = \ln K - \frac{\varepsilon Z_B \mu_0{}^* \cos \theta}{DkT r_0{}^2}. \tag{2.40}$$

Hence

$$\ln \frac{f_A f_B}{f_X} = \ln \frac{C_X}{C_A C_B} - \ln \frac{C_X{}^0}{C_A{}^0 C_B{}^0}$$

$$= \frac{\varepsilon Z_B \cos \theta}{DkT r_0{}^2} [\mu_0{}^* - \mu^* \exp(-\varkappa a)(1 + \varkappa a)] \tag{2.41}$$

$$= \mu_0 \frac{\varepsilon Z_B \cos \theta}{DkT r_0{}^2} \left[\frac{(n^2 + 2)D}{2D + n^2} - \frac{(1 + \varkappa a)(n^2 + 2)D}{D(2 + 2\varkappa a + \varkappa^2 a^2) + n^2(1 + \varkappa a)}\right]$$

and substituting for μ_0 in terms of μ_0^* in Eq. (2.41) gives

$$\ln \frac{f_A f_B}{f_X} = \frac{\varepsilon Z_B \mu_0^* \cos \theta}{DkTr_0^2} \left[\frac{D\varkappa^2 a^2}{2D\left(1 + \varkappa a + \dfrac{\varkappa^2 a^2}{2}\right) + n^2(1 + \varkappa a)} \right] \cdot \quad (2.42)$$

But

$$k' = k_0' \frac{f_A f_B}{f_X} \quad (2.43)$$

and

$$\ln k' = \ln k_0' + \ln \frac{f_A f_B}{f_X}, \quad (2.44)$$

therefore, from Eqs. (2.42) and (2.43),

$$\ln k' = \ln k_0' + \frac{\varepsilon Z_B \mu_0^* \cos \theta}{DkTr_0^2} \left[\frac{D\varkappa^2 a^2}{2D\left(1 + \varkappa a + \dfrac{\varkappa^2 a^2}{2}\right) + n^2(1 + \varkappa a)} \right],$$

$$(2.45)$$

which can be written

$$\frac{\ln k' - \ln k_0'}{\varepsilon Z_B \mu_0^* \cos \theta / 2DkTr_0^2} = W = \frac{\varkappa^2 a^2}{1 + \varkappa a + \dfrac{\varkappa^2 a^2}{2} + \dfrac{n^2}{2D}(1 + \varkappa a)} \cdot \quad (2.46)$$

For ion–ion reactions a reference point defined by the double transition $\varkappa = 0$, $D = \infty$ has proven successful in correlating the dielectric dependence of k'. In the case of ion–dipole reactions such a double transition results in a change in k' with $1/D$ opposite in sense to the change in k' with kappa. Data, however, indicate that in both cases the change in k should be in the same sense. For ion–ion reactions the electrostatic attractions or repulsions tend toward zero as D increases indefinitely. In the case of a dipole the external moment increases with increasing D, and though $\lim_{D=\infty} \mu^*$ exists, the point $D = \infty$ is not an adequate reference point since it accentuates rather than eliminates the effect of D. Amis and Jaffé assumed, therefore, that $\varkappa = 0$ was an adequate point of reference, and that Eq. (2.45) expressed the relationship of k' to both D and $\varkappa$.

These authors introduced dimensionless variables in order to put Eq. (2.45) into a form to be tested for the dependence of k' upon D. Let

$$\varkappa^2 = \lambda^2 / D, \quad (2.47)$$

where λ has the dimension of cm^{-1} but is free of D. Then if

$$S = \lambda a = \lambda r_0 \qquad (2.48)$$

and

$$W' = \frac{(\ln k' - \ln k'_{\varkappa=0})\,(2kT)}{\varepsilon Z_B \mu_0 {}^* \lambda^2 \cos \theta}, \qquad (2.49)$$

the general relation

$$W' = \frac{1}{D^2}\; \frac{1}{1 + S/D^{1/2} + S/2D} \qquad (2.50)$$

results. Equation (2.50) indicates that, in the limit $\lambda = 0$ and therefore $\varkappa = 0$, W' will vary as $1/D^2$. The increase of W' with decreasing dielectric constant will become less at higher concentrations, and the increase will depend on the parameter S which from Eq. (2.47) is proportional to the square root of the ionic strength.

This equation was applied by Amis and Jaffé to the inversion of sucrose by hydrochloric acid and the mutarotation of glucose by hydrochloric acid and found to reproduce the data well with reasonable values of the parameters r_0, $\mu_0{}^*$, and n when W' was plotted versus $1/D^2$. The acid-catalyzed inversion of sucrose, used in testing the predictions of Eq. (2.50), was studied (9) in water–dioxane solvents, and the acid-catalyzed mutarotation of glucose was investigated by Dyas and Hill (10) in water–methanol solvents. The acid-catalyzed inversion of sucrose in water–ethanol solvents was shown by energy considerations to be in opposition to the theory of electrostatics and was not applied in testing the predictions of Eq. (2.50).

It might be pointed out that the sign of the ion has to be taken into account in fitting Eq. (2.50) to data. This is not foreseen from electrostatics and probably arises from chemical rather than electrical properties of the system. It is found that much kinetic data give positive slopes when $\ln k'$ is plotted versus $1/D$ for positively charged ions reacting with dipolar molecules as required by Eq. (2.50); while these plots, in the case of negatively charged ions, give negative slopes, again as required by Eq. (2.50).

The Laidler-Eyring theory, Eq. (2.9), obviates these phenomena by having the valence of the ion appear as the square. The difficulty here is that the $\ln k'$ versus $1/D$ plots should always have positive slopes, which is contrary to much experimental data, unless $r_{M*} < r_A$ when the slope would be negative. It seems that r_{M*} would have little chance of being less than r_A, even including the influence of solvation, since r_{M*} is the radius of the ion plus the molecule plus any attached solvent and r_A is the radius of only the molecule plus any attached solvent. It would appear that the ion plus the molecule would be larger than the molecule alone, and that the ion–molecule combination would be more extensively solvated than the molecule if only because of the greater polarizing effect of the charged ion–molecule as compared to the dipolar molecule alone. Furthermore, r_{M*} would have to be less than r_A, in general, only for negative ions reacting with dipolar molecules. Such a phenomenon would seem to be merely coincidental.

The Amis Modification of the Amis-Jaffé Equation

Since there was some uncertainty about the extrapolation to a reference state of dielectric constant in the Amis-Jaffé approach to the theory of ion–dipolar molecule reactions, Amis (11) derived the dielectric constant dependence of k' using Coulombic energy considerations. This approach is limited to dielectric constant effects but is simple and can be applied to various electrically charged (12) and electrically unsymmetrical combinations of reactants. The complex formation–activity coefficient approach is, perhaps, still best for elucidating the salt effect.

E_c, the mutual potential (potential energy) between an ion and a dipole, is given by

$$E_c = -\frac{Z\mu_0 \cos\theta}{Dr^2}, \tag{2.51}$$

where higher powers than the first of the distance of separation of centers of charge in the dipole are neglected. The difference in Coulombic energy, ΔE_c, of the ion–dipole at two different dielectric constants D_1 and D_2 is

$$\Delta E_c = -\frac{Z\varepsilon\mu_0}{r^2}\left(\frac{1}{D_2} - \frac{1}{D_1}\right). \tag{2.52}$$

In order to specify quantitatively the effect of Coulombic energy on specific reaction rates, consider a reaction at dielectric constant D_2 and let its energy requirements be E_{D_2}. For the same reaction at dielectric constant D_1 let E_{D_1} represent the energy of activation. Then

$$E_{D_2} = E_{D_1} + \Delta E_c. \tag{2.53}$$

Now the specific velocity constant at dielectric constant D_2 is given by

$$k'_{D_2} = Z \exp\left(-\frac{E_{D_2}}{RT}\right) \tag{2.54}$$

and

$$\ln k'_{D_2} = \ln Z - \frac{E_{D_2}}{RT} = \ln Z - \frac{E_{D_1}}{RT} - \frac{\Delta E_c}{RT} \qquad (2.55)$$

but

$$\ln k'_{D_1} = \ln Z - \frac{\Delta E_{D_1}}{RT} \qquad (2.56)$$

hence,

$$\ln k'_{D_2} = \ln k'_{D_1} + \frac{Z\varepsilon\mu_0}{kTr^2}\left(\frac{1}{D_2} - \frac{1}{D_1}\right). \qquad (2.57)$$

If a reference dielectric constant D_2 is taken as infinite in magnitude so that all electrostatic effects between reactants vanish and if D_1 is taken as any dielectric constant D, we have, from Eq. (2.57),

$$\ln k_\infty' = \ln k_D' - \frac{Z\varepsilon\mu_0}{kTr^2D} \qquad (2.58)$$

or

$$\ln k_D' = \ln k_\infty' + \frac{Z\varepsilon\mu_0}{kTr^2D}. \qquad (2.59)$$

According to Eq. (2.59), if the charge sign of the ion is taken into account, a plot of $\ln k_D'$ vs $1/D$ should be a straight line of positive slope if $Z\varepsilon$ is positive and of negative slope if $Z\varepsilon$ is negative. These predictions are in harmony with those of the Amis-Jaffé equation. The specific velocity constants should preferably be those at zero ionic strength. From the slopes of the straight lines reasonable values of r should be calculable. Probably the better moment at zero ionic strength, in Eq. (2.59), would be μ_0^* which is related to μ_0 by the Onsager relationship, Eq. (2.31), when $\varkappa = 0$, namely,

$$\mu_0^* = \frac{\mu_0(n^2 + 1)D}{2D + n^2}. \qquad (2.60)$$

Equation (2.59) would then become

$$\ln k'_{\substack{D=D \\ \varkappa=0}} = \ln k'_{D=\infty} + \frac{Z\varepsilon\mu_0^*(2D + n^2)}{D^2kTr^2(n^2 + 2)}. \qquad (2.61)$$

This equation contains parameters μ_0^*, r, D, and n, and assumptions would have to be made concerning them since their values are ordinarily not recorded in the literature. One might assume reasonable values of r and n and plot

$$\ln k'_{\substack{D=D \\ \varkappa=0}} \text{ versus } \frac{Z\varepsilon(2D + n^2)}{D^2 kTr^2(n^2 + 2)} ,$$

and if a straight line with a slope that gives reasonable values of μ_0^* is obtained, theory is coinciding with experiment.

Quinlan and Amis (13) incorporated the simple expression Eq. (2.59) for the dielectric constant dependence of ion–dipolar molecule reaction rates into Eq. (2.45) and obtained

$$\ln k' = \ln k_{\substack{\varkappa=0 \\ D=\infty}} + \frac{Z\varepsilon\mu_0}{DkTr_0^2} + \frac{Z\varepsilon\cos\theta}{DkTr_0^2}\left(\mu_0^* - \frac{\mu^*(1 + \varkappa r_0)}{\exp(\varkappa r_0)}\right), \quad (2.62)$$

where the second term on the right gives the dielectric constant dependence of the rate and the last term on the right gives the ionic strength dependence of the rate. This equation is similar in appearance and function to the Brønsted-Christiansen-Scatchard equation, which can be obtained by substituting the value of $\log k'_{\varkappa=0}$ from Eq. (1.14) for $\log k_0'$ in Eq. (1.100). These latter equations are given in Chapter I.

Table I summarizes the application of this equation to both the acid and alkaline hydrolyses of esters in various mixed solvents at various temperatures. Only one datum is seen to be in real conflict with theory.

In Fig. 1 is shown the results of the application of Eq. (2.62) to the data on the rates of the sucrose inversion by hydronium ion and of the reaction of diacetone alcohol with hydroxide ion (12). As a first approximation the macroscopic dielectric constants of the mixed solvent pairs and the specific velocity constants unextrapolated to zero ionic strength were used. The solvents in the case of sucrose inversion (9) were water and water–dioxane at 41°C, and the solvents in the case of diacetone alcohol–hydroxide ion reaction (14) were water and water–ethanol at 25°C.

The value of the slope of the straight line for the sucrose inversion using Landt's (15) datum of 3.4 Debye units (D. U.) for the moment μ_0 of sucrose gave a value of r of 5.5 Å as the distance of approach for hydronium ion and sucrose to react. This compares favorably with 3.5 Å found by Amis and Jaffé (16). Using μ_0 as 4.0 D. U. for the moment of diacetone alcohol as did Amis, Jaffé, and Overman (17), the value for r is found to be 6.0 Å.

According to electrostatic requirements, the mechanism of the sucrose inversion is one involving a positive ion and a dipolar molecule and can be written

$$
\text{(structure of sucrose)} + H_3O^+ \underset{}{\overset{(1)}{\rightleftharpoons}}
$$

$$
\underset{R}{\overset{\overset{\displaystyle H^+}{\overset{|}{O}}}{\diagdown}}\underset{R'}{\diagup} + H_2O \overset{(2)}{\rightleftharpoons}
\begin{array}{c}
CHO \\
| \\
H-C-OH \\
| \\
HO-C-H \\
| \\
H-C-OH \\
| \\
H-C-OH \\
| \\
CH_2OH \\
\text{glucose}
\end{array}
+
\begin{array}{c}
CH_2OH \\
| \\
C=O \\
| \\
HO-C-H \\
| \\
H-C-OH \\
| \\
H-C-OH \\
| \\
CH_2OH \\
\text{fructose}
\end{array}
\qquad (2.63)
$$

TABLE I

VALUES OF PARAMETER r FOR BOTH THE ACID AND ALKALINE HYDROLYSES OF ESTERS

Ester	Solvent	Temp. (°C)	$r \times 10^8$ (cm)
(A) Acid hydrolysis			
Ethyl acetate	Water and water–dioxane	35.0	9.08
		45.0	9.18
		55.0	9.50
Ethyl acetate	Water and water–acetone	35.0	5.21
		45.0	6.19
		55.0	Unreasonably small
Methyl propionate	Water and water–acetone	25.13	3.43
		35.21	3.42
		45.48	3.01
(B) Alkaline hydrolysis			
Ethyl acetate	Water and water–ethanol	0.00	0.94
		9.80	0.93
		19.10	0.98
Ethyl acetate	Water and water–acetone	0.00	2.0
		15.87	1.8
		25.10	1.5
Methyl propionate	Water and water–acetone	15.00	1.4
		25.00	1.4
		35.03	1.3

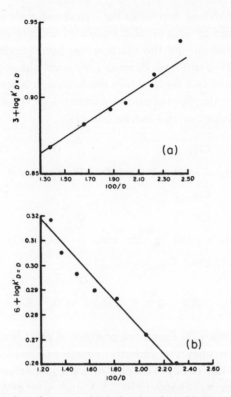

FIG. 1. (a) The reaction of sucrose with hydronium ion. (b) The reaction of diacetone alcohol with hydroxide ion. Dots are experimental points; straight lines through data yielded slopes from which the r parameters for the reactions were calculated.

The electrostatics of the diacetone alcohol–hydroxide reaction require a rate-controlling step involving a dipolar molecule and a negative ion. Such a mechanism is depicted below.

$$
\begin{array}{c}
\underset{\|}{O} \qquad \overset{OH}{|} \\
CH_3\overset{\|}{C}-CH_2-\overset{|}{\underset{|}{C}}-CH_3 + OH^- \underset{CH_3}{\overset{(1)}{\rightleftharpoons}} CH_3-\overset{O}{\overset{\|}{C}}-CH_2-\overset{O^-}{\underset{CH_3}{\overset{|}{C}}}-CH_3 + H_2O
\end{array}
$$

$$(2)$$

$$
^-CH_2-\overset{O}{\overset{\|}{C}}-CH_3 + CH_3-\overset{O}{\overset{\|}{C}}-CH_3
$$

$$(3) \quad H_2O$$

$$
CH_3-\overset{O}{\overset{\|}{C}}-CH_3 + OH^-
$$

$$(2.64)$$

Either the first or second step seems the logical one, since these steps involve negative ions reacting with dipolar molecules and are independent of the acetone concentration, and the reaction has been proven independent of this quantity (*18*). Frost and Pearson (*19*) select the second step as rate controlling. Nelson and Butler's (*20*) mechanism is where B as a negative ion would satisfy the electrostatic requirements of the kinetics, but there are certain objections to this mechanism (*19*).

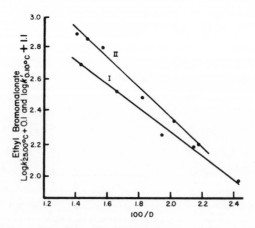

Broach and Amis (*21*) found the reaction of ethyl bromomalonate with thiosulfate ion to obey the dielectric constant as well as the salt effect requirements of Eq. (2.62). The plot of log $k'_{D=D}$ versus $1/D$ is given in Fig. 2. The r-value was chosen to be 2.5 Å, and μ_0 for ethyl bromomalonate calculated from the slopes of the straight lines are 5.45 and 4.60 Debye

FIG. 2. Plot of log k' vs $100/D$ for the ethyl bromomalonate–thiosulfate reaction. Curve I, 25.00°C; curve II, 0.10°C. Dots arc experimental data; slopes of straight lines were used to calculate r parameters.

units at $0.10°$ and $25.00°C$, respectively. The average value of μ_0 was thus 5.03 D. U., which is certainly a reasonable value.

From the electrostatics, the rate-governing step of the reaction involved a dipolar molecule and a bivalent negatively charged ion. Amis (*22*) concluded that the reaction is a substitution nucleophyllic reaction of the second order and that the probable mechanism is

$$C_2H_5OOCHCBrCOOC_2H_5 + S_2O_3^= \rightleftharpoons$$

$$\overset{\displaystyle S_2O_3^=}{\underset{\displaystyle Br}{\overset{\vdots}{\underset{\vdots}{C_2H_5OOCHCCOOC_2H_5}}}} \tag{2.66}$$

$$C_2H_5COOCHC(S_2O_3)\,^-COOC_2H_5 + Br^-.$$

Nolan and Amis (*23*) applied Eq. (2.62) to the dielectric constant dependence of the rates of hydrolyses of ethyl-α-haloacetates by negative hydroxyl ion in water and water–ethanol solvents at $15°$, $20°$, and $25°C$. The $\log k'$ versus $1/D$ curves were negative in slope, as required by theory, and only slightly curved. The range of dielectric constant is from 82.0 for water at $15°C$ to 26.7 for 91.77 weight percent ethanol at $30°C$. This is a wide range for data to approximate theory. Assuming at $25°C$ dipole moments of 2.09, 2.64, and 2.64 D. U. for ethyl fluoroacetate, ethyl chloroacetate, and ethyl bromoacetate, respectively, r values for the three esters were calculated to be 1.39, 1.35, and 1.30 Å, respectively. These are reasonable results. The mechanism involving the dipolar esters and negative hydroxyl ion was written

$$R\!-\!\underset{\overline{O}-R'}{\overset{\overline{O}|}{C}} + [\overline{O}\!-\!H]^- \rightleftharpoons \left[R\!-\!\underset{|\underline{O}-R}{\overset{|\overline{O}|}{C}}\!-\!\overline{O}H \right]$$

$$R\!-\!\underset{\underline{O}-H}{\overset{\overline{O}|}{C}} + [\overline{O}\ \ R']^-. \tag{2.67}$$

From the number and variety of reactions, at different temperatures, and in different solvents that conform with the requirement of Eq. (2.62), the validity of the assumptions made and the mathematical procedure used in its derivation seem strongly justified.

The Laidler-Landskroener Equation

Laidler and Landskroener (24) used Kirkwood's (25) and Kirkwood and Westheimer's (26) equation for the activity coefficient f_i of an arbitrary distribution of charges embedded in a sphere of radius b and dielectric constant D_i, submerged in a medium of dielectric constant D, and referred to a standard solvent of dielectric constant D_0 to derive a general equation for the rate of reaction between particles of charges z_A and z_B to yield a complex of charge $z_A + z_B$. The Kirkwood equation is

$$\ln f_i = \frac{1}{2kT} \sum_{n=0}^{\infty} \frac{(n+1)Q_n}{D_i b_i (2n+1)} \left[\frac{D_i - D}{(n+1)D + nD_i} - \frac{D_i - D_0}{(n+1)D_0 + nD_i} \right]$$

(2.68)

The principal requirement is that $D_i \ll D$, the actual dielectric constant of the medium. D_i was chosen as 2. Taking the standard state as that of a gas with $D_0 = 1$, $Q_0 = z_i^2 \varepsilon^2$, and $Q_1 = G_1 \varepsilon^2$, Eq. (2.68) can be written

$$\ln f_i = \frac{1}{2kT} \left[\frac{Q_0}{2b_i} \left(\frac{2-D}{D} - 1 \right) + \frac{Q_1}{b_i^3} \left(\frac{2-D}{2D+2} - \frac{1}{4} \right) \right]$$

$$= \frac{z_i^2 \varepsilon^2}{2kTb_i} \left(\frac{1-D}{D} \right) + \frac{3G_1 \varepsilon^2}{8kTb_i^3} \left(\frac{1-D}{D+1} \right).$$

(2.69)

Expanding the term $(1-D)/(D+1)$ yields

$$\frac{1-D}{D+1} = \left(\frac{1}{D} - 1 \right) \left(1 - \frac{1}{D} + \frac{1}{D^2} - \frac{1}{D^3} + ... \right).$$

(2.70)

This becomes, for D large,

$$\frac{1-D}{1+D} = \left(\frac{1}{D} - 1 \right) \left(1 - \frac{1}{D} \right) = \frac{2}{D} - \frac{1}{D^2} - 1 \simeq \frac{2}{D} - 1,$$

(2.71)

and this substituted into Eq. (2.69) gives

$$\ln f_i = \frac{z_i^2 \varepsilon^2}{2kTb_i} \left(\frac{1}{D} - 1 \right) + \frac{3G_i \varepsilon^2}{8kTb_i^3} \left(\frac{2}{D} - 1 \right).$$

(2.72)

Now for the reaction

$$A + B \; \rightleftarrows \; X \; \longrightarrow \; \text{Products},$$

$$\ln k' = \ln k_0' + \ln \frac{f_A f_B}{f_X}.$$

(2.73)

In this equation k' is the specific velocity constant for the reaction in a medium of dielectric constant D, k_0' is the specific velocity constant in a medium of reference dielectric constant D_0, and the f's are the activity coefficients of the respective species. In these considerations D_0 is taken as unity. Substitution of the respective f's from Eq. (2.72) into Eq. (2.73) yields, for the influence of dielectric constant on reaction rates,

$$\ln k' = \ln k_0' + \frac{\varepsilon^2}{2kT} \left(\frac{1}{D} - 1 \right) \left(\frac{z_A^2}{b_A} + \frac{z_B^2}{b_B} - \frac{(z_A + z_B)^2}{b_X} \right)$$

$$+ \frac{3\varepsilon^2}{8kT} \left(\frac{2}{D} - 1 \right) \left(\frac{G_A}{b_A^3} + \frac{G_B}{b_B^3} - \frac{G_X}{b_X^3} \right). \qquad (2.74)$$

The terms other than the G's have already been explained or are self-evident. G_A, G_B, and G_X are complex functions of the charges and structures of the respective species. In order to determine the charges, distances, and Legendre polynomials which make possible the evaluation of the G-factors, definite models of the reactant and complex species have to be assumed.

According to Eq. (2.74) a plot of $\log k'$ versus $1/D$ should yield a straight line of slope S, given by the expression

$$S = \frac{\varepsilon^2}{2.303(2kT)} \left[\frac{z_A^2}{b_A} + \frac{z_B^2}{b_B} - \frac{(z_A + z_B)^2}{b_X} \right.$$

$$\left. + \frac{3}{2} \frac{G_A}{b_A^3} + \frac{G_B}{b_B^3} - \frac{G_X}{b_X^3} \right]. \qquad (2.75)$$

For a univalent ion reacting with a dipolar molecule, $z_A = \pm 1$, $z_A^2 = +1$, $z_B = 0$, $G_A = 0$, and G_B is negligible. Under these circumstances Eq. (2.75) becomes

$$S = \frac{\varepsilon^2}{2.303(2kT)} \left(\frac{1}{b_A} - \frac{1}{b_B} - \frac{3}{2} \frac{G_X}{b_X^3} \right)$$

$$= \frac{\varepsilon^3}{9.212kTb_A} \left(\frac{2b_X^3 - 2b_Ab_X^2 - 3b_AG_X}{b_X^3} \right). \qquad (2.76)$$

Landskroener and Laidler point out that in order to evaluate b_X and G_X a model for the activated complex has to be constructed. Their models for the acid and base hydrolyses of esters are as follows:

(a)   (b)

where diagram (a) is the model for the acid hydrolysis and diagram (b) that for the base hydrolysis.

The model for the acid hydrolysis yields $G_X = 9.1 \times 10^{-14}$ sq cm and if b_A is taken as 1.7×10^{-8} cm, Eq. (2.76) becomes

$$S = \frac{1.07 \times 10^4}{b_X{}^3 T} (2b_X{}^3 - 3.4\, b_X - 47.31). \qquad (2.77)$$

The model for the base hydrolysis gives $G_X = 5.4 \times 10^{-16}$ sq cm, and if b_A is chosen as 1.4×10^{-8} cm, Eq. (2.76) becomes

$$S = \frac{1.30 \times 10^4}{b_X{}^3 T} (2b_X{}^3 - 2.8\, b_X{}^2 - 22.68). \qquad (2.78)$$

To test their theory Landskroener and Laidler plotted $\log k'$ versus $1/D$ for both the acid and base hydrolyses of esters. The slopes of the straight lines obtained were used to calculate b_X from Eq. (2.77) in the case of acid hydrolysis and from Eq. (2.78) in the case of basic hydrolysis. These values of b_X should be of the order of magnitude of a molecular dimension in the cases of conformity of theory with experiment. Table II lists the values of b_X for both acid and basic hydrolyses of esters. These values should be compared to the values of r for ester hydrolyses calculated using Eq. (2.62) and listed in Table II.

Laidler and Landskroener believe their equation to be more flexible and general and to take more proper cognizance of the charge distribution in the reactants and activated complex than in the instances of previous treatments of ion–dipolar molecule reactions. However their treatment requires a definite model of the activated complex for a reaction and conse-

TABLE II

VALUES IN ANGSTROMS OF b FUNCTION IN BOTH ACID AND BASE HYDROLYSES OF ESTERS

Ester	Solvent	Ionic strength	Temp. (°C)	Slope	b_X
		Acid hydrolysis			
Ethyl acetate	Dioxane–water	0.05	35.0	— 11	3.3
	Acetone–water	0.05	35.0	— 34	3.0
Methyl propionate	Acetone–water	0.02	35.3	— 31	3.0
		Base hydrolysis			
Ethyl acetate	Dioxane–water	0.05	25	— 24	2.5
	Acetone–water	0.05	25	— 32	2.4
	iso-PrOH–water	0.05	25	— 50	2.3
	tert-BuOH–water	0.05	25	— 61	2.2
	n-PrOH–water	0.05	25	— 74	2.1
	EtOH–water	0.05	25	— 96	2.0
	MeOH–water	0.05	25	—155	1.8

quently is rather difficult to apply. Also, model (b) for basic hydrolysis contains water in the complex, and there is no experimental evidence that water is involved in this process.

The Solvent and Energy of Activation

Equation (2.59) was derived by considering the effect of dielectric constant on the energy of activation for the reaction between an ion and a dipolar molecule, which effect can be calculated as a Coulombic energy of activation represented in Eq. (2.52).

Nathan and Watson (27) represented, by the empirical equation

$$E_s = E_u + K_1\mu + K_2\mu^2 ,$$ (2.79)

the energy of activation for a series of reactions between a common ion and aromatic molecules separately substituted with groups of dipole moment μ. In Eq. (2.79), E_s is the energy of activation of the substituted molecule and E_u the energy of activation of the unsubstituted molecule with the common ion.

Moelwyn-Hughes (28) ignores the quadratic term in μ and writes the energy of activation in terms of two separate linear equations:

$$E_s = E_u + K_m\mu$$ (2.80)

and

$$E_s = E_u + K_p\mu .$$ (2.81)

In these equations the gradient K_m refers to the *meta-* and the gradient K_p to the *para-*substituted molecules. The kinetics of somewhat similar molecules with an ion of opposite sign, studied by Ingold and Nathan (29) and by Evans, Gordon, and Watson (30), exhibit a similar linear relationship but with slopes of opposite sign.

Moelwyn-Hughes, in explaining these facts in terms of electrostatic principles, resolves the total energy of activation E into these components

$$E = E_0 + E_e(r_1, \theta_1, D_1) + E_e(r_2, \theta_2, D_2) ,$$ (2.82)

where E_0 is all the energy of activation except the two electrostatic contributions, $E_e(r_1, \theta_1, D_1)$ is the electrostatic energy arising from the interaction between the reactant ion and the dipole of the group which it is attacking, and $E_e(r_2, \theta_2, D_2)$ is the electrostatic energy arising from the interaction of the ion at the distance of reaction and the second dipole situated at some distance from the seat of action in the dipole. $E_e(r_1, \theta_1, D_1)$, which may be assumed to be constant for all members in the series, may be combined with E_0 to give E_u and, dropping the subscript, the equation

$$E_s = E_u + E_e(r, \theta, D)$$ (2.83)

can be written, where r is the distance between the center of the attacking ion and the center of the nonreacting dipole when the critical complex is formed between the ion and the dipole, θ is the angle between the line of centers and the polar axis of the substituted dipolar molecule, and D is the dielectric constant of the intervening medium, which, for the cases being considered, is principally the benzene ring. But

$$E_e(r, \theta, D) = -\frac{Nz\mu \cos \theta}{Dr^2} , \qquad (2.84)$$

and from Eqs. (2.83) and (2.84)

$$E_s = E_\mu - \frac{Nz\mu \cos \theta}{Dr^2} \qquad (2.85)$$

which for head-on alignment of ion and dipole becomes

$$E_s = E_\mu - \frac{Nz\mu}{Dr^2} . \qquad (2.86)$$

Thus the magnitude and sign of the constants in Eqs. (2.80) and (2.81) are explained. According to Eq. (2.86) the energy of activation plotted against the dipole moment of dipolar molecules reacting with ions should be linear with a negative slope when the reacting ion is positive in charge and with a positive slope when the ion is negative in charge. Compare these predictions with those of the Amis-Jaffé and the Amis theories.

Moelwyn-Hughes applies the theory to the alkaline hydrolyses of para-substituted ethyl benzoates (28, 29) and to the hydrogen ion catalyzed enolization of para-substituted acetophenones. Both plots of E_s versus μ were linear, but the slope of the plot for alkaline hydrolyses was the steeper of the two. This indicates that the hydroxyl ion which attacks the carbon atom of the carboxyl group in the saponification reaction gets nearer to the center of the substituted dipole than does hydronium ion which attacks the oxygen atom of the carboxyl group in the enolization reaction. Moelwyn-Hughes illustrates by means of indicated dipoles of substituted molecules how substitution in the benzene ring affects the velocities of chemical reactions.

However, as to the influence of solvent on reaction rates we are more concerned with the evaluation of the energy of activation E_u of the reaction between an ion and an unsubstituted polar molecule. Moelwyn-Hughes approaches the problem from the standpoint of Ogg and Polanyi's (31) adaptation of London's treatment to the problem of substitution, with optical inversion, at a saturated carbon atom. A mechanistic picture can be represented as follows:

$$(2.87)$$

where the solvated ion Y^-, at infinite distance from the dipolar-substituted methane molecule in the ground electronic and vibrational state, approaches the dipolar molecule in the direction indicated because of the nature of the ion–dipole interaction. In the critical complex the hydrogen and carbon atoms are coplanar. The energy of the covalent links are represented by the Morse function, and the energy of the interactions of the ions and dipoles are represented by the Born-Heisenberg function. The ion–dipole interactions include those between the ion and the solvent as well as those between the ion and reacting molecule. The computed values of the energies of activation for the iodide ion–methyl fluoride and the iodide ion–methyl bromide reactions were calculated to be 31.6 and 26.5 kcal/mole, respectively. The energy of the latter reaction is now known to be less than 26.5 kcal/mole. Experimentally, it has been found to be 18.3, 18.3, and 14.3 kcal/mole in in methanol, water, and acetone, respectively. In the complete methyl halide series reacting with hydroxyl ion, the observed energies of activation are not indicative of the strengths of the covalent bonds which are being broken. Glew and Moelwyn-Hughes (32) found that E_A was less for the methyl fluoride than for the methyl iodide reaction. Moreover Moelwyn-Hughes (28) points out that calculations of complete energy surfaces show that the energy of activation in the gas phase is zero for the reaction of the hydroxyl ion with alkyl halides. The solvent seems to greatly retard these reactions, so that extremely rapid reactions in the gas phase become measurable in solution. The hypothesis that the energy of activation applies, in the case of anions reacting with dipolar molecules, to the escape of the ion from its solvent sheath is explored by the above author. In this case, the reorientation and rearrangement of the solvent molecules solvating the ion would account for most of the energy of activation, and its magnitude would depend on the intrinsic properties of the ion and solvent, though it would be modified by the polar molecule with which the ion reacts during or following its escape from the solvation shell.

Let c be the number of molecules of solvent normally coordinated with each ion or molecule of solute, and let Φ be the energy of one solute–solvent interaction. The concentration n^* of solute molecules which are deficient by one solvent molecule can be written approximately in terms of the Boltzmann equation:

$$\frac{n^*}{n} = \frac{\exp\left[-(c-1)\,\Phi/kT\right]}{\exp\left(-c\Phi/kT\right)} = \exp\left(\Phi/kT\right) = \exp\left(-E'/kT\right), \quad (2.88)$$

where n is the total concentration of solute and E' is the energy required to remove one solvent molecule from the solvation sheath of one solute molecule. If reaction occurs upon collision of an ion deficient by one solvent molecule of solvation with a solute molecule deficient by one solvent molecule of solvation, the reaction rate according to Moelwyn-Hughes is

$$-\frac{dn}{dt} = Zn_1^* n_2^* = Zn_1 n_2 \exp\left[-(E_1' + E_2')/kT\right], \qquad (2.89)$$

where Z is the collision frequency and the energy of activation is

$$E = E_1' + E_2'. \qquad (2.90)$$

E_1' and E_2' have to be evaluated from known properties of the reactants and presumably of the particular solvent. Such calculations have been encouraging according to Moelwyn-Hughes, who explains, on the basis of this hypothesis, the difference between the energies of activation of hydrogen ion and hydroxyl ion catalysis of esterification as mainly due to the difference in the heats of hydration of these ions. The difference in these energies of activation is roughly equal to the difference in the heats of hydration divided by the coordination number.

Thus the energies of reorientation and rearrangement of solvent molecules around reactant particles and the heat of solvation of reactant species are important in explanations of rates of reaction.

Such a mechanism would undoubtedly evince large specific effects arising from selective solvation of either or both reactant species by one component of a mixed solvent.

As pointed out by Moelwyn-Hughes (28) ion–dipolar molecule reactions show more regularity than other types of reactions in solution. The rates in solution or certain second-order ion–dipolar molecule reactions may differ greatly depending on the ion and dipole reactants selected and on the solvent used. In such cases the Arrhenius frequency factor A in the equation for the second-order velocity constant, namely, $k_2' = A \exp(- E_A/RT)$, may remain practically constant, and the difference in rates arise principally from variations in E_A. The above author cites the case of the study of catalytic chlorination of ethers of the type ROC_6H_4X by Brynmor Jones (33) in which the rates, depending on the nature of R and X, varied at 20°C by a factor of 3300. These differences were accounted for by a variation in E_A of 4800 cal while A remained constant. Since E_A is dependent so markedly and in so many aspects on the solvent, we can see why rates would depend strongly on the solvent.

Benson (34) points out that the real difficulty in applying the simple electrostatic models to ion-dipolar molecule reactions is that the Coulombic term is of the same order of magnitude as the difference in free energies of hydration of the ion A^{z_A} and the transition complex X^{z_A}. He suggests that a fairer model for the formation of the transition complex might result from considering it as resulting from the displacement in the solvent sheath of the ion of a solvent molecule of dipole moment μ_s by a reactant solute molecule of dipole moment μ_B.

Neglecting dipole–dipole interactions this procedure would give, for the free energy of formation of the transition state,

$$F_\mu = \frac{z_A \varepsilon \mu_B \cos \theta}{D r_B^2} \left(1 - \frac{\mu_s}{\mu_B} \frac{r_B^2}{r_s^2} \right). \tag{2.91}$$

From this equation the effect of the dielectric constant is tied up with μ- and r- the dipole moment and the distance of closest approach of a solvent molecule to the ion. For $\mu_s/r_s^2 = \mu_B/r_B^2$ there would be no effect of dielectric constant on the rate since F_μ would be zero while depending on which is greater, opposite effects may be predicted.

It would seem, to the present author, that such a model would give rise to pronounced specific effects in mixed solvents, if there were selective solvation by one of the solvent components.

REFERENCES

1. K. J. Laidler and H. Eyring, *Ann. N. Y. Acad. Sci.* **39**, 303 (1940).
2. E. Hückel, *Physik. Z.* **26**, 19 (1935).
3. J. G. Kirkwood, *J. Chem. Phys.* **2**, 351 (1934).
4. P. Debye and J. McAulay, *Physik, Z.* **26**, 22 (1925).
5. C. C. J. Fontein, *Rec. Trav. Chim.* **47**, 635 (1928).
6. E. S. Amis and G. Jaffé, *J. Chem. Phys.* **10**, 598 (1942).
7. L. C. Bateman, M. G. Church, E. D. Hughes, C. K. Ingold, and N. A. Taher, *J. Chem. Soc.* **1940**, 979.
8. L. Onsager, *J. Am. Chem. Soc.* **58**, 1486 (1936).
9. E. S. Amis and F. C. Holmes, *J. Am. Chem. Soc.* **63**, 2231 (1941).
10. H. E. Dyas and D. G. Hill, *J. Am. Chem. Soc.* **64**, 236 (1942).
11. E. S. Amis, *J. Chem. Educ.* **30**, 351 (1953).
12. E. S. Amis, *J. Chem. Educ.* **29**, 337 (1952).
13. J. E. Quinlan and E. S. Amis, *J. Am. Chem. Soc.* **77**, 4187 (1955).
14. G. Akerlof, *J. Am. Chem. Soc.* **48**, 3046 (1926); **49**, 2960 (1927); **50**, 1272 (1928).
15. E. Landt, *Naturwissenshaften* **22**, 809 (1934).
16. E. S. Amis and G. Jaffé, *J. Chem. Phys.* **10**, 646 (1942).
17. E. S. Amis, G. Jaffé, and R. T. Overman, *J. Am. Chem. Soc.* **66**, 1823 (1944).
18. L. P. Hammett, "Physical Organic Chemistry," p. 344. McGraw-Hill, New York, 1940.
19. A. A. Frost and R. G. Pearson, "Kinetics and Mechanism," p. 283-96. Wiley, New York, 1953.
20. W. E. Nelson and J. A. V. Butler, *J. Chem. Soc.* **1938**, 957.
21. W. J. Broach and E. S. Amis, *J. Chem. Phys.* **22**, 39 (1954).
22. E. S. Amis, *Anal. Chem.* **27**, 1672 (1955).
23. G. J. Nolan and E. S. Amis, *J. Phys. Chem.* **65**, 1556 (1961).
24. K. J. Laidler and P. A. Landskroener, *Trans. Faraday Soc.* **52**, 200 (1956).
25. J. G. Kirkwood, *J. Chem. Phys.* **2**, 351 (1934).
26. J. G. Kirkwood and F. H. Westheimer, *J. Chem. Phys.* **6**, 506 (1938).
27. W. S. Nathan and H. B. Watson, *J. Chem. Soc.* **1933**, 217; **1933**, 890.
28. E. A. Moelwyn-Hughes, "Physical Chemistry," Chapter XXIV. Pergamon Press, New York, 1957.
29. C. K. Ingold and W. S. Nathan, *J. Chem. Soc.* **1936**, 222.
30. D. P. Evans, J. J. Gordon, and H. B. Watson, *J. Chem. Soc.* **1937**, 1430.
31. R. A. Ogg, Jr. and M. Polanyi, *Trans. Faraday Soc.* **31**, 482 (1935).
32. D. N. Glew and E. A. Moelwyn-Hughes, *Proc. Roy. Soc.* (*London*) **A211**, 254 (1952).
33. B. Jones, *J. Chem. Soc.* **1928**, 1006; **1928**, 3073.
34. S. W. Benson, "The Fundations of Chemical Kinetics," Chapter XV. McGraw-Hill, New York, 1960.

CHAPTER III

SOLVENT EFFECTS ON DIPOLAR MOLECULE-DIPOLAR MOLECULE REACTIONS

THEORETICAL

Introduction

The electrical forces between dipolar molecules, while smaller than those between either ions and ions or between ions and dipolar molecules, are nevertheless significant and must be accounted for in any comprehensive treatment of factors governing rates of reaction between electrically unsymmetrical molecules. Since there are dipole–dipole interactions, the solvent, through its dielectric constant, will influence the forces which such molecules exert upon each other, and hence their ability to contact each other and undergo chemical change. This chapter will deal with some approaches that have been made to account theoretically for the influence of the solvent, in so far as its dielectric constant is concerned, on the rates of reaction between dipolar molecules. Because of their relatively small magnitude, dielectric constant effects on the electrostatics of reactions between dipolar molecules will tend to be more easily obscured by specific solvent and structural effects, which in themselves may be difficult or impossible to elucidate.

Laidler and Eyring's Equation

In the case of dipolar substances van der Waals forces cannot, except as a very rough approximation, be neglected. Laidler and Eyring (1), however, believed that dipolar forces are somewhat stronger than nonelectrostatic ones in the case of strongly polar molecules, and treated certain reactions involving fairly strong dipoles in terms of electrostatic dipolar forces. To do this they used Kirkwood's (2) expression for the free energy of transfer of a strong dipole of moment μ from a vacuum to a medium of dielectric constant D. For a symmetrical charge distribution within a molecule, this expression for a molecule of radius a is

$$\Delta F = kT \ln \beta = -\frac{\mu^2}{a^3} \left[\frac{D-1}{2D+1} \right]. \tag{3.1}$$

Including a nonelectrostatic term, this becomes

$$\Delta F = kT \ln \beta = -\frac{\mu^2}{a^3} \left[\frac{D-1}{2D+1} \right] + \Phi. \tag{3.2}$$

Bell (3) has shown that for low dielectric constant media Eq. (3.1) cannot be applied. Laidler and Eyring point out, however, that the expression is very useful in dealing with fairly concentrated solutions of polar molecules in which electrostatic forces are predominant. Martin and co-workers (4–8) have plotted the logarithms of the activity coefficients versus $(D-1)/(2D+1)$ for various polar substances in solution and obtained satisfactory straight lines of reasonable slopes. Deviation from linearity were found only at low dielectric constants.

Using the bimolecular reaction

$$A + B \longrightarrow M^* \longrightarrow X + Y \tag{3.3}$$

and applying Eq. (3.2), the equation for the specific velocity constant,

$$\ln k' = \ln \left(\varkappa \frac{kT}{h} K_0^* \right) - \frac{1}{kT} \frac{D-1}{2D+1} \left[\frac{\mu_A^2}{a_A^3} + \frac{\mu_B^2}{a_B^3} - \frac{\mu_{M^*}^2}{a_{M^*}^3} \right]$$

$$+ \frac{\Phi_A + \Phi_B - \Phi_{M^*}}{kT}, \tag{3.4}$$

can be obtained. In this equation, D is the dielectric constant of the final solution formed, and for dilute solutions D is effectively the dielectric constant of the pure solvent.

Provided the nonelectrostatic terms are negligibly small, a plot of $\ln k'$ versus $(D-1)/(2D+1)$ should give a straight line. Laidler and Eyring plot the above function using the dielectric constants of the solvents for quaternary ammonium salt formations at 29°C in alcohol–benzene mixtures, for the acid hydrolysis of ethyl orthoformate [$H_2O + HC(OEt)_3$], the alkaline hydrolysis of ethyl benzoate [$H_2O + PhCOOEt$], and the water hydrolysis of tertiary butyl chloride [$H_2O + (CH_3)_3CCl$], the last three in ethanol–water mixtures. These authors argue that, from the standpoint of the influence of the solvent, ester hydrolysis should be regarded as a reaction between the two dipoles, water and ester.

The plot of the data for the quaternary ammonium salt formation in alcohol–benzene mixtures is shown in Fig. 1. The plots are linear as expected

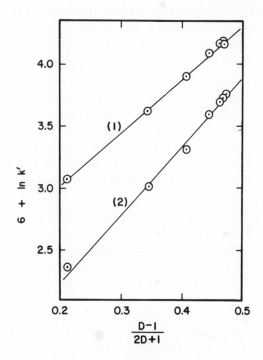

FIG. 1. The logarithm of the specific velocity constant versus D-1/$2D$ + 1 for quaternary ammonium salt formations in ethanol–benzene at 29°C. (1) Triethylamine and benzyl bromide. (2) Pyridine and benzyl bromide. The ⊙ symbols represent experimental data. The lines through the data are straight as expected by theory.

from theory. For the same reactions in benzene–nitrobenzene solvents, however, similar plots show considerable curvature. In solutions rich in nitrobenzene the reactions are much faster than theory predicts. Specific solvent effects are common in quaternary ammonium salt formation reactions. For example, the triethylamine–ethyl iodide reaction, studied by

Grimm, Ruf, and Wolff (9) in a variety of solvents, gave rates and dielectric constants that are not even in the same order. Menschutkin (10) did point out that the reaction does tend to be accelerated by solvents of high dielectric constant, so that there is some correlation between the two. Laidler and Eyring explained the deviations in terms of the nonelectrostatic contributions Φ, to which the most important contribution is probably the solvating power of the solvents. In nitrobenzene the quaternary salts are soluble and ionized (11, 12), while in benzene they are only slightly soluble. The solutions of amines and iodides probably do not show large deviations from ideality. Stearn and Eyring (13) show that the activated complex in quaternary salt formations, while not ionic, has a large dipole moment and therefore approaches the salt in general properties. Thus, in nitrobenzene the complex is perhaps selectively solvated by this highly polar component of the solvent; this results in a low activity coefficient of the complex and hence a high velocity of reaction. In terms of the Φ-functions, Φ_A and Φ_B may be sensibly normal in nitrobenzene-containing solvents, while Φ_M is very small due to solvation and the rate of reaction rapid. On the other hand, benzene may have the opposite effect to nitrobenzene. The product of the reaction in benzene is low, and hence the solubility of the activated complex in benzene is, perhaps, low also. Thus β_{M^*} and Φ_{M^*} would be large and the reaction slow in benzene. Even in benzene, however, the solvent must effect some type of stabilization of the partially ionized complex, since the Menschutkin reaction will not take place in the gas phase.

For the above theory to be applied widely and quantitatively, data on activity coefficient of reactants and complexes are required. Such data, especially in the case of the complex, are meager.

The Hydrolyses of Esters

Laidler and Eyring apply dipolar molecule–dipolar molecule rate theory to acid and alkaline hydrolyses of esters on the basis of the argument that, from the standpoint of solvent influence, an ester hydrolysis should be considered as a reaction between dipolar molecules, the ester and water molecules. They contend that, if the reactions were between ions and neutral molecules, the rate of reaction might be expect to be retarded by ionizing solvents. They quote literature (14, 15) to show that both the acid and alkaline hydrolyses are accelerated by such solvents. They substantiate their theory by the mechanisms of the two reactions based on the scheme proposed by Lowry (16). For acid hydrolysis the mechanism is

$$\text{HOH} + \text{R}-\overset{\overset{\text{O}}{\|}}{\text{C}}-\text{OR}' + \overset{\downarrow}{\text{H}^+} \longrightarrow \text{R}-\overset{\overset{-\text{O}}{|}}{\underset{\underset{+\text{OH}}{|}}{\text{C}}}-\overset{\overset{\text{H}}{|}}{\text{O}}-\text{R} \longrightarrow \text{RCOOH} + \text{HOR}' + \text{H}^+,$$

$$\text{(3.5)}$$

$$+\ddot{\text{H}}$$

and for the alkaline hydrolysis

$$\text{HO}^- + \text{R}-\overset{\overset{\text{O}}{\|}}{\text{C}}-\text{OR}' + \overset{\downarrow}{\text{H}}\text{OH} \longrightarrow \text{R}-\overset{\overset{-\text{O}}{|}}{\underset{\underset{+\text{OH}}{|}}{\text{C}}}-\overset{\overset{\text{H}\ \ddot{\,}\,\text{OH}^-}{|}}{\text{O}}-\text{R}' \longrightarrow \text{RCOOH} + \text{HOR}' + \text{OH}^-,$$

$$\text{(3.6)}$$

where the arrows represent the movements of electrons.

For the acid hydrolysis the equation for the specific velocity constant is written

$$k' = \varkappa \, \frac{kT}{h} \, K_0{}^* \, \frac{\beta_{\text{H}_2\text{O}}\beta_{\text{ester}}\beta_{\text{H}^+}}{\beta_{\text{complex}}}, \tag{3.7}$$

and the actual rate is written

$$v = \varkappa \, \frac{kT}{h} \, K_0{}^* \, \frac{\beta_{\text{H}_2\text{O}}\beta_{\text{ester}}\beta_{\text{H}^+}}{\beta_{\text{complex}}} \, C_{\text{H}_2\text{O}}C_{\text{ester}}C_{\text{H}^+}, \tag{3.8}$$

and the corresponding equations for alkaline hydrolysis are

$$k' = \varkappa \, \frac{kT}{h} \, K_0{}^* \, \frac{\beta_{\text{OH}^-}\beta_{\text{ester}}\beta_{\text{H}_2\text{O}}}{\beta_{\text{complex}}} \tag{3.9}$$

and

$$v = \varkappa \, \frac{kT}{h} \, K_0{}^* \, \frac{\beta_{\text{OH}^-}\beta_{\text{ester}}\beta_{\text{H}_2\text{O}}}{\beta_{\text{complex}}} \, C_{\text{OH}^-}C_{\text{ester}}C_{\text{H}_2\text{O}}. \tag{3.10}$$

Laidler and Eyring attribute the catalyzing effect of the ions to the involvement of the ions in the reactions, although they point out that the ions may have some effect on the activity coefficients of the activated complexes. The solvent effect is explained on the hypothesis that the complex is more polar than the reactants, even though one of the reactants is an ion. The complex is not considered as a spherical ion, and the hydrogen or hydroxyl ion which is about to split off from the decomposing complex is probably fully charged in the complex. When the complex is formed an extra separation of charge takes place with practically no increase of the radius of the hydrogen or hydroxyl ion. Thus an ionizing, high dielectric constant, solvent would promote the formation of the more polar complex, thus increasing its concentration and speeding up the reaction rate.

It is hard to reconcile the above explanation of Laidler and Eyring with the previously quoted statement by these authors that, from the standpoint of solvent influence, an ester hydrolysis reaction should be considered as a reaction between two dipoles, the ester and the water molecule. This statement hardly coincides with the explanation based on the complex's being more polar than the reactants, one of which is an ion.

In Chapter II, it will be recalled, ester hydrolyses were treated as ion–dipolar molecule reactions. In the case of alkaline hydrolysis, in these correlations of empirical data and theory, the rate-controlling step would be that between a negatively charged ion and a dipolar molecule, though the over-all mechanism might involve several steps. These steps could be (17)

$$\tag{3.11}$$

From the standpoint of dielectric constant on the rate, either step 1 or some other step involving the reaction of a dipolar molecule with a negatively charged ion is rate controlling.

For the dielectric constant effect on the acid hydrolysis of esters, Hockersmith and Amis (18) use the mechanism proposed by Bender as satisfactory in correlating experiment with theory. This mechanism is

$$\tag{3.12}$$

The first step would be rate controlling. Any similar mechanism in which the rate-controlling step is the reaction between a positive ion and a dipolar molecule would be satisfactory.

The Amis Equation

Amis (19) assumed that for dipole–dipole interactions the dielectric constant's influence on the rate is given by the following equations, where all powers higher than the second of distances of separation of charge centers in dipolar molecules have been neglected. For the dipolar reactants the Coulombic energy E_c is

$$E_c = \frac{2\mu_1\mu_2 \cos\theta_1 \cos\theta_2}{Dr^3} + \frac{\mu_1\mu_2 \sin\theta_1 \sin\theta_2}{Dr^3}. \tag{3.13}$$

and for head-on alignment

$$E_c = \frac{2\mu_1\mu_2}{Dr^3}. \tag{3.14}$$

The difference in Coulombic energy at two different dielectric constants D_1 and D_2 is, from this restricted equation,

$$\Delta E_c = \frac{2\mu_1\mu_2}{r^3}\left(\frac{1}{D_2} - \frac{1}{D_1}\right), \tag{3.15}$$

and the energy of activation for the dipole–dipole reaction at dielectric constant D_2 can be related to the energy of activation at dielectric constant D_1 by the equation

$$E_{D_1} = E_{D_2} + \Delta E_c. \tag{3.16}$$

Now, using the energy per mole,

$$k'_{D_2} = Z \exp\left(-\frac{E_{D_2}}{RT}\right) \tag{3.17}$$

and

$$\ln k'_{D_2} = \ln Z - \frac{E_{D_2}}{RT} \tag{3.18}$$

$$= \ln Z - \frac{E_{D_1}}{RT} - \frac{\Delta E_c}{RT}, \tag{3.19}$$

where ΔE_c is the Coulombic energy per mole. But

$$\ln k'_{D_1} = \ln Z - \frac{E_{D_1}}{RT}. \tag{3.20}$$

Therefore,

$$\ln k'_{D_2} = \ln k'_{D_1} - \frac{\Delta E_c}{RT}, \tag{3.21}$$

and substituting ΔE_c from Eq. (3.15) into Eq. (3.21) yields

$$\ln k'_{D_2} = \ln k'_{D_1} - \frac{2\mu_1\mu_2 N}{RTr^3}\left(\frac{1}{D_2} - \frac{1}{D_1}\right). \tag{3.22}$$

If dielectric constant D_2 is taken as a reference value of infinite magnitude and D_1 is taken as any dielectric constant D, we have, from Eq. (3.22) and remembering that $N/R = 1/k$.

$$\ln k_\infty' = \ln k_D' + \frac{2\mu_1\mu_2}{kTDr^3}, \tag{3.23}$$

or the velocity constant k_D' at any dielectric constant is related to the velocity constant k_∞' at the infinite or reference value of the dielectric constant by the equation

$$\ln k_D' = \ln k_\infty' - \frac{2\mu_1\mu_2}{kTDr^3}. \tag{3.24}$$

In these equations k is the Boltzmann gas constant and μ_1 and μ_2 are the dipole moments of the respective dipolar reactants.

Equation (3.24) would have a linear relationship when $\ln k_D'$ is plotted versus $1/D$, so long as the approximations made in this derivation are valid, and so long as the simple moments *in vacuo* of the dipoles can be used. The slope of the plot should be negative and should permit the calculation of a reasonable value of r, the distance of approach for the two dipoles to react. This distance $r = r_A + r_B$ should be the sum of the radii of the dipoles and should be a molecular dimension.

Figure 2 shows a plot of $\log k_D'$ versus $1/D$ for the reaction between the dipolar reactants water and tertiary butyl chloride in water and water–ethanol of various compositions at 25°C. The data are those of Hughes (20). The scatter of points is perhaps due to lack of precision. Equating the slope of the line to $- 2\mu_1\mu_2/kTr^3$, as required by Eq. (3.24), the value of r is found to be 0.75 Å. The dipole moments for water and tertiary butyl chloride were taken as 1.84 and 2.10 D. U. (21), respectively.

The value of r is low, perhaps by a factor of 4 or 5, but the value found is gratifying considering the approximations made in deriving Eq. (3.24), its simplicity, and the ease of visualizing its predictions.

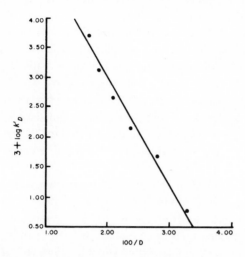

FIG. 2. The reaction of tertiary butyl chloride with water. Dots represent experimental data; slope of straight line through data was used to calculate the reaction parameter, r.

From these electrostatic considerations the mechanism must show the slow step as a dipolar molecule reacting with a dipolar molecule and could be step 1 given in the following mechanism:

$$CH_3-\underset{\underset{CH_3}{|}}{\overset{\overset{CH_3}{|}}{C}}-Cl + HOH \underset{(1)}{\rightleftarrows} CH_3 \underset{\underset{CH_3}{|}}{\overset{\overset{CH_3}{|}}{C}}{}^+ + Cl^-H_2O$$

$$(2)\downarrow\uparrow H_2O$$

$$CH_3-\underset{\underset{CH_3}{|}}{\overset{\overset{CH_3}{|}}{C}}-OH_2{}^+$$

$$(3)\downarrow H_2O$$

$$CH_3-\underset{\underset{CH_3}{|}}{\overset{\overset{CH_3}{|}}{C}}-OH + H_3O^+$$

The plot of the data by Laidler and Eyring, according to the requirements of Eq. (3.4), did not show markedly better precision than that in Fig. 2. A quantitative test of the predictions of Eq. (3.4) would be more difficult to make, however, since the moments and radii of both reactants and complex would have to be known.

Entelis et al. (22) studied, using spectrophotometry, the hydrolysis of diacid chlorides of phthalic and terephthalic acids in water–dioxane mixtures. For both reactions the rate was first order with respect to each reagent and second order over-all. The rates increased with the increase of the polarity of the medium according to the Kirkwood relationship for reactions of two dipolar molecules. From the agreement of data with theory, dipole moments of 6.85 and 6.95 D. U. were calculated for the activated complexes in the case of the diacid chlorides of phthalic and terephthalic acids, respectively.

Moelwyn-Hughes (23) points out that, if the energy of activation between two dipolar molecules contains a Coulombic term $E_c = \mu_1\mu_2/Dr^3$, the energy of activation should vary linearly as $\mu_1\mu_2$ for a series of chemically comparable reactions measured in a common solvent at the same temperature, and the apparent energy of activation should vary as $(1 - LT)/D$ for a given reaction measured in a variety of solvents of different dielectric constants at a fixed temperature. He states that both these and other consequences of electrostatic theory have been confirmed for reactions between molecules in solution. The magnitude of these effects helps in establishing the extent of Coulombic energy in the formation of the critical complex and in interpreting the other components of the energy of activation. It is this contribution to the energy of activation that we have quantitatively related to the rates of dipolar molecule–dipolar molecule reactions in Eq. (3.24).

Benson (24) believes that in the case of dipolar molecule–dipolar molecule reaction the electrostatic interactions between the reactant species are of the order of the van der Waals forces in magnitude. He thinks it is grossly oversimplifying matters to neglect the van der Waals interactions in the case of such reactants and to throw the entire behavior of solutions on the influence of the dielectric constant. He questions whether the correlation between the dielectric constant of the solvent and the rate of reaction between dipolar molecules can throw much light either on the transition state complex or the theory of solutions. Benson suggests that a detailed molecular model for the solution using only near-neighbor interactions would be no more complex and would probably yield more interesting information. For such a model the only parameters would be the dipole moments and the radii of the solute and solvent species.

REFERENCES

1. K. J. Laidler and H. Eyring, *Ann. N. Y. Acad. Sci.* **39**, 303 (1940).

2. J. G. Kirkwood, *J. Chem. Phys.* **2**, 351, (1934).

3. R. P. Bell, *Trans. Faraday Soc.* **31**, 1557 (1935).

4. A. R. Martin and B. Collie, *J. Chem. Soc.* **1932**, 2658.

5. A. R. Martin, *Trans. Faraday Soc.* **30**, 759 (1934).

6. A. R. Martin and C. M. George, *J. Chem. Soc.* **1933**, 1413.

7. A. R. Martin, *Trans. Faraday Soc.* **33**, 191 (1937).

8. A. R. Martin and A. C. Brown, *Trans. Faraday Soc.* **34**, 742 (1938).

9. H. G. Grimm, H. Ruf, and H. Wolff, *Z. physik. Chem.* **B13**, 301 (1931).

10. N. Menschutkin, *Z. physik. Chem.* **6**, 41 (1890).

11. N. Bjerrum and E. Jozefawicz, *Z. physik. Chem.* **A159**, 194 (1932)

12. P. Walden, *Bull. Acad. Sci. Petersburg* **7**, 427 (1913).

13. A. E. Stearn and H. Eyring, *J. Chem. Phys.* **5**, 113 (1937).

14. H. S. Harned and N. T. Samaras, *J. Am. Chem. Soc.* **54**, 9 (1932).

15. R. A. Fairclough and C. N. Hinshelwood, *J. Chem. Soc.* **1937**, 538.

16. See W. A. Waters, "Physical Aspects of Organic Chemistry," Chapter 13. Van Nostrand, New York, Princeton, New Jersey, 1950.

17. E. S. Amis, *Anal. Chem.* **27**, 1672 (1955)

18. J. L. Hockersmith and E. S. Amis, *Anal. Chim. Acta* **9**, 101 (1953).

19. E. S. Amis, *J. Chem. Educ.* **30**, 351 (1953).

20. E. D. Hughes, *J. Chem. Soc.* **1935**, 255.

21. R. H. Wiswall, Jr., and C. P. Smyth, *J. Chem. Phys.* **9**, 356 (1941).

22. S. G. Entelis, R. P. Tiger, E. Ya. Nevel'skii, and I. V. Epel'baum, *Izv. Akad. Nauk SSSR, Otd. Khim. Nauk* **1963**, 245.

23. E. A. Moelwyn-Hughes, "Physical Chemistry," Chapter XXIV. Pergamon Press, New York, 1957.

24. S. W. Benson, "The Foundation of Chemical Kinetics," Chapter XV. McGraw-Hill, New York, 1960.

CHAPTER IV

THE INFLUENCE OF THE SOLVENT ON ELECTRON EXCHANGE REACTIONS

THEORETICAL

Introduction

In organic reactions ionic mechanisms generally involve the transfer of atoms or ions in the individual steps. Some of these steps, depending on the charge of the species transferred, may correspond to oxidation–reduction. In these cases there is no exchange of charge due to electron transfer.

Oxidation–reduction reactions, such as

$$2Fe^{3+} + Sn^{2+} \rightleftarrows 2Fe^{2+} + Sn^{4+} ,$$

which occur in the field of inorganic chemistry are ordinarily referred to as electron-exchange or electron-transfer reactions.

Theoretical equations have been derived to explain the effect of charge and size of the reacting ions and the temperature and dielectric constant of the medium upon the specific velocity constant of electron-exchange reactions between ions. These theories include those of R. J. Marcus, B. J. Zwolinski, and H. Eyring (1, 2); J. Weiss (3); R. A. Marcus (4a, b, c); and Edward S. Amis (5).

The Marcus, Zwolinski, and Eyring Theory

Marcus, Zwolinski, and Eyring (1) have proposed an electron-tunneling hypothesis for electron-exchange reactions. The possibility of the penetration of energy barriers by the electron will be mentioned again later. This penetration of energy barriers by the electron is referred to as tunneling. When, in the electron-exchange reactions, the reactants and products are identical, the standard free energy change is zero. Usually radioactive tracers are used to follow these reactions, since the tracer concentration in the two valence states can be determined, and hence the rate of exchange between the two states can also be determined.

Zwolinski, Marcus, and Eyring (2) give a more detailed discussion of the electron-tunneling hypothesis. These authors believe that the application of the Franck-Condon principle to the slow readjustment of the solvation sphere of two ions which have exchanged an electron is based on a misinterpretation of the principle. They likewise question that it is this readjustment of hydration spheres which requires energy of activation. Their opinion is that it is the rearrangement of electronic structure of the interacting complex ions which requires considerable energy and which they designate as rearrangement free energy ΔF_r. The rearrangement of electronic structures is necessitated perhaps since the principle of conservation of energy requires energetic equality of the electronic states of the activated complex during electron transfer. The rearrangement is also necessitated from the Franck-Condon principle that, since the nuclei do not move during the electronic transition, the electronic states of the reactants must be made equal before electron transfer can occur, since the transitions are found to be radiationless. The value for ΔF_r will be much greater for ions of greater dissimilarity of structure in the coordination shells as $Fe(aq)^{2+}$ and $Fe(aq)^{3+}$ and less for ions of greater similarity in these shells as $Fe(CN)_6^{3-}$ and $Fe(CN)_6^{4-}$. The values of ΔF_r of the hydration shells of negative ions is so small that the electron transfer is very, and in many cases immeasurably, fast.

Rabinowitch (6) has dealt with the subject of electron-transfer spectra and their photochemical effects. These absorption bands arise from the reduction of a cation and the oxidation of an anion, as in the case of the iron-halogen complexes. The spectra of all complex ions investigated in the far ultraviolet give bands which from their intensity and position, can be interpreted as electron-transfer bands. While these types of electron-transfer processes are not radiationless, the processes involve transfer of electrons within complexes and not between two entities only instantaneously in reaction configuration. The two entities may be of the same charge sign.

Complexing can be seen to have two effects in oxidation–reduction reactions, or electron-transfer reactions of the type being discussed here. The charge product of two reactant ions may be reduced and the barrier to their approach decreased by using a complexing agent of opposite sign. Complexing may also affect the activation free energy of rearrangement, this effect being greater when both ions are complexed. The effect essentially determines if it will be very fast or very slow, corresponding to similarity or dissimilarity of the complex structure of the two ions.

Two possible paths are indicated for electron-exchange reactions. One has a low energy of activation and a negative entropy of activation, that is, a low frequency constant. The other has a higher energy of activation and a positive entropy of activation, that is, a high frequency constant. Some reactions may proceed by either path.

Marcus, Zwolinski, and Eyring explain these data on the basis of media effects and the hypothesis of electron tunneling. The media effects considered include dielectric constant, ionic strength, electrostatic charge effects, and the effects of solvation.

The electron-tunneling hypothesis assumes that a determining factor in regulating electron-exchange reactions in solutions is the probabilty for the penetration of an electron through a barrier. The specific velocity constant k' was estimated using the equation

$$k' = \varkappa_e \varkappa_a \frac{kT}{h} \exp\left(\frac{\varDelta S}{R}\right) \exp\left(-\frac{\varDelta H}{RT}\right) = \varkappa_e k_0' , \qquad (4.1)$$

where $\varkappa_e$ is the product of the average chance of the electron penetrating the barrier per electronic vibration into the electronic vibrations and k_0' in the number of ionic collisions represented by the product of the remaining terms between the equality signs in Eq. (4.1). $\varkappa_a$ is the ordinary atomic or nuclear transmission coefficient, k the Boltzmann constant, h Planck's constant, T the absolute temperature, R the molar gas constant, $\varDelta S$ the entropy of activation, and $\varDelta H$ the enthalpy of activation.

A whole set of values of the potential energy of the transition electron, varying with the potential along the electronic coordinate, exists for each point along the atomic reaction coordinate. One seeks the best distance along both coordinates where there will be a maximum rate of electron transfer. To find the best distance, one maximizes the rate constant so that it represents the best possible compromise between close approach of the reactant ions with high energy of activation and farther distances of approach of reactant ions with accompanying large resistance to electron transfer.

If it is assumed that only two factors contribute to the free energy of activation of k_0' namely, $\varDelta F_{\text{rep}}$ arising from the Coulombic interaction of the charges on two reacting ions, and $\varDelta F_r$ representing the energy required for the rearrangement of the coordination and solvation shells of the reacting ions then, since $\varkappa_a$ is probably close to unity, the apparent over-all free energy of activation ΔF_{app} is

$$\Delta F_{\text{app}} = - RT \ln \varkappa_e + \Delta F_{\text{rep}} + \Delta F_r . \qquad (4.2)$$

ΔF_{rep} is difficult to calculate since it includes the activity coefficients of ions and the activated complex which, in turn, represent the contribution of external factors to the free energy of activation. Marcus, Zwolinski, and Eyring tried taking ΔF_{rep} as constant. For collisions that make important contributions to the electron-transfer process, the apparent free energy ΔF_{app} will be a minimum.

An explicit expression for V, the height of the electronic potential barrier, has to be found before the limiting value of ΔF_{app} can be formulated. A triangular shape was assumed for the electronic potential barrier, and the probability for an electron penetrating this barrier was represented by the Gamow factor (7). This factor can be written

$$\varkappa_e = \exp \left[- \frac{8\pi}{3h} r_{ab} [2m(V - W)]^{1/2} \right], \qquad (4.3)$$

where V is the height of the electron barrier, W is the kinetic energy of the electron, r_{ab} is the width of the barrier at the height of penetration, m is the mass of the electron, and the other terms have been defined. The shape of the barrier will no doubt be between the two extremes of triangular and rectangular barriers. The choice of a triangular barrier simplifies algebraic manipulations.

The height of the electronic barrier was estimated using a smoothed potential function based on a simple one-dimensional electrostatic model. The model assumed the cations of respective charges $n_a\varepsilon$ and $n_b\varepsilon$ fixed at some distance r_{ab} with the electron of charge at a distance x from the charge $n_a\varepsilon$ and at a distance $(r_{ab} - x)$ from the cation of charge $n_b\varepsilon$. All are immersed in a medium of dielectric constant D. This model is given in Fig. 1.

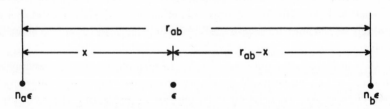

FIG. 1. Assumed model for the two cations and the electron.

The potential V' is given by

$$V' = \frac{\varepsilon^2 n_a n_b}{D r_{ab}} - \frac{\varepsilon^2 n_a}{Dx} - \frac{\varepsilon^2 n_b}{D(r_{ab} - x)}, \tag{4.4}$$

and the height of the potential barrier V is

$$V = V_m - V_0 \tag{4.5}$$

where V_0 is the zero-point energy of the electron and V_m is the maximum potential energy for the system consisting of the exchange electron interacting with the two cations. These potential energies are illustrated in Fig. 2.

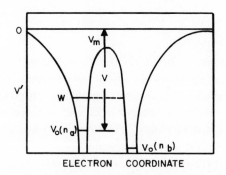

FIG. 2. Schematic model of the potential function.

To maximize with respect to x, one sets the derivative dV'/dx from Eq. (4.4) equal to zero. This gives

$$\frac{dV'}{dx} = 0 = \frac{\varepsilon^2 n_a}{Dx^2} - \frac{\varepsilon^2 n_b}{D(r_{ab} - x)^2}, \tag{4.6}$$

and therefore

$$\frac{r_{ab}^2 - 2r_{ab}x + x^2}{x^2} = \frac{n_b}{n_z} = \gamma^2. \tag{4.7}$$

Solving for the maximum value of x or x^*, one obtains

$$x^* = \frac{r_{ab} \pm r_{ab}\gamma}{1 - \gamma^2} = \frac{r_{ab}}{1 + \gamma}. \tag{4.8}$$

The negative sign in the numerator of Eq. (4.8) is used in the final step since $\gamma^2 = n_b/n_a$ is greater than unity and x would be negative if the positive sign were used.

Substituting the value of x^* from Eq. (4.8) into Eq. (4.4) gives the maximum potential V_m as

$$V_m = - \frac{\varepsilon^2}{Dr_{ab}} f(n), \qquad (4.9)$$

where the factor $f(n)$ is given by

$$f(n) = n_a[(1 + \gamma)^2 - n_b]. \qquad (4.10)$$

It is assumed that the zero-point energy is given by

$$V_0 = - \frac{\varepsilon^2 Z^*}{r_0}, \qquad (4.11)$$

where Z^* is the positive charge on the central atom of the complex ion and r_0 is the radius of the classical orbit of the electron which is transferring. Equation (4.11) applies to the complex ion whose central coordinated atom or ion has the smallest ionization potential and is, therefore, in general the lower valence form of the two ions. Thus $Z^* = n_a$ when all the coordinating groups about the central ion are neutral.

It is assumed that the radius of the electronic orbit is given by

$$r_0 = n^{*2}Q_0 \qquad (4.12)$$

where n^* is the effective principal quantum number as given by Rice (8) and Q_0 is the Bohr radius.

Therefore from Eqs. (4.5), (4.9), and (4.11), we have

$$V = V_m - V_0 \frac{\varepsilon^2 Z^*}{r_0} - \frac{\varepsilon^2 f(n)}{Dr_{ab}}. \qquad (4.13)$$

The kinetic energy of the electron is, from the virial theory,

$$W = \frac{\varepsilon^2 Z^*}{2r_0}, \qquad (4.14)$$

and from Eqs. (4.3) (4.13), and (4.14) we can obtain the following expression for the transmission coefficient $\varkappa_e$:

$$\varkappa_e = \exp - \left\{ \frac{8\pi}{3h} r_{ab} \left[2m\varepsilon^2 \left(\frac{Z^*}{2r_0} - \frac{f(n)}{Dr_{ab}} \right) \right] \right\}. \qquad (4.15)$$

From Eqs. (4.1), (4.2), and (4.15), the specific rate constant k' for the electron-exchange reaction is

$$
k' = \frac{kT}{h} \exp\left\{ - \frac{8\pi}{3h} r_{ab} \left[2m\varepsilon^2 \left(\frac{Z^*}{2r_0} - \frac{f(n)}{Dr_{ab}} \right) \right]^{1/2} \right.
$$
$$
\left. - \frac{\Delta F_r}{RT} - \frac{\varepsilon^2 n_a n_b}{RTDr_{ab}} \right\},
\tag{4.16}
$$

where it is assumed that the ordinary atomic or nuclear transmission coefficient $\varkappa_a$ is unity and the Coulombic free energy of activation ΔF_{rep} of Eq. (4.2) is equal to $\varepsilon^2 n_a n_b / Dr_{ab}$ of Eq. (4.16). The rest of the terms in Eq. (4.16) have been defined.

The above formulation of the rate constant expresses the competition between the "easy" path of low repulsive energy leading to the greater tunneling distance for the electron and the "hard" path of close approach where $\varkappa_e$ has an increased value approaching unity. At some definite value of r_{ab} these two tendencies will balance each other and, thus, result in a critical or "best" interionic distance for which the rate of electron exchange is a maximum. This resistance to electron exchange prevents electron tunneling from occurring at great distances.

The critical value of the interionic distance in the activated state is found by finding the extremal value for the specific rate constant with respect to r_{ab}. To do this, Marcus, Zwolinski, and Eyring define the following dimensionless parameters:

$$
\alpha = \frac{64\pi^2 m\varepsilon^2 r_0 Z^*}{9h^2},
\tag{4.17}
$$

$$
b = \frac{128\pi^2 m\varepsilon^2 r_0 f(n)}{9h^2 D},
\tag{4.18}
$$

$$
c = \frac{\varepsilon^2 n_a n_b}{kTDr_0},
\tag{4.19}
$$

$$
d = \frac{1}{r_0 kT} \times \frac{d\Delta F_r}{dx} \approx 0,
\tag{4.20}
$$

and also the normalized variables

$$
y = \frac{k'h}{kT},
\tag{4.21}
$$

$$
x = \frac{r_{ab}}{r_0}.
\tag{4.22}
$$

Using these parameters and variables, Eq. (4.16) can be written in logarithmic form:

$$- \ln y = (ax^2 - bx)^{1/2} + \int a\,dx + \frac{c}{x}. \tag{4.23}$$

Differentiating with respect to x and setting the resulting expression equal to zero, one obtains after factoring

$$a^{1/2} (1 - b/2ax)/(1 - b/ax)^{1/2} = c/x^2. \tag{4.24}$$

Expanding the quadratic term and retaining only the first two terms of the expansion, Eq. (4.24) simplifies to

$$x^2 = c/a^{1/2}. \tag{4.25}$$

Substituting the values of x, c, and a given above into Eq. (4.25) yields

$$r_{ab}^{*2} = \frac{3\varepsilon \, n_a n_b h r_0^{1/2}}{8\pi \, kDT(mZ^*)^{1/2}}. \tag{4.26}$$

To apply the theory, one uses Eq. (4.12) to calculate r_0 employing the principal quantum number as given by Rice for the electron in the atom concerned and the radius of the first Bohr orbit, namely, 0.529 Å. The factor Z^* is chosen for the positive charge on the central atom of the complex ion, the central atom chosen being the one of the lowest ionization potential, and hence usually refers to the lowest valence or least ionized of the exchanging ionic species. The quantity Z^* is given by the total positive valence n_a, of the central atom of the complex ion reduced by the sum of the negative valences of the coordinating groups. Thus if all the coordinating groups are neutral, $Z^* = n_a$.

From the positive charges on the two cations, n_a and n_b, and the values of r_0 and Z^*, r_{ab}^* is calculated from Eq. (4.26). From the ratio n_a/n_b, γ is calculated and then $f(n)$ is obtained from Eq. (4.10).

With r_0, r_{ab}^*, Z^* and $f(n)$ known, if k' is measured, then the various free energies contributing to the rate can be found from Eq. (4.16) for a solvent of known dielectric constant at a known temperature. Or if ΔF_r is assumed (ΔF_r was taken as 8.1 kcal/mole by Marcus, Zwolinski, and Eyring) then, from known values of r_0, r_{ab}^*, Z^*, and $f(n)$, rate constants in solvents of known dielectric constants at known temperatures may be calculated from Eq. (4.16) and compared with measured rate constants.

Plots of ΔF (total) calculated or of $\ln k'$ calculated versus r_{ab}, Z^*, $n_a n_b$,

r_0, or $1/D$ should yield characteristic curves. The measured values of $\ln k'$ versus $1/D$ should coincide with the calculated values of $\ln k'$ versus $1/D$ if the right values of the parameters have been used in calculating the values of k'.

Marcus, Zwolinski, and Eyring state that the dielectric constant relationship is particularly important since it provides a convenient experimental approach for testing the correctness of the proposed model by studying the kinetics of an exchange reaction in media of varying dielectric constants. Table I contains the observed value of ΔF (total), as well as the values of r_{ab}^*, $-RT\ln \varkappa_e$, and ΔF_{rep} as presented by Marcus, Zwolinski, and Eyring.

TABLE I

COMPARISON OF CALCULATED AND EXPERIMENTAL DATA FROM EQ. (4.16)

Reaction	Reference	r_{ab}^* (Å)	$-RT\ln \varkappa_e$	ΔF_{rep}	ΔF (total) calc	ΔF (total) obs
Co(en)_3^{2+}–Co(en)_3^{3+}	(9)	5.9	4.38	4.32	16.8	23.5
Tl^{1+}–TlOH^{2+}	(10)	3.3	2.63	2.57	13.3	23.9
VOH^{2+}–VO^{2+}	(11)	4.4	4.00	3.86	16.0	17.2
Fe^{2+}–Fe^{3+}	(12)	6.0	4.37	4.25	$16.7^{\,a}$	16.7
Fe^{2+}–Fe(OH)^{2+}	(12)	4.9	3.57	3.47	15.1	12.2
Fe^{2+}–FeCl^{2+}	(12)	4.9	3.57	3.47	15.1	15.3
Fe^{2+}–FeCl_2^{1+}	(12)	3.4	2.48	2.50	13.1	15.1
Ce^{3+}–Ce^{4+}	(13)	9.3	5.52	5.48	19.1	18.0

^{a} Fitted value

The rearrangement free energy ΔF_r was taken a 8.1 kcal/mole. This value was that found for the ferrous–ferric reaction. All energy values were in kcal/mole for $D = 78$.

Cohen et al. (14) studied the isotopic exchange reaction between Np(V) and Np(VI) ions at 0°C and in ethylene glycol–water and in sucrose–water media of varying composition and containing perchloric acid. The principal features of the experimental procedure were given by Cohen et al. (15). The exchange reaction is first order in each of the metal ions and zero order in hydrogen ion. The reaction is therefore second order over-all. The rate R is given by the expression

$$R = k[\text{Np(V)}]\,[\text{Np(VI)}]. \tag{4.27}$$

By methods described above, r_{ab}^* and k' were calculated using Z^* as 5, ΔF_r as 8.1 kcal/mole, D as 88.3 for water at 0°C. and varying values for n_a and n_b. The results are reproduced in Table II. From the table it is seen that the values $n_a = 1.59$, $n_b = 2.69$, $r_{ab}^* = 5.04$ Å, and $\Delta F_r = 8.1$ kcal/mole give a value of k' of 20.8 liters mole^{-1} sec^{-1}, which approximates the experimental value of 19.5 liters mole^{-1} sec^{-1}.

By proper choice of the parameters any one data value of k', at any dielectric constant used in these experiments, could be reproduced by Eq. (4.16). Table II however shows the extreme sensitivity of k' to the values chosen for the adjusted parameters. The lower rearrangement free

TABLE II

VALUES OF r_{ab}^* AND k' [a]

n_a	n_b	r_{ab}^*	k' (calc) (liters mole^{-1} sec^{-1})
1	2	3.51	795
1.5	2.6	4.90	34.9
1.59	2.69	5.04	20.8
2	3	6.07	2.45
3	4	8.59	0.00794
3	4	8.59	15.1 [b]

[a] Calculated for $D = 88.3$, $Z^* = 5$, and $\Delta F_r = 8.1$ kcal/mole at 0°C. The experimental value of k' in 0.1 M HClO$_4$ was 19.5 liter mole^{-1} sec^{-1} at 0°C.

[b] $\Delta F_r = 4.1$ kcal/mole.

energy leads to a reasonable value of the tunneling distance r_{ab}^* if the electron-transfer process is to occur without interference of the two ions with each others primary hydration shells. For the hydration shells of NpO$_2^+$ and NpO$_2^{2+}$ to remain undisturbed, the radius of the activated complex must exceed 7.5 Å, as can be seen by allowing the reasonable values of 1Å as the radius of each of the Np(V) and Np(VI) ions and 5.25 Å for the diameters of two water molecules. On the other hand, 3+ and 4+ are probably too large for n_a and n_b respectively, since the Np-O bonds are somewhere between single and double bonds (15–17).

The Np(V)–Np(VI) electron-exchange rate is out of harmony with the electron hypothesis in relaion to the effect of dielectric constant upon the rate. This is clearly seen in Fig. 3.

The solid line represents the theoretical predictions using $n_a = 1.59$ Å, $n_b = 2.69$ Å, and $r_{ab} = 5.04$ Å. The slope of the theoretical line in the region of the dielectric constant used experimentally is -2.0. The experimental data represented in the figure by circles for the ethylene glycol–water solvent and by squares for the sucrose–water solvent give a least square straight line with a slope of 0.08 ± 0.13 standard deviation. The rate of exchange is therefore independent of the gross dielectric constants of the solutions for this reaction and for the experimental conditions which prevailed. Larger values of n_a, n_b, and r_{ab} enhance the divergence between theory and experi-

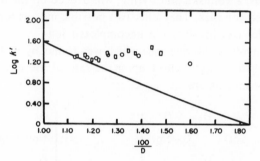

FIG. 3. Effect of dielectric constant on the rate of the Np(V)–Np(VI) exchange; $t = 0°$, $HClO_4 = 0.106\ M$: —, theoretical curve; ○, water–ethylene glycol, Np(V) = Np(VI) $\sim 1.41 \times 10^{-5}\ M$; □, water–sucrose, Np(V) = Np(VI) $\sim 1.87 \times 10^{-5}\ M$.

ment since larger values of these parameters increase the slope of the theoretical $\log k'$ versus $1/D$ plot, while the slope of the data plot is unaffected of course. The data for any particular solvent can be made to conform to the theory by proper choice of parameters.

The above results suggest that perhaps the exchange does not occur by a direct electron exchange but by an atom transfer. However, the lack of agreement of the data with theory is not proof that there is not a direct electron exchange. Platzman and Franck (18) point out that the Christiansen-Scatchard treatment is valid only for large distances. Polarization phenomena markedly alter the dielectric properties of the medium; consequently, as pointed out by Platzman and Franck and by Amis and Jaffé (19), the dielectric constant of the medium loses significance at the distance involved in the formation of the activated complex. Weiss (3), in his treatment of the electron-transfer process, has tried to take this factor into account. He believes that only the interaction energy in the final state, that is, in the activated complex, enters into the activation energy. In this case, only the optical dielectric constant (the square of the refractive index) and not the macroscopic dielectric constant should be considered. What is particularly needed, however, is a knowledge of the dielectric properties of the medium in high potential fields.

Amis (5) explains the results by an electron transfer using a bridge mechanism and by assuming that there is selective solvation of the activated complex by the higher dielectric component of the solvent medium under conditions of complete dielectric saturation. These theories will be discussed more fully later.

Cohen et al. (20) point out that the results of any calculations, such as those of Marcus, Zwolinski, and Eyring, will depend on the model assumed for the activated complex. They offer two possible models for the activated complex. Model (A) involves the uncomplexed ion.

Model (B) involves chloride ions which was found by Cohen et al. (20) to have a marked catalytic effect upon the rate of the electron–exchange reaction. These models are

$$\left[{}^{-}O{-}Np^{4+}\underline{\quad\quad}{}^{-}O \overset{\overset{\displaystyle H}{\overset{|}{\diagdown}}}{\underset{\underset{\displaystyle H}{\diagup}}{\quad}} \overset{\overset{\displaystyle O^{-}}{\overset{|}{}}}{\underset{\underset{\displaystyle O}{|}}{O{-}Np^{3+}}} \right]^{3+}$$

(A)

$$\left[\overset{\overset{\displaystyle O^{-}}{\overset{|}{}}}{\underset{\underset{\displaystyle O^{-}}{|}}{Np^{3+}}} \cdots Cl{-} \overset{\overset{\displaystyle O^{-}}{\overset{|}{}}}{\underset{\underset{\displaystyle O^{-}}{|}}{Np^{3+}}} \right]^{2+}$$

(B)

If the bridging is through water molecules, no rearrangement in the hydration spheres would be involved in model (A). In model (B), however, water molecules would be unfrozen in the formation of the intermediate. Direct electron transfer presumably could take place in either case; however, there would be a marked difference in both energy and entropy values in the two cases. Cohen, Sullivan, and Hindman believe, therefore, that it would appear naive to attempt to argue about the possible mechanism from the energetics alone.

The Theory of Weiss

It is assumed, from earlier works (21–23) on electron-transfer processes in gaseous systems, that electron-transfer processes can take place by (i) nonadiabatic process at the crossing point (or in regions of close approach) of the potential curves of the initial and final state or by (ii) a tunnel effect between electronic levels of equal energy.

For nonadiabatic electron transfer the interaction energy $U_{if}(R)$ between the colliding systems as a function of the separation R is given by the equation

$$U_{if} \approx (E_1^\infty - E_2^\infty) + (U_i - U_f),\tag{4.28}$$

where $(E_2^\infty - E_1^\infty) = \Delta E$ is the energy difference at infinite separation of the reaction particles and U_i and U_f are the interionic energies in the initial and final states, respectively. In the case of electron-transfer processes between ions, U_{if} decreases exponentially with distance and can therefore be neglected in comparison with long-range electrostatic interactions (Coulombic forces) (21). Thus U_{if} in Eq. (4.28) becomes zero and

$$\Delta E = (E_2^\infty - E_1^\infty)(U_i - U_f).\tag{4.29}$$

Thus the crossing point is at the distance R_0, where ΔE is of the order of the difference of the interaction energies between the initial and final states.

From Landau's work, Weiss deduced the expression for the cross section for the nonadiabatic process between particles of reduced mass M to be

$$\sigma(e) = \frac{2^{5/2}\pi^2 U_{if}^2 R_0^2 M^{1/2}}{\hbar \left|\dfrac{d}{dR}(U_i - U_f)\right|_{R=R_0}} \frac{[E - \Delta U]^{1/2}}{E},\tag{4.30}$$

E is the total energy and ΔU is the potential energy at the crossing point $(R = R_0)$. $[E - \Delta U]^{1/2}$ is proportional to the relative velocity.

If the electron process is exothermic $[\Delta U(R_0) < 0]$ then even for relatively low values of E, $\sigma(e)$ may assume appreciable values. Thus the temperature dependence of the cross section would arise only through the velocity $(V \propto T^{1/2})$.

If the process is endothermic $[\Delta U(R_0) > 0]$, $\sigma(e)$ will have only positive values for $E > \Delta U(R_0)$. In this case the cross section $\sigma'(e)$ as a function of temperature is given by the equation

$$\sigma'(e) \simeq \sigma(e) \times \exp(-\Delta U/kT).\tag{4.31}$$

Thus the endothermicity (ΔU) assumes the role of an activation energy.

In the case of electron transfer by tunnel effect, electron transfer can occur only between levels of equal energy. The energy difference between the electronic levels of the initial and final state $[\Delta U(R')]$ at the distance R' where the electron transfer takes place is again the energy of activation, and the rate to a first approximation is given by the equation

$$\text{rate} \propto \Gamma \times \exp(-\Delta U/kT).\tag{4.32}$$

For endothermic processes this exponential dependence of the rate is supported by Bell (24) and by Hellman and Syrkin (25). Γ is the Gamow factor and is a function of the width (R') of the potential barrier (26). Γ is given to a first approximation by the expression

$$\Gamma \propto \exp\left\{-\frac{\sqrt{2m}}{\hbar}\int_{R_1}^{R_2}\sqrt{(U_0-E)}dR\right\} \propto \exp\left\{-\frac{\sqrt{2m}}{\hbar}\sqrt{(U_0-E)}R'\right\} \tag{4.33}$$

where U_0 is the height of the rectangular barrier and E the total energy of the electron mass m.

Relevant to either of the above theories are ΔE, the energy difference of the electronic levels at infinite separation, and ΔU, the energy difference at a distance r'. ΔU differs from ΔE by the interaction energy of the reacting particles.

At equilibrium the energy difference (ΔH) between the initial and final states is given by

$$\Delta H = \Delta I + \Delta S, \tag{4.34}$$

where

$$\Delta I = \Sigma - I. \tag{4.35}$$

Σ is the electron affinity of the electron acceptor and I is the ionization potential of the electron donor referred to vacuum. ΔS is the difference between the total hydration energies of the initial (S_i) and final (S_f) states.

$$\Delta S = S_i - S_f. \tag{4.36}$$

To evaluate ΔE, which is relevant to the actual electron transfer, the Franck-Condon principle must be considered. When this is done, the first terms of Eq. (4.34) remain unaltered for monatomic ions. For diatomic or polyatomic ions or molecules this is not true since the "vertical" ionization potential or electron affinity, respectively, must be taken, because only those transitions have high probability where the electron is transferred, without change of internuclear distances.

With respect to the solvation term (ΔS) a somewhat similar situation exists since, again due to the Franck-Condon principle, the orientation of the water molecules remains practically fixed during electron transfer. Only the smaller difference of "electronic hydration" energies ($\Delta S^{(e)}$) have to be considered. These arise from the *electronic polarization* of the solvent around the charged particles, thus

$$\Delta S^{(e)} = S_i^{(e)} - S_f^{(e)}, \tag{4.37}$$

whereas the total hydration energy (S) is given by

$$S = S^{(e)} - S^{(d)}, \tag{4.38}$$

where $S^{(d)}$ is the hydration energy due to the orientation of the water-molecules. The Born formula gives for the "electronic" energy of hydration ($S^{(e)}$ of an ion of charge Ze)

$$S^{(e)} \simeq \frac{Z^2 e^2}{2r_i} \left(1 - \frac{1}{n^2} \right), \tag{4.39}$$

where r_i is an ionic radius and n^2 is the square of the refractive index (optical dielectric constant) for the electronic frequency being considered. $S^{(d)}$ is given in a similar way by

$$S^{(d)} = -\frac{Z^2 e^2}{2r_i} \left(\frac{1}{n^2} - \frac{1}{D} \right), \tag{4.40}$$

where D is the static dielectric constant of the medium.

The field due to the water dipoles surrounding the ion produces still a further energy term, which, since these dipoles remain practically fixed during the electron transfer, can be regarded roughly as a small spherical double layer with the ion in the center. According to Gurney (27) the transfer of an electron to or from the inside of such a double layer is accompanied by a change in the potential energy of the electron.

Since the spherical layer is positive outward around a positive ion, work has to be performed in passing an electron through the double layer to the central ion. In the case of a negative ion the spherical layer is negative outward, and work is required to remove an electron from the inside and bring it through the double layer.

Schottky and Rothe (28) would consider the charge of the dipoles on the inside to be neutralized by the polarization of the ion in the center, which would give the change in the potential energy ($\Delta\varphi$) in passing through the double layer to be

$$\Delta\varphi = 2\pi\sigma\mu. \tag{4.41}$$

However Weiss gives the change in potential energy for this process to be twice that given by Eq. (4.41), namely,

$$\Delta\varphi = 4\pi\sigma\mu, \tag{4.42}$$

where $\sigma = Z\varepsilon/4\pi\bar{r}^2$ is the surface charge density and μ is the dipole moment of the double layer. $Z\varepsilon$ is the charge of the central ion and $\bar{r}^2$ is the mean radius of the spherical double layer. The change of potential energy then, when an electron is conveyed through this double layer, is given by

$$\Delta\varphi = \pm \frac{Z\varepsilon\mu}{n^2\bar{r}^2}, \tag{4.43}$$

where n^2 is the optical dielectric constant of the medium. If the negative sign holds for the removal of an electron the positive sign holds for the addition of an electron.

If the spherical double layer is considered as a small spherical condenser with spheres of radius a and b and a charge of $Z\varepsilon$, the potential difference ΔV in a medium of dielectric constant n^2 is given by

$$\Delta V = \frac{Z\varepsilon}{n^2} \frac{b - a}{ab}, \tag{4.44}$$

which becomes identical with Eq. (4.43) when $b - a = S$, $\varepsilon S = \mu$, $a \sim b \sim \bar{r}$, and $\varepsilon\,\Delta V = \Delta\varphi$.

The difference in potential energy of the electron in the initial and final states is given by

$$\Delta\varphi = \Delta\varphi_i - \Delta\varphi_f, \tag{4.45}$$

and the total energy difference ΔE at infinite separation between the electronic levels becomes

$$\Delta E = \Delta I + \Delta S^{(e)} + \Delta\varphi. \tag{4.46}$$

From Eq. (4.43) $\Delta\varphi$ is estimated assuming that, in the electron-transfer process, the electron has to pass through the first coordination sphere of water dipoles which surround the ion. Thus, normally, electron transfer would occur virtually between these coordination spheres.

Much more energy is required to transfer the electron outside the Debye sphere (29, 30) to the region where the solvent has its normal dielectric constant. The critical radius inside which water no longer has its true dielectric properties extends quite far beyond the first coordination sphere of water dipoles.

Bernal and Fowler (31) have estimated the energy of hydration of ions. These authors disregard the intermediate very complex region of the hydration layer which adjoins the coordination sphere and assume a sphere of sharp discontinuity of radius r_z. Inside the sphere the energy is that of ionic coordination, and outside the sphere is that of ordinary water. Within this approximation Eq. (4.43) with $\bar{r} = r$ gives the energy to transport the electron through the coordination sphere of water dipoles.

However, to remove the electron from the (virtual) coordination sphere to the region outside the Debye sphere, in the case of an ion of central charge Ze, would require the energy $\Delta\psi$ given by the equation

$$\Delta\psi = -\frac{Z\varepsilon^2}{r_z}\left(\frac{1}{n^2} - \frac{1}{D}\right). \tag{4.47}$$

This energy difference is based on the assumption that for $r > r_z$ the electron is in a field in which the potential is of the form

$$\psi_r \simeq \frac{Z\varepsilon}{r}\left(\frac{1}{n^2} - \frac{1}{D}\right). \tag{4.48}$$

The displacement polarization in a dielectric medium, as suggested by Landau (32) and by Mott and Gurney (33) is similar to this expression. Platzman and Franck (34) have used this potential in their theoretical discussion of the absorption spectra of halide ions in solution.

This may be of some importance in photo-excited electron transfer; however, in thermal-electron-transfer processes, the electron may only have to pass through the first coordination sphere of water dipoles.

Approaching reacting particles will, in general, exert forces on each other in the initial as well as in the final states. Cross sections for nonadiabatic as well as tunneling effects may be relatively large compared with molecular dimensions. Therefore, in these cases, particularly in the case of ions where Coulombic forces and other electrostatic interactions are present, short-range interactions may be neglected.

The energy difference $[\Delta U(R')]$, which in the case of endothermic reactions may coincide with the energy of activation, is given by

$$\Delta U(R') = \Delta E + U_f(R'). \tag{4.49}$$

Only the interaction energy of the final state is evident. The interaction energy (U_i) cancels out since, if energy is supplied to bring the particles to the required distance (r') in the initial state, an equal amount of energy is gained in the subsequent stages or vice versa. $\Delta U(R')$ is thus the total energy difference between the initial state of infinite separation and the final state where the separation is R'. Thus R' is the separation of the particles in the transition state.

On the basis of the simple theoretical considerations presented above, Weiss drew certain conclusions, particularly with regard to the transition state for the actual electron-transfer process.

The gain in Coulombic energy when electron transfer leads to the formation of ions of opposite sign should reduce the energy difference ΔU. This would partly compensate for the loss of part of the hydration energies arising from the orientation of water molecules.

On the account of Coulombic repulsion, reactions between ions of like sign are hindered. Debye (35) and others have treated the effect of electrostatic attraction and repulsion on the number of collisions between ions.

A new factor acting in a somewhat similar way arises in the Weiss theory. This factor arises from a gain in Coulombic energy in the final state which preferably occurs in reactions between ions of opposite charge or in processes between charged or between charged and uncharged particles.

For electron transfer by nonadiabatic processes, the Weiss theory applies only when Eq. (4.28) leads to a positive and real value of R_0, where R_0 is then generally of the form

$$R_0 \propto \frac{\varepsilon^2}{|\Delta E|}. \tag{4.50}$$

For large values of R_0, the interaction energy $U_{if}(R_0)$ may be very small, and then the actual cross section of the electron-transfer process would not be appreciable. For R_0 relatively small (of the order of molecular dimensions) it will enter the region of short-range repulsive forces and thus prevent the nonadiabatic electron-transfer reaction altogether.

If Eq. (4.28) results in an imaginary value of R_0 there is, according to Landau (21, 22), crossing in the near imaginary, that is the curves only approach each other closely.

From the physical considerations given above, Weiss gives several examples of pictures of the transition state for the electron-transfer process. For the reaction

$$A^+ + B^- \longrightarrow \underset{\text{transition state}}{(A^+ \cdots B^-)} \longrightarrow (A \cdots B) \longrightarrow A + B,$$

$\Delta I = E_B - I_A$ (E_B, the electron affinity of B; I_A, ionization potential of A),

$$\Delta S^{(e)} \simeq \frac{\varepsilon^2}{2} \left[\frac{1}{r_A^+} + \frac{1}{r_B^-} \right] \left(1 - \frac{1}{n^2} \right) \quad (r_A^+, r_B^-, \text{ ionic radii})$$

$$\Delta \varphi \;\simeq\; \frac{\varepsilon \mu_W}{n^2} \left[\frac{1}{\bar{r}_{A^+}^2} + \frac{1}{\bar{r}_{B^-}^2} \right] \quad (\bar{r}_{A^+}^2,\; \bar{r}_{B^-}^2,\; \text{mean radii of the hydration dou-}$$

ble layer of the respective ions)

$$U_i \;\simeq\; -\frac{\varepsilon^2}{DR'}, \qquad U_f \simeq 0,$$

$$\Delta E \;=\; E_B - I_A + \frac{\varepsilon^2}{2} \left[\frac{1}{r_{A^+}} + \frac{1}{r_{B^-}} \right] \left(1 - \frac{1}{n^2} \right)$$

$$+ \frac{\varepsilon \mu_W}{\hbar^2} \left[\frac{1}{\bar{r}_{A^+}^2} + \frac{1}{\bar{r}_{B^-}^2} \right].$$

If $\Delta E \sim \Delta U > 0$, the actual electron transfer is endothermic.

However, for an exothermic process, $R \propto -\varepsilon^2/\Delta E$, which is positive only if $\Delta E < 0$.

For the reaction

$$A^+ + B \quad \longrightarrow \quad (A^+ \cdots B) \quad \longrightarrow \quad (A^{2+} \cdots B^-) \quad \longrightarrow \quad A^{2+} + B^-,$$

$$\Delta A \;=\; I_{A^+} - E_B \qquad (I_{A^+},\; \text{second ionization potential of A}),$$

$$\Delta S^{(e)} \;=\; \frac{\varepsilon^2}{2} \left[\frac{1}{r_{A^+}} - \frac{4}{r_{A^{2+}}} - \frac{1}{r_{B^-}} \right] \left(1 - \frac{1}{n^2} \right) \quad (r_{A^+}\; r_{A^{2+}},\; r_{B^-},\; \text{ionic radii})$$

$$\Delta \varphi \;\simeq\; \frac{\varepsilon \mu_W}{n^2 \bar{r}_{A^+}^2},$$

$$U_i \;\simeq\; 0 \text{ (neglecting dipole interactions)},$$

$$U_f \;\simeq\; -\frac{2\varepsilon^2}{DR'},$$

$$\Delta U \;\simeq\; \Delta E - \frac{2\varepsilon^2}{DR'}.$$

The Coulombic interaction in the final state would reduce the energy difference between the initial and final states in the case of an endothermic process ($\Delta E < 0$).

For the reaction

$$A^+ + B^+ \quad \longrightarrow \quad \underset{\text{transition state}}{(A^+ \cdots B^+)} \quad \longrightarrow \quad (A^{2+} \cdots B) \quad \longrightarrow \quad A^{2+} + B,$$

$$\Delta I = I_A - I_B,$$

$$\Delta S^{(e)} \simeq \frac{\varepsilon^2}{2} \left[\frac{1}{r_{A+}} + \frac{1}{r_{B+}} - \frac{4}{r_{A^{2+}}} \right] \left(1 - \frac{1}{n^2} \right),$$

$$\Delta\varphi \sim 0,$$

$$U_i \simeq \frac{\varepsilon^2}{DR'},$$

$$U_f \sim 0 \quad \text{(neglecting dipole forces)}.$$

For the reaction

$$A^- + B \longrightarrow \underset{\text{transition state}}{(A^-\cdots B)} \longrightarrow (A\cdots B^-) \longrightarrow A + B^-,$$

$$\Delta I = E_A - E_B \quad (E_A, E_B, \text{ electron affinities of } A \text{ and } B, \text{ respectively}),$$

$$\Delta S^{(e)} \simeq \frac{\varepsilon^2}{2} \left[\frac{1}{r_{A-}} - \frac{1}{r_{B-}} \right] \left(1 - \frac{1}{n^2} \right),$$

$$\Delta\varphi \simeq \frac{\varepsilon\mu_W}{Dr_{A-}^2},$$

$$U_i \simeq -\frac{\varepsilon^2\mu_A}{DR'^2}, \quad U_f \simeq -\frac{\varepsilon^2\mu_B}{DR'^2} \qquad \begin{array}{l} (\mu_A, \mu_B, \text{ permanent dipole mo-} \\ \text{ments of the molecules } A \text{ and} \\ B, \text{ respectively)}. \end{array}$$

For the reaction

$$A^- + B^- \longrightarrow \underset{\text{transition state}}{(A^-\cdots B^-)} \longrightarrow (A^{2-}\cdots B) \longrightarrow A^{2-} + B,$$

$$\Delta I = E_B - E_A^- \quad (E_A^-, \text{ second electron affinity of } A),$$

$$\Delta S^{(e)} \simeq \frac{\varepsilon^2}{2} \left[\frac{1}{r_{A-}} + \frac{1}{r_{B-}} - \frac{4}{r_{A^{2-}}} \right] \left(1 - \frac{1}{n^2} \right),$$

$$\Delta\varphi = 0,$$

$$U_i \simeq \frac{\varepsilon^2}{DR'}, \qquad U_f \simeq 0 \quad \text{(neglecting dipole forces)}.$$

Since the dielectric constant of the medium enters into the expressions for $\Delta S^{(e)}$ and U_f, the magnitudes of ΔE and ΔU depend upon the nature of the solvent. For example, the formation of an ion from a neutral particle or of an ion of higher charge should be favored in a medium of high optical dielectric constant (high refractive index) and vice versa.

In the discussion thus far it has been assumed that ions are monoatomic.

In general, intercalation of the coordination shell should lead to an increased ionic radius, which in the first instance should reduce the interaction with the solvent.

In dealing with complex ions one should distinguish between two cases: (1) where, under equilibrium conditions, the structure of the two forms of different valency, that is, of the two states differing by one electron, is essentially the same; and (2) where there is a difference in structure and bonding of the coordination shells belonging to the two states of different valency.

In the latter case, since the electron transfer normally takes place without change of the internuclear distance, the electron transfer will lead to a non-equilibrium condition in the coordination shell. In this instance the "vertical" ionization potentials and electron affinities, respectively, have to be taken into account. This may cause an increase in ΔI and consequently in ΔU. This should result in a decrease in the rate of the electron transfer according to Eqs. (4.30) and (4.31).

However the situation is again somewhat similar to monoatomic ions if the two states of the different valency of the complex ion differ essentially only in their charge. This is apt to be the case if a system of highly conjugated double bonds with loosely bound π-electrons compose the coordination shell. There may be considerable shielding of the positive charge of the central metal ion by the electronic cloud of the π-electrons. This should lead to a further reduction of the electrostatic interaction with the solvent.

This latter group apparently includes most of the metal porphyrin complexes and thus includes also the biologically important haematin compounds, for example, the iron porphyrins, which act as prosthetic groups of the cytochrome enzymes and of the enzymes peroxidase and catalase (36).

In the haemoproteins the metal ion is in the middle of the porphyrin molecule. The four central nitrogen atoms of the molecule are bound to the central ferrous or ferric ion. The disk-shaped haemin result. At right angles from the plane of the disk, the two remaining valencies are probably bound to the protein component (36, 37).

Cytochromes through oscillation between the ferrous and ferric state can act as carriers in biological oxidation–reduction systems. This leads to a transfer of electrons from the direction toward the molecular oxygen from the hydrogen-transferring enzymes.

The potential differences between adjacent carriers must not be too small since the "carrier" chain transfers the oxidation toward the substrate. However, as indicated previously, with decreasing energy difference between two adjacent states, the rate of the electron-transfer process should increase. To a marked extent both of these conditions are apparently fulfilled in biological oxidation–reduction chains where the potential difference between adjacent chains is of the order of 100 mV and $kT/e \sim 27$ mV at 37°C.

The weak interaction of these complex metal ions with the aqueous medium explains, at least partly, these relatively small energy and entropy differences. This may be one explanation of why metal ions and other ionic species which occur in biological oxidation–reduction systems form relatively large complex ions with extended systems of conjugated double bonds surrounding the central ion.

Electron-transfer processes between like ions are studied using radioactive isotopes. Libby (38) has proposed a qualitative theory of these electron-exchange processes. Libby concluded that electron transfer should be favored if there is a "configurational symmetry" between the exchanging ions, particularly if the hydration spheres of the ions are "shared" to a marked extent, since the dissimilarities would be reduced in magnitude in this case.

The theory of Weiss has certain features in common with that of Libby. The two theories differ in the general treatment and in the conclusions as is illustrated by the following discussion of electron-transfer processes between isotopic ions.

For the reaction

$$Fe^{2+} + *Fe^{3+} \longrightarrow (Fe^{2+}\cdots*Fe^{3+}) \longrightarrow (Fe^{3+}\cdots*Fe^{2+}) \longrightarrow Fe^{3+} + *Fe^{2+},$$

where $*Fe$ represents the radioactive isotope, because of the Franck-Condon principle, the electron transfer will result in the formation of a ferric ion and a ferrous ion with practically identical hydration atmospheres. In this and similar cases, the ionization potentials and the "electronic" energies of hydration show no difference. Thus,

$$\Delta I \sim 0, \qquad \Delta S^{(e)} \sim 0.$$

There is an energy difference arising from the transfer of the electron through the double layer of oriented water molecules surrounding the ions.

According to Eq. (4.43) this energy difference is given by

$$\Delta\varphi = \frac{3\varepsilon\mu_W}{n^2\bar{r}^2} - \frac{2\varepsilon\mu_W}{n^2\bar{r}^2} = \frac{\varepsilon\mu_W}{n^3\bar{r}^2} , \qquad (4.51)$$

and the total energy difference for the electron-transfer process becomes

$$\Delta U(R') = \frac{\varepsilon\mu_W}{D\bar{r}^2} + \frac{6\varepsilon^2}{DR'} , \qquad (4.52)$$

which, as pointed out above, would be identified with an energy of activation.

This total energy difference can be estimated as follows. Taking $\mu_W = 1.8 \times 10^{-18}$ esu, $n^2 = 2$, $\bar{r} = (0.85 \text{ Å} + 0.88 \text{ Å}) = 1.73 \text{ Å}$, where the value of $\bar{r}$ is found by increasing the Goldschmidt radius of ferrous iron by 0.85 suggested by Latimer, Pitzer, and Slansky (39) as giving an "effective" ionic radius of positive ions in aqueous solutions. This gives 0.83 eV for the first term in Eq. (4.52). For the second term Weiss chooses R' as 5 Å and D as 80, the static dielectric constant of liquid water. The value of the second term in Eq. (4.52) thus becomes 0.2 eV. Therefore

$$\Delta U \sim (0.83 + 0.2) = 1.03 \text{ eV} ,$$

which is about twice the value of 0.43 eV as observed experimentally by Silverman and Dodson (40). However, the estimated value of $\bar{r}$ may be too low, and it is possible that only half the dipole layer surrounding the ion is effective because, on the inside, the charge of the dipoles is neutralized by the ion in the center. These considerations would reduce the final term to 0.43 eV or less. In the second term on the right of Eq. (4.52), D would, due to dielectric saturation, be more nearly 30 or 40 or even less rather than 80, the value used in the calculation. This may be partly compensated for by a R' value larger than the 5 Å that was used. It is doubtful that R' would be greater than 10 Å, since beyond this distance the Gamow factor would be rather small. In estimating the second term it might have to be remembered that in an electrolyte the electric potential at a distance r from an ion varies not with $1/r$ but rather like $\exp(-\varkappa r/r)$, because the diffuse ionic layer surrounding an ion shields the charge on the ion. The factor $1/\varkappa$ represents the "thickness" of this ionic layer. The Debye-Hückel theory permits the calculation of this thickness. Consideration of this shielding effect will lead to marked reduction of the second term. Thus a final value of 0.4 to 0.5 eV may be obtained for ΔU.

Silverman and Dodson found an acceleration by OH^- and Cl^- ions of the electron exchange in the ferrous–ferric system. They conclude that in these cases the electron transfer takes place between the Fe^{2+} ion and the respective ion pairs, $FeOH^{2+}$ and $FeCl^{2+}$. These ion pairs are present in the equilibria

$$Fe^{3+} + OH^- \rightleftharpoons Fe(OH)^{2+},$$

$$Fe^{3+} + Cl^- \rightleftharpoons FeCl^{2+}.$$

Weiss represents the actual electron transfers by the processes

$$Fe^{2+} + {}^*FeOH^{2+} \longrightarrow (Fe^{2+}\cdots OH^- - {}^*Fe^{3+}) \longrightarrow (Fe^{3+}\cdots OH^- -$$
$${}^*Fe^{2+}) \longrightarrow Fe^{3+}OH^- + {}^*Fe^{2+}, \tag{4.53}$$

$$Fe^{2+} + {}^*FeCl^{2+} \longrightarrow (Fe^{2+}\cdots Cl^- - {}^*Fe^{3+}) \longrightarrow (Fe^{3+}\cdots Cl^- -$$
$${}^*Fe^{2+}) \longrightarrow Fe^{3+}Cl^- + {}^*Fe^{2+}. \tag{4.54}$$

According to the Franck-Condon principle these processes should proceed without appreciable change of interionic distance between OH^- or Cl^- and the metal ion. Weiss believes that ${}^*FeOH^+$ and ${}^*FeCl^+$ are first produced in a "non-equilibrium" state and should be followed by

$$ {}^*FeOH^+ \longrightarrow {}^*Fe^{2+} + OH^-, \tag{4.55}$$

$$ {}^*FeCl^+ \longrightarrow {}^*Fe^{2+} + Cl^-. \tag{4.56}$$

Electron transfer, according to Eq. (4.24) or Eq. (4.25), results in a change of total hydration energies of magnitude

$$\Delta S^{(e)} \simeq \frac{\varepsilon^2}{2} \left[\frac{2\times 4}{r_{(2+)}} - \frac{9}{r_{(3+)}} - \frac{1}{r_{(+)}} \right] \left(1 - \frac{1}{n^2} \right)$$
$$\simeq -\frac{\varepsilon^2}{r} \left(1 - \frac{1}{n^2} \right); \tag{4.57}$$

$\Delta\varphi \sim 0$ in these cases since both the electron donor and the electron acceptor have the same positive charge initially. $\Delta I \neq 0$, however, because the electron affinity of $FeOH^{2+}$ or $FeCl^{2+}$ is evidently different from that of the simple Fe^{3+} ion. The following sequence of reactions, where X^- is the monovalent anion, makes possible the calculation of ΔI:

$$FeX^{2+} \longrightarrow Fe^{3+} + X^- \qquad -D_{(2)} \tag{4.58}$$

$$Fe^{3+} + e \longrightarrow Fe^{2+} \qquad +I_{Fe^{2+}} \tag{4.59}$$

$$Fe^{2+} + X^- \longrightarrow FeX^+ \qquad +D_{(1)} \tag{4.60}$$

$$FeX^{2+} + e \longrightarrow FeX^+ + I_{Fe^{2+}} + D_{(1)} - D_{(2)} \tag{4.61}$$

and

$$\Delta I \simeq [E_d - E_d'] \simeq \frac{\varepsilon^2}{n^2 r_X}. \tag{4.62}$$

In these equations, E_d and E_d' represent the dissociation energies of FeX^{2+} and FeX^+, respectively, for the same interionic distance r_X. The total energy difference is given by the expression

$$\Delta U(R') = \Delta S^{(e)} + \frac{\varepsilon^2}{n^2 r_X} + \frac{3\varepsilon^2}{DR'}. \tag{4.63}$$

As indicated by Eq. (4.57), $\Delta S^{(e)}$ will generally be negative. However, the last two terms in Eq. (4.63) will always be positive. For the same separation R' the Coulombic repulsion is less in this instance than for monatomic ions. Thus the ions might approach more closely and cause the width R' of the potential barrier to be smaller. An increase in transmission coefficient for the "tunnel" effect should result. This might explain the greater nonexponential (temperature-independent) factor observed in this case by Silverman and Dodson (40) as compared to the exchange reaction between ordinary ferrous and ferric ions.

The electron exchange between Eu^{II} and Eu^{III} ions in the presence of Cl^- ions studied by Meier and Garner (41) and that between $Ce^{III}-Ce^{IV}$ ions in the presence of F^- ions by Hornig and Libby (42) probably involve similar mechanisms.

The slow exchange reaction observed by Lewis et al. (43) between cobaltous and cobaltic hexamine ions could have resulted from the fact that the structure and the bonding of the coordination shells (at equilibrium) are different in the two valence states. This would cause ΔI to be positive and would thus increase the energy difference between the initial and final states.

These conclusions are in agreement with those of Brown (44) and also harmonize with the results of Adamson (45). Brown pointed out that there is a relatively large difference in the length of the $CO^{2+}-N$ bond and in the $Co^{3+}-N$ bond in the equilibrium state.

Lewis and co-workers (43) found that molecular oxygen catalyzes the slow electron transfer between $[Co(NH_3)_6]^{2+}$ and $[Co(NH_3)_6]^{3+}$. Weiss (46) suggested oxidation–reduction processes of the type

$$Me^{2+} + O_2 \longrightarrow (Me^{3+} \cdot O_2^-) \longrightarrow Me^{3+} + O_2^-, \tag{4.64}$$

$$Me^{3+} + O_2^- \longrightarrow Me^{2+} + O_2 \tag{4.65}$$

to explain this oxidation.

For slow normal electron transfers such processes may be important. Weiss (46) had suggested that the uncatalyzed electron-exchange reaction between ferrous and ferric ions was catalyzed to some extent by molecular oxygen. However, Silverman and Dodson (40) have established that the electron reaction was too fast to be brought about by molecular oxygen catalysis. That oxygen catalysis is not important in this reaction has been experimentally confirmed by Eimer et al. (47).

Electron-exchange reactions would be fast, according to the above theory, when, in the two different valence states, the equilibrium structure of the coordination shell is the same. Under these conditions the coordination shell may cause a shielding of the charge of the central metal ion and thus bring about a greater effective "ionic radius." Both of these effects should produce a reduced electrostatic interaction with the solvent.

The electron reaction (45, 48)

$$MnO_4^{2-} + {}^*MnO_4^- \;\; \underset{\longleftarrow}{\overset{\longrightarrow}{}} \;\; MnO_4^- + {}^*MnO_4^{2-} \tag{4.66}$$

and the fast electron exchange reaction (49)

$$FeCy_6^{4-} + {}^*FeCy_6^{3-} \;\; \underset{\longleftarrow}{\overset{\longrightarrow}{}} \;\; FeCy_6^{3-} + {}^*FeCy_6^{4-},$$

where Cy stands for the cyanide radical, apparently belong to this class, as do the rapid electron-exchange reactions of certain metal porphyrin complexes, e.g.,

$$Co^{II}(Por) + {}^*Co^{III}(Por) \;\; \underset{\longleftarrow}{\overset{\longrightarrow}{}} \;\; Co^{III}(Por) + {}^*Co^{II}(Por). \tag{4.67}$$

Here (Por) represents tetraphenyl porphyrin radicals. Dorough and Dodson (50) studied this reaction.

As a first approximation in all these cases $\Delta E \sim 0$, $\Delta I \sim S^{(e)} \sim \Delta \varphi \sim 0$, so that $\Delta U(R') \sim 0$ are very small. There then remains only the (electrostatic) interaction in the final state (U_f), which in these cases is also expected to be relatively small.

The R. A. Marcus Theory: The Theory of Little Overlap of Electronic Orbitals

R. A. Marcus (4a) describes a mechanism for electron-exchange reactions in which there is very little spatial overlap of the electronic orbitals of the two reacting species in the activated complex. In the resulting intermediate state X* the electrical polarization of the solvent does not have the usual value appropriate for the given ionic charges, that is, it does not have an equilibrium value. Marcus used an equation developed by himself (*4a*) for the electrostatic free energy of nonequilibrium states to calculate the free energy of all possible intermediate states. Using the calculus to minimize the free energy subject to certain restraints the properties of the most probable state was determined. The electrostatic contribution to the free energy of formation of the intermediate state from the reactants, ΔF, is thereby obtained in terms of known quantities, such as ionic radii, charges, and the standard free energy of reaction.

The intermediate state X* can return to reactants or go to a state X, the ions of which are characteristic of the products. When the latter process is more probable than the former, the over-all reaction rate is simply the rate of formation of the intermediate state, namely, the collision number in solution multiplied by $\exp(-\Delta F/RT)$. Marcus states that on the basis of this theory no arbitrary parameters are needed to obtain reasonable agreement between calculated and experimental results.

The reaction scheme is as follows, with A and B denoting the reactants involved in the electronic transition,

$$A + B \underset{k_{-1}'}{\overset{k_1'}{\rightleftarrows}} X^*, \qquad\qquad I$$

$$X^* \underset{k_{-2}'}{\overset{k_2'}{\rightleftarrows}} X, \qquad\qquad II$$

$$X \xrightarrow{k_3'} \text{products}. \qquad\qquad III$$

Since we are interested in calculating only the over-all forward reaction, the reverse of step (III) can be neglected. The rate constant for the over-all backward reaction can be calculated from the rate constant of the over-all forward reaction and the equilibrium constant for the over-all reaction.

If the observed rate constant of this reaction sequence is k_{bi}' and C's denote concentrations, the over-all rate of the reaction sequence is $k_{bi}'C_A C_B$. According to step (III) the rate is also given by $k_3'C_X$, therefore

$$k_{bi}'C_A C_B = k_3'C_X, \qquad\qquad (4.68)$$

and the steady-state equations for the concentrations of X* and X, namely, C_{X^*} and C_X, are, respectively,

$$\frac{dC_{X^*}}{dt} = 0 = k_1'C_A C_B - (k_{-1}' + k_2')\,C_{X^*} + k_{-2}'\,C_X \qquad (4.69)$$

and

$$\frac{dC_X}{dt} = 0 = k_2' C_{X^*} - (k_{-2}' + k_3')C_X . \tag{4.70}$$

Solving these two equations simultenously for C_X and introducing this value for C_X into Eq. (4.68) we find

$$k_{bi}' = k_1'/[1 + (1 + k_{-2}'/k_3')k_{-1}'/k_2')]. \tag{4.71}$$

For the case where forward step (II) is more probable or about as probable as the reverse step (I), then

$$k_{bi}' \cong k_1' . \tag{4.72}$$

It is to be remembered that the electrical polarization of the solvent at each point is not in electrostatic equilibrium with the electrical field produced by ionic charges, that the overlap of the electronic orbitals of the two reacting particles is small in the activated complex, and that the thermodynamic function of such systems must be calculated using the method of Marcus (*4a*). The most probable pair of intermediate states which constitute the activated complex is found by minimizing the free energy of formation of X* from the reactants subject to the restriction that X* and X have the same total energy.

The model assumes that the reactants are spheres of radii a_1 and a_2 and that each reactant particle in turn is surrounded by a concentric spherical region of saturated dielectric having a radius a. Outside these spherical regions the medium is dielectrically unsaturated. For monatomic ions a is generally assumed to equal the sum of the crystallographic radius and the diameter of a solvent molecule.

It was found, after much mathematical manipulation that k_{bi}' was given by the collision number in solution, Z, multiplied by $\exp(-\Delta F/RT)$, that is,

$$k_{bi}' = Z \exp(-\Delta F/RT) , \tag{4.73}$$

where

$$Z = \sqrt{\frac{8\pi RT}{\mu}}\, r_g^2 \left(\frac{V}{V_f}\right). \tag{4.74}$$

In Eq. (4.74), R is the gas constant per mole, μ is the reduced molar mass of the two reactants, T is the absolute temperature, r_g is the distance between the centers of gravity of the two reactant particles, V is free volume in the standard state, and V_f is the free volume under experimental conditions of either reactant. The free volumes of the two reactants are assumed to be equal in any state.

It was shown that the excess free energy of activation ΔF is given by the equation

$$\Delta F = m^2\lambda + e_1{}^*e_2{}^*/Dr , \qquad (4.75)$$

where

$$2m + 1 = -\frac{[\Delta F^0 + T\,\Delta S_e + (e_1e_2 - e_1{}^*e_2{}^*)/Dr]}{\lambda} , \qquad (4.76)$$

$$\lambda = \left(\frac{1}{2a_1} + \frac{1}{2a_2} - \frac{1}{r}\right)\left(\frac{1}{D_{\text{op}}} - \frac{1}{D}\right)(\Delta e)^2, \qquad (4.77)$$

and

$$\Delta e = e_1 - e_1{}^* = e_2{}^* - e_2 . \qquad (4.78)$$

In these equations $e_1{}^*$ and $e_2{}^*$ are the charges of the reactants and e_1 and e_2 are those of the products. The effective radii of the reactants are a_1 and a_2. The free energy change of the complete reaction sequence (I) through (III) is ΔF^0, and ΔS_e is the corresponding electronic contribution to the standard entropy of reaction. ΔS_e arises from any change in the electronic degeneracy accompanying the formation of products in the reaction. Depending on the reaction, it is either zero or negligible and can be ignored. The distance r between the centers of reactants in the complex should be chosen so as to maximize the over-all expression for the reaction rate [Eq. (4.73)]. The optical dielectric constant D_{op} is the square of the refractive index. D is the static dielectric constant.

For isotopic exchange reactions, the products in the elementary electron-transfer step [sequences (I) through (III)] are chemically indistinguishable from the reactants though they are distinguishable in some physical properties such as radioactive behavior. In such reactions $\Delta F^0 = 0$, $\Delta S_e = 0$, $e_1{}^* = e_2$, and $e_2{}^* = e_1$. If then $r = 2a_1 = 2a_2 = 2a$, as it will in this case, then, in Eq. (4.76), $e_1e_2 - e_1{}^*e_2{}^* = 0$, and from Eq. (4.76) $m = -1/2$. From Eq. (4.77),

$$\lambda = \frac{1}{2a}\left(\frac{1}{D_{\text{op}}} - \frac{1}{D}\right)(\Delta e)^2 \qquad (4.79)$$

and, from Eq. (4.75),

$$\Delta F = \frac{\lambda}{r} + \frac{e_1{}^*e_2{}^*}{D\,(2a)} = \frac{1}{2a}\left[\frac{e_1{}^*e_2{}^*}{D} + \frac{(\Delta e)^2}{4}\left(\frac{1}{D_{\text{op}}} - \frac{1}{D}\right)\right]. \qquad (4.80)$$

Therefore, knowing the radii of the ions (assumed equal), the charges on the reactant ions, the difference between the charge of a reactant and product ion, and the static and optical dielectric constants of the medium, ΔF can be calculated from Eq. (4.80). Then calculating the collision frequency in solution and multiplying this by exp $(- \Delta F/RT)$, one obtains k'_{bi}.

The a values were calculated as illustrated in the case of iron ions. The radius of a ferrous ion is the sum of its crystallographic radius (0.75 Å) and the diameter of a water molecule (2.76 Å) or 3.51 Å. In a similar way a for the ferric ion is found to be 3.36 Å. The mean of the two values gives $a = 3.44$ Å.

The refractive indexes of water at various temperatures are recorded in the literature, and from their squares the optical dielectric constants of water at various temperatures are obtained. Wyman's (51) values of the static dielectric constant of water were used. Then ΔF was calculated using Eq. (4.80).

From a measured value of k'_{bi}, ΔF was calculated, using Eq. (4.73), taking $Z = 10^{16}$ cc mole^{-1} sec^{-1}. This was called ΔF experimental or ΔF_{expt}.

Similar calculations were made for various other ion species. Also, experimental values of ΔF from measured values of k'_{bi} for various ion species were substituted into Eq. (4.80), and corrected values of a (a_{corr}) were obtained.

The results of these calculations are recorded in Table III.

Marcus thinks that, in the light of the accuracy of prediction of free energies of activation in general, the agreement between the calculated and experimental values of free energy given in Table III is encouraging, since

TABLE III

EXCESS FREE ENERGY OF ACTIVATION AND RADII OF IONS FOR ISOTOPIC EXCHANGE REACTIONS

Reaction	Temp. (°C)	ΔF_{expt} (kcal/mole)	ΔF_{calc} (kcal/mole)	a_{cryst} (Å)	a_{corr} (Å)
Fe^{2+}–Fe^{3+}	0	16.3	9.8	3.44	2.07
Co^{2+}–Co^{3+}	0	16.4	9.9	3.41	2.06
$Fe(CN)_6^{4-}$–$Fe(CN)_6^{3-}$	4	12.7	10.1	4.5	3.59
$Mo(CN)_8^{4-}$–$Mo(CN)_8^{3-}$	2	<12.6	9.5	4.8	>3.60
MnO_4^{2-}–MnO_4^{-}	1	12.8	9.1	2.9	2.06
Tl^{4+}–Tl^{3+}	24.9	18.8	—	—	—
$Os(bipy)_3^{2+}$–$Os(bipy)_3^{3+}$	5	14.7	—	—	2.43

no adjustable parameters were introduced into the calculations. However, in the presentation of the electron-tunneling hypothesis, it was seen that both the ionic charges and ionic radii were considered as adjustable parameters. It is seen that Marcus' calculations show a wide divergence between the crystallographic and what Marcus termed the correct value of a. Marcus' theory, like the electron-tunneling hypothesis, indicates a strong dependence of the free energy [Eq. (4.80)] and hence of the rate upon the static dielectric constant of the medium. However, it was shown that, for the Np(V)–Np(VI) electron-exchange reaction at least, the rate was sensibly independent of the static dielectric constant in mixed solvents.

Marcus (52) has applied his theory of oxidation–reduction reactions to the rates of organic redox reactions.

The Theory of Libby

a. Introduction

Libby (53) points out that electron exchange between aqueous ions has been clearly demonstrated in many cases and some of the characteristics of the kinetics defined. Measurable rates of exchange occur between simple ions, such as ferrous and ferric (54), cerous and ceric (55), and eurapous and europic (56). Immeasurably fast reactions occur between certain coordinated ions like manganate and permanganate (57, 58) and ferrocyanide and ferricyanide (58–62). Catalysis of the exchange between simple ions by the chloride ion, which is apparently first power in the chloride concentration, has been observed (53, 55).

These results may be explained on the basis of the ideas of Franck (63a) based on the Franck-Condon (63b) principle. Essentially Franck assumes that the hydration spheres of ions are unable to move in the time required for the electron to transfer. This makes necessary the movement of hydration energy from one site to another and constitutes a barrier inhibiting the exchange. For simple ions, the hydration atmospheres probably involve considerable difference in energy and geometric arrangement. In the case of large coordinated ions like manganate and permanganate and ferro- and ferricyanide there is, in the electron-transfer process, due to the symmetry of the oxygens and the cyanides, little movement likely of the oxygen atoms or cyanide groups beyond that of the amplitude of the zero-point motion. Also, these larger ions have smaller energies of hydration. Therefore, in the case of these highly symmetric but firmly coordinated ions, the barrier is greatly reduced.

Libby explains the catalysis by small negative ions as probably due to the formation of a linear complex with the negative ion between the two exchanging positive ions. The complex is assumed to be small enough so that the hydration spheres of the two positive ions share several water molecules, and thus cause a reduction in the height of the reorientation barrier.

He divides the problem into two parts. Part 1 deals with the quantum mechanics of the electron-exchange reaction as treated in vacuum using hydrogen molecule ion as a model, since a solution of this case is possible. Part 2 involves the application of the Franck-Condon principle to aqueous solutions.

If it is assumed that the two most important constituent states of the molecule are those in which the electron is either on the one atom or the other. It can be shown (64) that the energy of separation of the two states for this molecule, derivable from any hydrogen-like wave function, can be calculated from the frequency with which the electron interchanges from one atom to the other divided by Planck's constant. From Fig. 4 the exchange frequency is given by

$$h\nu = E_{aa} \frac{2S}{1 - S^2} - E_{ab} \frac{2S}{1 - S^2} \, , \tag{4.81}$$

where

$$E_{aa} = \int \frac{\psi^2 a}{r_b} \, d\tau \, , \tag{4.82}$$

$$E_{ab} = \int \frac{\psi_a \psi_b}{r_a} \, d\tau \, , \tag{4.83}$$

and

$$S = \int \psi_a \psi_b \, d\tau \, . \tag{4.84}$$

One can derive, using $1s$ wave functions,

$$\nu = \nu_0 E^{-R} \left(\frac{2}{R} - \frac{4}{3} R \right) \tag{4.85}$$

in which $\nu_0 = e^2/a_0 = 6.58 \times 10^{15}$ sec^{-1} and R is measured in units of the Bohr radius, 0.5282 Å.

It was shown by Libby that even the $1s$ wave function gave very appreciable rates of exchange at distances of 10 Å and greater.

Libby used the generalized hydrogen molecule ion in which the two positive nuclei have charges of $+ Ze$ to explain the exchange frequencies of various $3d$ wave functions. These exchange frequencies were desired since most of the ions concerned in electron-exchange studies have incomplete $3d$ shells which are probably involved in the actual exchange.

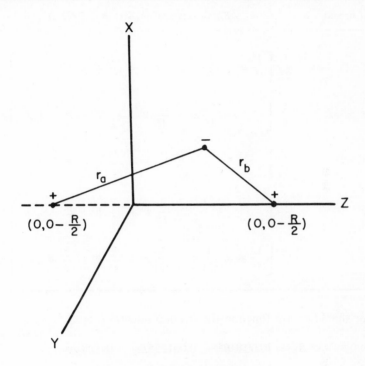

FIG. 4. Hydrogen molecule ion.

Table IV gives the exchange integrals E_{ab} for the five $3d$ wave functions. The z-axis is, as shown in Fig. 4, the axis of the molecule, and the usual nomenclature is used in labeling the five orbitals. As is to be expected, the z^2 orbital has the largest value of the exchange integral, and the discussion is therefore limited to this orbital.

E. P. Wolfgarth (65) used the wave function

$$\psi = A\mathrm{r}^2 \exp\left(-d\mathrm{r}^2\right), \tag{4.86}$$

which is spherically symmetrical, in discussing the exchange integral for the $3d$ states. In this equation A and d are constants. This gives E_{ab} under the label w in Table IV. Libby believes that this could have been used as a sort of average for the $3d$ electrons, but the points he makes in the following presentation seem not to be decisively dependent on this choice.

TABLE IV

Exchange Integrals for $3d$ Wave Functions

$3d$ orbital	E_{ab} (in units of $h\nu_0 Z^2$; $B = RZ/3a_0$)
z^2	$\dfrac{1}{27}\left[\dfrac{1}{15}B^5 - \dfrac{1}{5}B^4 - 3B^3 \quad + \dfrac{2}{5}B^2 + 3B \quad + 3 \right]e^{-B}$
yz or xz	$\dfrac{1}{9}\left[\quad - \dfrac{1}{15}B^4 - \dfrac{2}{15}B^3 + \dfrac{1}{2}B^2 + B \quad + 1 \right]e^{-B}$
$x^2 - y^2$	$\dfrac{1}{9}\left[\qquad\qquad \dfrac{1}{15}B^3 + \dfrac{2}{5}B^2 + B \quad + 1 \right]e^{-B}$
xy	$\dfrac{1}{12}\left[\qquad\qquad \dfrac{1}{15}B^3 + \dfrac{2}{5}B^2 + B \quad + 1 \right]e^{-B}$
w	$\left[\dfrac{1}{625}B^5 + \dfrac{6}{625}B^4 + \dfrac{6}{121}B^3 + \dfrac{7}{45}B^2 + \dfrac{1}{3}B + \dfrac{1}{3}\right]e^{-B}$

For the $3d_{z^2}$ wave function the overlap integral S is

$$S_{z^2} = [0.0040B^6 - 0.001555B^5 - 0.0556B^4 - 0.09523B^3 + 0.2381B^2 + B + 1]\,e^{-B}, \tag{4.87}$$

which gives, for S_{z^2} when B (units $RZ/3a_0$) is 10, a value of 0.15; when B is 15, a value of 0.015; and when B is 20, a value of 0.00049.

The discussion concerns exchange at distances corresponding to $B = 10$ or larger, it is clear that S^2 in the denominator of Eq. (4.81) can be dropped. Even for closer distances the error will not be greater than an order of magnitude.

The Coulombic integral E_{aa} is

$$E_{aaz^2} = Z^2\nu_0 h\left[\left(\frac{30}{B^5} + \frac{4}{3B^3} + \frac{1}{3B}\right) - 128\,\frac{\exp(-2B)}{B^2}\right]. \tag{4.88}$$

This equation is valid for value of B of 5 or greater, since some terms involving $\exp(-2B)$ have been neglected. Finally, substitution in Eq. (4.81) yields, for the exchange frequency, the equation

$$v_{z^2} = Z^2 v_0 \left[\exp\left(-B\right) \left(-0.00227B^2 + 0.0138B^4 + 0.196B^3 \right.\right.$$

$$-0.0976B^2 + 0.0283B + 0.0918 - 2.03\,\frac{1}{B} - 54.5\,\frac{1}{B^2}$$ (4.89)

$$+ 17.0\,\frac{1}{B^3} + 60\,\frac{1}{B^4} + 60\,\frac{1}{B^5} \left.\right) - \exp\left(-3B\right)\left(1.024B^{10}\right.$$

$$\left.\left. - 400B^9\right) \right].$$

The values of v/v_0 from Eq. (4.89) ranged from 3.4×10^{-3} at B (units $ZR/3a_0$) = 10 to -4.4×10^{-9} at $B = 30$. Minus values for v were obtained for B values larger than about 11, but the signs were dropped since the sign of the energy separating the two states does not affect the frequency.

b. *The Franck-Condon Principle and the Hydration Atmospheres*

Franck (*62*) stated that electron transfer in aqueous solutions should be inhibited since the heavy-water molecules constituting the hydration atmosphere of the ions would require a relatively longer time of movement as compared to the transit time of the electron.

If, as in the hydrogen molecule ion, the electron transfer occurs before the movement of the hydration atmosphere, the relative time should be roughly equal to the inverse square root of the mass ratio of the water molecule and the electron, or about 200. The extra energy of hydration at the site of one of the electron-exchanging ions, the ferrous ion in the case of electron transfer between ferrous and ferric ions, must leak across to the original site of the other ion by the slow type of collisional process which accomplishes heat conduction. There is no net energy change in the reaction. Since there is a difference in the rates of movement of the electron and the heat, the electron must transfer against a barrier comparable in magnitude to the energy involved in the subsequent slow reorientation of the water molecules to harmonize with the new charge situation.

The magnitude of the barrier can be calculated as follows: Let the electrical charges which appertain to the ions be dissolved in the solution at infinite distance from each other. The charges are then assembled on a sphere of radius r, equal to the ionic radius and in a medium of dielectric constant D, which for water at 25°C is 78.5. The classical expression $Z^2e^2/2Dr$, where Z is the charge on the ion and e is the electronic charge, gives the work of assembly. The energy barrier ΔH^* in the case under consideration is, therefore,

$$\Delta H^* = \frac{e^2}{2Dr} \; [(Z + 1)^2 - Z^2] = \frac{e^2}{2Dr} \; [2Z + 1] , \qquad (4.90)$$

where Z is the charge on the ion having the smaller charge. For $r = 2$ Å and $D = 80$, the barrier for bi- and trivalent ions would be only 5.4 kcal/mole. But, as Libby points out, for distances as small as ionic radii, D is not so large, and the difference in hydration energies will be considerably larger, perhaps two or three times the lower limit set by the value of 80 for D.

The electron in transit has no time for solvation. Further, the electron polarizability of the aqueous medium as measured by the square of the refractive index will be effective and, therefore, about a two fold reduction of the ionization potential will be expected. This ionization potential is not too dissimilar from that for the gaseous problem discussed in the previous section. Thus this lowering of the energy required to remove the electron from the donor ion partially counterbalances the obstructive action of the molecules and ions that lie between the two exchanging ions. This makes the model of the gaseous hdyrogen molecule ion somewhat more applicable to the solution case.

In the case of exchanging ions surrounded by fixed coordination spheres, as in $Fe(CN)_6^{4-}$ and $Fe(CN)_6^{3-}$ or in MnO_4^- or MnO_4^{2-}, the Franck-Condon principle probably interposes a much smaller boundary since considerable probability of spatial orientation exists in the ground state due to sufficiently large amplitudes of the zero-point vibrations arising from the similarities of the geometries of the ions in the two valence states. Also, the large sizes of the ions reduce the hydration energies to small values where the hydration barrier will not be serious as is seen from Eq. (4.90). Libby therefore states the principle that electron exchange can be catalyzed by complexing the exchanging ions in such a way that the complexes are symmetrical providing their geometries are identical to within the vibration amplitudes involved in zero-point motion. This symmetry principle seems to explain the rapid electron exchange between ions such as the iron cyanides and the manganate–permanganate ions.

Exchange between asymmetrical complexes involving significant energy differences is strongly inhibited by the symmetry principle. Libby believed that optical isomers would exchange even though their geometries are different.

Small negative ions would, on the basis of the Franck-Condon principle, produce by electrostatic effects a symmetric stable configuration involving reduced distances between the exchanging ions with considerable sharing of hydration spheres and a consequent diminution of dissimilarities. This would produce a catalytic effect of the small, negative ion, probably proportional to the first power of the concentration at low concentrations (63b). Such an effect of the chloride ion is found in the europous–europic case (55). There may also be a catalytic effect of the small ion due to the formation of stable complexes with the positive ions. In the case of cyanide ion with the ferrocyanide–ferricyanide pair and in other similar cases there may be catalysis both by electrostatic and complex–formation effects. Perchlorate ion, on the other hand, is very large and should have little catalytic effect, and it is found that iron exchange occurs most slowly in perchlorate media.

Since the reduction in ionization energy by the electronic dielectric constant partially offsets the obstructing effect of coordination shell and the intervening water molecules, Libby believed the gaseous calculation, presented in the previous section, not too inappropriate for aqueous solutions and thought it reasonable to suppose that the electron wave function reaches through several layers of solution, as far as exchange and oxidation–reduction reactions are concerned. The interaction energies required are small, the unit corresponding to only 27.1 eV, so that rapid exchange reactions may result from interaction energies of less than a microvolt. Even though the coordination sphere is of an insulating nature, large complex ions with a reducing metallic ion in the center can be rapid reducing agents.

The mechanism for electron-exchange reactions, whereby the electron is solvated followed by its subsequent transfer from the solvent to the receiving ion, is believed by Libby to be of no great importance, since electrons should reduce water and hence would not be solvated to any extent by it and since the process would involve an energy barrier equal to the difference between the electronic solvation energy and the ionization potential of the donor ion, which would perhaps be several volts.

All oxidation–reduction reactions in aqueous media may conform with the considerations described above. For example, electrodes may be reversible if both ions involved are complexed in such a way that both valence states have the same coordination spheres. As examples of these, the manganate–permanganate and ferro- and ferricyanide half-cells may be cited. The part played by the Franck-Condon principle might not be minor in the electrode reactions themselves. Just what its effects may be is by no means obvious. The problem of overvoltage might be related to this phenomenon.

Excited end products, which satisfy the symmetry principle and which later lose their excitation energy and revert to final end products, might be formed in ordinary exothermic oxidation–reduction reactions. The energy of the exothermic process might the used to cross the Franck-Condon barrier. Heat might be liberated at both the site of the reduced and of the oxidized ion, whereas the over-all heat is zero for exchange reaction, while hydration energy must move from the oxidized to the reduced ion. Only at the site of the oxidized ion is heat generated.

The Theory of Amis: The Model

The model proposed (66, 67) may be a spherical-, spheroidal-, dumbbell-, or capsular-shaped region, preferably the latter. The two reactant ions together with their solvent sheaths occupy the ends of the capsule. A bridge arrangement composed of a water molecule, a hydrogen ion, an oyxgen molecule, the electron in intermediate position, or overlapping electronic orbitals is contained in the center of the capsule. This capsular region contains, in a condition of complete dielectric saturation, the component of the mixed solvent of the highest dielectric constant, which is preferentially solvating the ionic reactants. Thus, since the region contains only the solvent component of highest dielectric constant under conditions of complete dielectric saturation, and since it is the dielectric constant in this region only which can influence the rate of reaction, the reaction rate will be independent of the composition of the solvent and of its gross properties, such as dielectric constant, refractive index, and viscosity. Rearrangement free energies will involve some electronic rearrangement on the part of the participating species.

That preferential solvation of ions with respect to the more polar component of the solvent does occur has been discussed by Amis and co-workers (68).

If D' represents the dielectric constant of the higher dielectric component under conditions of complete dielectric saturation, this D' is the dielectric constant that should be used in Eq. (4.16). It should also be used for D in Eqs. (4.75) and (4.77). D_{op} in Eq. (4.77) would be the square of refractive index of the highest dielectric component of the solvent at complete dielectric saturation and should be independent, as mentioned before, of the gross composition of the solvent.

Hindman (69) and co-workers found for 0°C that the rate of electron exchange between Np(V) and Np(VI) depends on hydrogen, since the exchange reaction was gradually slowed down when ordinary water solvent was progressively replaced by deuterium oxide. They also found that at high acid concentration the rate of the Np(V) and Np(VI) electron reaction depends on the first power of the hydrogen ion concentration when the temperature was 5°C. The reaction was first order with respect to both Np(V) and Np(VI) concentrations as shown by Cohen et al. (70).

Suppose the model of the electron-exchange process be that of a hydronium ion of charge $+\varepsilon$, forming a bridge between ions of charges $z_a\varepsilon$ and $z_b\varepsilon$ for the transfer of the electron between the two ions.

There are two potential barriers to the transfer of the electron, one between the ion of charge $z_a\varepsilon$ from which the electron starts and the hydronium ion of charge ε and another between the hydronium ion and the ion of charge $z_b\varepsilon$ which is the destination of the electron. A diagrammatic sketch is given in Fig. 5, and a potential model showing the potential barriers is given in Fig. 6.

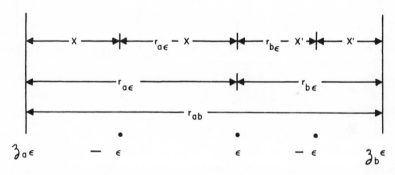

FIG. 5. Particle model for hydronium ion acting as bridge for the exchange of an electron between two ions.

It is realized, of course, that the electron cannot, as might be implied by Fig. 5, be simultaneously in the two positions between the ion of charge $z_a\varepsilon$ and the hydronium ion and between the hydronium ion and the ion of charge $z_b\varepsilon$. However, the total path will include these segments of path.

The problem resolves itself into finding V_m', V_m'', $V_0(z_a)$, $V_0(\varepsilon)$, $V' = V_m' - V_0(z_a)$, $V'' = V_m'' - V_0(\varepsilon)$, W', and W''. The transmission coefficient for the transfer of an electron from ion of charge $z_a\varepsilon$ to the hydronium ion of charge ε can be found from V' and W', and the transmission coefficient for the transfer of the electron from the hydronium of charge ε to ion of charge $z_b\varepsilon$ can be found from V'' and W''. The transmission coefficient for the total process is the product of the two transmission coefficients described and can be found by summing the two potential energies V' and V'' and the two work terms W' and W'' since these appear as exponents in the expression for the two partial transmission coefficients. The mathematical procedure will be similar to that used by Marcus, Zwolinski, and Eyring (*1, 2*).

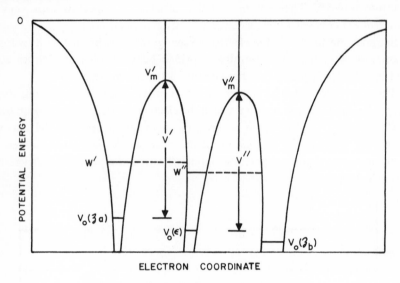

FIG. 6. Schematic model of potential function.

First let us find the transmission coefficient $\varkappa_\varepsilon'$ for the transfer of the electron from the ion of charge $z_a\varepsilon$ to the proton of charge ε. The transmission coefficient $\varkappa_\varepsilon''$ for the transfer of the electron from the proton to the ion of charge $z_b\varepsilon$ will then be found and the product $\varkappa_\varepsilon' \cdot \varkappa_\varepsilon''$ will yield $\varkappa_\varepsilon$ for the total process of transferring the electron from the ion of charge $z_a\varepsilon$ to the ion of charge $z_b\varepsilon$.

If the ion of charge $z_a\varepsilon$ is fixed at a distance $r_{a\varepsilon}$ from the hydronium ion of charge ε, and if the electron is at a distance x from the ion of charge $z_a\varepsilon$ then, using Coulomb's law for point charges, the potential energy V' is given by the expression

$$V' = \frac{z_a\varepsilon^2}{D'r_{a\varepsilon}} - \frac{z_a\varepsilon^2}{D'x} - \frac{\varepsilon^2}{D'(r_{a\varepsilon} - x)} , \qquad (4.91)$$

where D' is the dielectric constant of the higher dielectric constant component of the solvent under conditions of complete dielectric saturation.

The ion of charge $z_a \varepsilon$ is the metal-containing cation of smaller ionization potential.

Maximizing with respect to x one obtains

$$x = \frac{r_{a\varepsilon}}{1 + \gamma} \tag{4.92}$$

where $\gamma^2 = 1/z_a$. Substituting this value of x into Eq. (4.91) there results

$$V_m{}' = \frac{\varepsilon^2}{D' r_{a\varepsilon}} f'(z_a) . \tag{4.93}$$

In this equation

$$f'(z_a) = 2z_a{}^{1/2} + 1 . \tag{4.94}$$

In like manner,

$$V_m{}'' = - \frac{\varepsilon^2}{D' r_{b\varepsilon}} f''(z_b) , \tag{4.95}$$

where

$$f''(z_b) = 2z_b{}^{1/2} + 1 . \tag{4.96}$$

The total maximum potential V_m for the entire system is

$$V_m = V_m{}' + V_m{}'' = - \frac{\varepsilon^2}{D'} \left(\frac{f(z_a)}{r_{a\varepsilon}} + \frac{f(z_b)}{r_{b\varepsilon}} \right). \tag{4.97}$$

The total zero-point energy V_0 is given by the equation

$$V_0 = V_0(z_a) + V_0(\varepsilon) = - \frac{z_a{}^* \varepsilon^2}{r_{0a}} - \frac{z_\varepsilon \varepsilon^2}{r_{0\varepsilon}} = - \left(\frac{z_a{}^*}{r_{0a}} + \frac{z_\varepsilon{}^*}{r_{0\varepsilon}} \right) \varepsilon^2, \tag{4.98}$$

$z_a{}^*$ is the charge on the central atom of the complex ion whose central coordinating atom or ion has the smallest ionization potential and $z_\varepsilon{}^*$ is the charge on the hydronium ion and is taken to be unity. The quantities are related theoretically to the square of the effective principal quantum number of the central atom of the two ions and to the Bohr radius. In this treatment these quantities will be taken merely as adjustable parameters.

In like manner the total work will be given by

$$W = W' + W'' = \left(\frac{z_a{}^*}{2r_{0a}} + \frac{z_\varepsilon{}^*}{2r_{0\varepsilon}} \right) \varepsilon^2. \tag{4.99}$$

Therefore the total height of the potential barrier V can be found from $V_m - V_0$ which gives, from Eqs. (4.97) and (4.98),

$$V = V_m - V_0 = \left(\frac{z_a{}^*}{r_{0a}} + \frac{z^*}{r_{0\varepsilon}} \right) \varepsilon^2 - \frac{\varepsilon^2}{D'} \left(\frac{f'(z_a)}{r_{0\varepsilon}} + \frac{f''(z_b)}{r_{b\varepsilon}} \right), \quad (4.100)$$

and

$$V - W = \left(\frac{z_a{}^*}{r_{0a}} - \frac{z_\varepsilon{}^*}{r_{0\varepsilon}} \right) \frac{\varepsilon^2}{2} - \frac{\varepsilon^2}{D'} \left(\frac{f'(z_a)}{r_{a\varepsilon}} + \frac{f''(z_b)}{r_{b\varepsilon}} \right). \quad (4.101)$$

Now let

$$f(z) = \tfrac{1}{2} \left[f'(z_a) + f'(z_b) \right] = z_a^{1/2} + z_b^{1/2} + 1 . \quad (4.102)$$

To reduce the number of parameters and simplify calculations we shall make two assumptions. These are

$$r_{0a} = r_{0\varepsilon} = r_0 , \quad (4.103)$$

$$r_{a\varepsilon} = r_{b\varepsilon} = \tfrac{1}{2} r_{ab} . \quad (4.104)$$

The assumption represented by Eq. (4.104) is probably not in too great error if the charges $z_a{}^*$ and $z_b{}^*$ on the central atoms of the two electron-exchanging complex ions do not differ greatly. However, the assumption represented by Eq. (4.103) is probably much more questionable. Using these assumptions,

$$V - W = \frac{(z_a{}^* + z_\varepsilon{}^*)}{2r_D} \varepsilon^2 - \frac{4\varepsilon^2}{D'r_{ab}} f(z) . \quad (4.105)$$

We now write the equation for the total transmission coefficient:

$$\varkappa_\varepsilon = \varkappa_\varepsilon' \varkappa_\varepsilon'' = \exp \left[- \frac{8\pi}{3h} r_{ab}(2m(V - W))^{1/2} \right] \quad (4.106)$$

$$= \exp \left\{ - \frac{8\pi}{3h} r_{ab} \left[2m \left(\frac{(z_a{}^* + z_b{}^*)}{2r_0} \varepsilon^2 - \frac{4\varepsilon^2}{D'r_{ab}} f(z) \right) \right]^{1/2} \right\} .$$

The electrostatic repulsion to the energy contribution ΔF_{rep} to the free energy of activation is

$$\Delta F_{\text{rep}} = - \frac{z_a \varepsilon^2}{kTD'r_{a\varepsilon}} - \frac{z_b \varepsilon^2}{kTD'r_{b\varepsilon}} - \frac{z_a z_b \varepsilon^2}{kTD'r_{ab}}$$

$$= - \frac{\varepsilon^2}{kTD'r_{ab}} \left[2(z_a + z_b) + z_a z_b \right] = - \frac{\varepsilon^2 f'(z)}{kTD'r_{ab}} , \quad (4.107)$$

where

$$f'(z) = [2(z_a + z_b) + z_a z_b] \,. \tag{4.108}$$

It will be assumed that there is no activation free energy of rearrangement of hydration shells during the rate-controlling process since the solvent shells will be assumed to be fixed during the instant of time required for the exchange process to take place. The expression for the specific velocity constant k' is, therefore,

$$k' = \frac{kT}{h} \exp \left\{ - \frac{8\pi}{3h} r_{ab} \left[2m\varepsilon^2 \left(\frac{(z_a{}^* + z_\varepsilon{}^*)}{2r_0} - \frac{2}{D'r_{ab}} f(z) \right) \right]^{1/2} \right.$$

$$\left. - \frac{\varepsilon^2 f'(z)}{kTD'r_{ab}} - \frac{\Delta F_r}{RT} \right\} \,. \tag{4.109}$$

In this equation, k is the Boltzmann gas constant, h is Planck's constant, m is the rest mass of the electron, ΔF_r is the electronic, as contrasted to the solvent, rearrangement free energy, and the other quantities have already been defined.

To find the interaction distance r_{ab} in the activated state it is necessary to find the extremal value for the specific velocity constant with respect to r_{ab}. Let us define the following dimensionless parameters:

$$a = 64\pi^2 m\varepsilon^2 r_0 (z_a{}^* + z_\varepsilon{}^*)/9h^2 \,, \tag{4.110}$$

$$b = 512\pi^2 m\varepsilon^2 r_{ab} f(z)/9h^2 D \,, \tag{4.111}$$

$$c = \frac{\varepsilon^2 f'(z)}{kTDr_0} \,, \tag{4.112}$$

$$\alpha = \frac{1}{r_0 kT} \times \frac{d\Delta F_r}{dx} \approx 0 \,, \tag{4.113}$$

and the normalized variables

$$y = k'h/kT \,, \tag{4.114}$$

$$x = r_{ab}/r_0 \,. \tag{4.115}$$

Using these parameters we can put Eq. (4.109) in the following logarithmic form

$$- \ln y = (ax^2 - bx)^{1/2} + c/x \,. \tag{4.116}$$

To maximize with respect to x the equation is differentiated with respect to x, and the results set equal to zero and solved for x. Then

$$-\frac{d \ln y}{dx} = \frac{(2ax - b)}{2(ax^2 - bx)^{1/2}} - \frac{c}{x^2} = 0. \qquad (4.117)$$

Factoring we have

$$a^{1/2}(1 - b/2ax)/(1 - b/ax)^{1/2} = \frac{c}{x^2}. \qquad (4.118)$$

Expanding the square root term by the binomial theorem and retaining only the first two terms of the expansion simplifies Eq. (4.118) to

$$x^2 = c/a^{1/2}. \qquad (4.119)$$

Substituting the values of a, c, and x from Eqs. (4.110), (4.112), and (4.115) into Eq. (4.119) gives

$$r_{ab}^2 = \frac{3hf'(z)\varepsilon r_0^{1/2}}{8\pi kTD'[m(z_a^* - z_\varepsilon^*)]^{1/2}}. \qquad (4.120)$$

Equation (4.109) will be tested by applying it to the data for the rate of the electron-transfer process between Np(V) and Np(VI). The dielectric constant D' will be taken as that of the highest dielectric component (water) of the mixed solvents at the temperature of the experiments, namely, 0°C. The dielectric constant of water at 0°C is 88.3. The actual dielectric constant in the capsular-shaped volume occupied by the reaction complex, although constant, will probably be somewhat less than 88.3, due to dielectric saturation of the included solvent. Since, however, the actual value of D' is not known, 88.3 will be used as a fair approximation. The quantities z_a^* and z^* will be taken as 5.0 and 1.0, respectively. The quantity z_a^* is the charge on the central atom of the complex pentavalent neptunium ion, and z^* is the charge on the hydronium ion. The quantity r_0 will be taken as a variable depending on the values of z_a, z_b, and r_{ab}. The latter quantity will be set at 10.0 Å. It would probably have been better to select a value of r_0 and calculated r_{ab}, but these are both really adjustable parameters, and r_{ab} = 10.0 Å seems a reasonable, though somewhat small, value for the distance between two neptunium ions surrounded by their water sheaths and separated by the hydronium ion. Cohen et al. (71) estimate the distance between two neptunium ions to exceed 7.5 Å without the presence of the hydronium ion bridge. This is equivalent to assigning a radius of 1 Å to both Np(V) and Np(VI) ions plus the diameters of two water molecules. In the present instance if we assume the diameter of the hydronium ion to equal to that of the water molecule and use the other dimensions given by Cohen, Sullivan, Amis and Hindman, the distance between the centers of neptunium ions will be 10.3 Å.

The rearrangement free energy of activation for electron-exchange reactions was chosen as 8.1 kcal/mole by Marcus, Zwolinski, and Eyring. Cohen, Sullivan, Amis, and Hindman estimated 4 kcal/mole as the rearrangement free energy of activation for the Np(V)–Np(VI) reaction since the symmetry of the ions would cause a lower rearrangement free energy to be required. There is no way, of which the author is aware, of calculating this quantity, since the electronic states of the species involved are not known for the activated state. It is conceivable that in the formation of the activated state the electron rearrangements of the particles are spontaneous and actually contribute energy to the process. The energy expended on the reaction species in bringing them into the activated configuration may make the normal electron configuration untenable and hence cause a spontaneous electron rearrangement in the species.

In Table V the results of the calculation using ΔF_r as 4 kcal/mole are tabulated. The value of k' is seen to be strongly dependent upon the values chosen for the parameters z_a, z_b, and r_0. The values of z_a and z_b are fairly reasonable. Especially are the values $z_a = 1.80$ and $z_b = 2.80$, which give the k' value of 19.2 liters mole^{-1} sec^{-1}, reasonable since the Np–O bonds are somewhere between single and double bonds (70–74), and also since the experimental value of k' is 19.5 liters mole^{-1} sec^{-1}. The only parameter

TABLE V

VALUES OF z_a, z_b, r_0, AND k' [a]

z_a	z_b	r_0 (Å)	k' (calc) (liters/mole^{-1} sec^{-1})
1.85	3.00	16.4	1.59
1.85	2.85	19.5	11.2
1.80	2.85	20.2	14.6
1.80	2.80	20.7	19.2
1.00	2.00	65.6	5260

[a] Calculated for $D' = 88.3$, $z_a^* = 5.0$, $z^* = 1.0$, and $r_{ab} = 10.0 \times 10^{-8}$ cm. $t = 0°C$, k'_{exptl} (0.1 M HClO$_4$) = 19.5 liters mole^{-1} sec^{-1}, $\Delta F_r = 4.0$ kcal/mole.

which assumes a value that is somewhat unreasonable is r_0. This parameter seems too large by a factor of 3 or 4.

In Table VI are listed values of z_a, z_b, r_0 and ΔF_r to give values of k' approaching the experimental value of 19.5 liters mole^{-1} sec.$^{-1}$. In these calculations the values of D', z_a^*, z_b^*, and r_{ab} were the same as for Table V.

It is seen that, as the value of ΔF_r is continually lowered, the value of r_0 grows more reasonable down to $\Delta F_r = -4.0$ kcal/mole. This improvement in r_0 is at some sacrifice in acceptable values of z_a and z_b though the extreme values used here, of $z_a = 2.90$ and $z_b = 3.86$, are not too unreasonable.

TABLE VI

VALUES OF z_a, z_b, r_0, AND ΔF_r TO GIVE VALUES OF k' APPROACHING 19.5 liters mole^{-1} sec^{-1} [a]

z_a	z_b	r_0 (Å)	ΔF_r (kcal/mole)	k' (calc) (liters/mole^{-1} sec^{-1})
1.80	2.80	20.7	4.0	19.2
2.35	3.39	11.1	0.0	19.7
2.90	3.86	6.87	—4.0	20.0

[a] Using $D = 88.3$, $z_a^* = 5.0$, $z^* = 1.0$, and $r_{ab} = 10.0 \times 10^{-8}$ cm. $t = 0°C$, k'_{exptl} (0.1 M HClO$_4$) = 19.5 liters/mole^{-1} sec^{-1}.

It thus appears that a rearrangement free energy of 4.0 kcal/mole is too great for the Np(V)–Np(VI) electron-exchange reaction.

The theory does obviate the difficulty noted by Cohen et al. (70) of giving a curve of negative slope when values of log k' are plotted as ordinates versus the reciprocal of dielectric constant. Since the composition of the solvent in the capsular volume of the complex and the condition of complete dielectric saturation of this solvent do not change with the composition of the gross solvent, log k' is independent of the reciprocal of the dielectric constant of the gross solvent as was observed experimentally. The theory also accounts for a first-order dependence of the rate of the Np(V)–Np(VI) reaction on the hydrogen ion concentration. The mixed solvent must have emphasized this acid-dependent path to the exclusion of the acid-independent path found for lower acid concentrations by Cohen et al. (70).

Further, the model presented here would account for the dependence of the rate on hydrogen, as has been observed (69). It is certainly possible, if deuterium were substituted for hydrogen in the hydronium ion of the model presented here, that this alteration of the nature of the bridge across which the electron travels would change the rate of the electron exchange between the two neptunium ions. Then too the solvent sheaths would become deuterated and, while the dielectric constant of H_2O and D_2O are very nearly the same (74), the masses would be altered and other isotope effects become operative.

It should be pointed out that Eq. (4.109) would hold for a system where the dielectric constant D' varied with the gross composition of the solvent; however, it is assumed here that D' is the dielectric constant of the higher dielectric component of the solvent under conditions of complete dielectric saturation and is therefore constant at all gross compositions of the solvent.

A similar procedure could be used to derive equations for bridges composed of water molecules, chloride ions, or other substances.

For example, if two positive ions A and C having charges $z_a \varepsilon$ and $z_c \varepsilon$, respectively, are separated by distance r_{ac}, and if a negative ion B of valence $z_b \varepsilon$ is placed between A and C at distance r_{ab} from ion A and r_{cb} from ion C, the equation for the rate of the electron-exchange reaction is

$$k' = \frac{kT}{h} \left\{ - \frac{8\pi}{3h} r_{ac} \left[2m \left(\frac{z_a{}^* + z_b{}^*}{2r_0} + \frac{4\varepsilon^2 f(n)}{dr_{ac}} \right) \right]^{1/2} \right.$$
$$\left. + \frac{\Delta F_r}{RT} + \frac{\varepsilon^2 f_r(n)}{kTDr_{ac}} \right\}, \tag{4.121}$$

where

$$f(n) = 2z_b + z_a z_b + z_c z_b - z_a - z_c \tag{4.122}$$

$$f_r(n) = 2z_b(z_a + z_c) - z_a z_c , \tag{4.123}$$

and

$$r_{ac}^2 = \frac{3h\varepsilon f_r(n)r_0^{1/2}}{8\pi [m(z_a + z_b)]^{1/2}kTD} , \tag{4.124}$$

and the meaning of the other terms can be surmised from previous treatment.

REFERENCES

1. R. J. Marcus, B. J. Zwolinski, and H. Eyring, *J. Phys. Chem.* **58**, 432 (1954).
2. B. J. Zwolinski, R. J. Marcus, and H. Eyring, *Chem. Rev.* **55**, 157 (1955).
3. J. Weiss, *Proc. Roy. Soc. (London)* **A222**, 128 (1954).
4a. R. A. Marcus, *J. Chem. Phys.* **24**, 966 (1956); **24**, 979 (1956).
4b. N. Sutin, *Ann. Rev. Nuclear Sci.* **12**, 285 (1962).
4c. J. Halpern, *Quart Rev.* **15**, 207 (1961).
5. E. S. Amis, Accademia Nazionale dei Lincei, Fondazione Donegani, Estratto dal Fascicolo "Cinetica Chimica" (II° Corso Estivo di Chimica, Varenna, 7–22 agosto 1957), p. 53.
6. E. Rabinowitch, *Rev. Mod. Phys.* **14**, 112 (1942).
7. N. F. Mott and I. N. Sneddan, "Wave Mechanics and Its Applications." Oxford Univ. Press (Clarendon), London and New York, 1948.
8. O. K. Rice "Electronic Structure and Chemical Binding" p. 96. McGraw-Hill, New York, 1940.
9. W. B. Lewis, C. D. Coryell, and J. W. Irvine, Jr., *J. Chem. Soc.*, Suppl. 2 1949, S 386.
10. G. Harbottle and R. W. Dodson, *J. Am. Chem. Soc.* **73**, 2442 (1951).
11. S. C. Furman and C. S. Garner, *J. Am. Chem. Soc.* **74**, 2333 (1952).
12. J. Silverman and R. W. Dodson, *J. Phys. Chem.* **56**, 846 (1952).
13. J. W. Gryder and R. W. Dodson, *J. Am. Chem. Soc.*, **73**, 2890 (1951).
14. D. Cohen, J. C. Sullivan, E. S. Amis, and J. C. Hindman, *J. Am. Chem. Soc.* **78**, 1543 (1956).
15. D. Cohen, J. C. Sullivan, and J. C. Hindman, *J. Am. Chem. Soc.* **76**, 352 (1954).
16. W. H. Zachariasen, "The Actinide Elements," pp. 784-785. McGraw-Hill, New York, 1954.
17. J. C. Eisenstein and M. H. L. Pryce, *Proc. Roy. Soc. (London)* **A229**, 20 (1955).
18. R. Platzman and J. Franck, *Z. Physik* **138**, 411 (1954).
19. E. S. Amis and G. Jaffé, *J. Chem. Phys.* **10**, 646 (1942).
20. D. Cohen, J. C. Sullivan, and J. C. Hindman, *J. Am. Chem. Soc.* **77**, 4964 (1955).
21. L. Landau, *Physik Z. Sowjet* **1**, 88 (1932).
22. L. Landau, *Physik Z. Sowjet* **2**, 46 (1932).
23. D. R. Bates and H. S. W. Massey, *Phil. Trans.* **A239**, 269 (1943).
24. R. P. Bell, *Proc. Roy. Soc. (London)* **A148**, 241 (1935).
25. H. Hellman and J. K. Syrkin, *Acta Physicochim.* **2**, 433 (1935).
26. G. Gamow, *Z. Physik* **51**, 205 (1928).
27. R. W. Gurney, "Ions in Solution." Cambridge Univ. Press, London and New York, 1936.
28. W. Schottky and H. Rothe, "Handbuch der Physik" (H. Geiger and K. Scheel, eds.), 1st ed., Vol. 13, p. 162. Springer, Berlin, 1928.
29. P. Debye, "Polare Molekeln." Hirzel, Leipzig, 1929.
30. C. K. Ingold, *J. Chem. Soc.* **1931**, 2179.

31. J. D. Bernal and R. H. Fowler, *J. Chem. Phys.* **1**, 515 (1933).

32. L. Landau, *Phys. Z. Sowjet.* **3**, 664 (1933).

33. N. F. Mott and R. W. Gurney, "Electronic Processes in Ionic Crystals." Oxford Univ. Press (Clarendon), London and New York, 1940.

34. R. Platzman and J. Franck. "L. Farkas Memorial Volume," p. 21. Research Council of Israel, Jesusalem, 1952.

35. P. Debye, *Trans. Am. Electrochem. Soc.* **82**, 265 (1942).

36. R. Lemberg and J. W. Legge, "Hematin Compounds and Bile Pigments." Oxford Univ. Press (Clarendon), London and New York, 1949.

37. H. Theorell, *Advances in Enzymol.* **7**, 265 (1947).

38. W. F. Libby, *J. Phys. Chem.* **56**, 863 (1952).

39. W. M. Latimer, K. S. Pitzer, and C. M. Slansky, *J. Chem. Phys.* **7**, 108 (1939).

40. J. Silverman and R. W. Dodson, *J. Phys. Chem.* **56**, 846 (1952).

41. D. J. Meier and C. S. Garner, *J. Phys. Chem.* **56**, 853 (1952).

42. H. C. Hornig and W. F. Libby, *J. Phys. Chem.* **56**, 869 (1952).

43. W. B. Lewis, C. D. Coryell, and J. W. Irvine, Jr., *J. Chem. Soc.*, Suppl. 2 **1949**, S 386.

44. H. C. Brown, *J. Phys. Chem.* **56**, 868 (1952).

45. A. W. Adamson, *J. Phys. Chem.* **56**, 858 (1952).

46. J. Weiss, *J. Chem. Phys.* **19**, 1066 (1951).

47. L. Eimer, A. Medalia, and R. W. Dodson, *J. Chem. Phys.* **20**, 743 (1952).

48. H. C. Hornig, G. L. Zimmerman, and W. F. Libby, *J. Am. Chem. Soc.* **78**, 3808 (1950).

49. J. W. Cobble and A. W. Adamson, *J. Am. Chem. Soc.* **72**, 2276 (1950).

50. G. D. Dorough and R. W. Dodson, Brookhaven Quart. Rept. July–September, 1951.

51. J. Wyman, Jr., *Phys. Rev.* **35**, 623 (1930).

52. R. A. Marcus, *J. Chem. Phys.* **26**, 872 (1957).

53. W. F. Libby, *J. Phys. Chem.*, **56**, 863 (1952).

54. J. Silverman and R. W. Dodson, Brookhaven Quart. Rep. BNL-93, p. 65, 1950.

55. J. W. Gryder and R. W. Dodson, *J. Am. Chem. Soc.* **71**, 1894 (1949).

56. D. J. Meter and C. S. Garner *J. Am. Chem. Soc.* **73**, 1894 (1951).

57. H. C. Hornig, G. L. Zimmerman, and W. F. Libby, *J. Am. Chem. Soc.* **72**, 3808 (1950).

58. N. A. Bonner and H. A. Potratz, *J. Am. Chem. Soc.* **73**, 1845 (1951).

59. R. C. Thomson, *J. Am. Chem. Soc.* **70**, 1045 (1948),

60. W. B. Lewis, ONR Tech. Rep. No. 19, M.I.T., 1949.

61. J. W. Cobble and A. W. Adamson, *J. Am. Chem. Soc.* **72**, 2276 (1950).

62. L. Eimer and R. W. Dodson, Brookhaven Quart. Rep. BNL-93, 1950.

63. (a) Cf. E. J. B. Wiley, "Collisions of the Second Kind, Arnold, London, 1937; and review by K. J. Laidler and K. E. Schuler, *Chem. Rev.* **48**, 153 (1951).

(b) W. F. Libby, *Abstr. Phys. and Inorg. Section, 115th Meeting Am. Chem. Soc.,* *San Francisco, Calif., March. 27–April 1,* 1949.

64. H. Eyring, J. Walter, and G. E. Kimball, "Quantum Chemistry," pp. 192–199. Wiley, New York, 1944.

65. E. P. Wolfgarth, *Nature* **163,** 57 (1949).

66. E. S. Amis, *J. Chem. Phys.* **26,** 880 (1957).

67. E. S. Amis, Accademia Nazionale dei Lincei, Fondazione Donegani, Estratto dal Fascicolo " Cinetica chimica" (II° Corso Estivo di Chimica - Varenna, 7–22 agosto 1957).

68. E. S. Amis, *J. Phys. Chem.* **60,** 428 (1956); N. G. Foster and E. S. Amis, *Z. physik. Chem.* [N. F.] **7,** 360 (1956); R. Whorton and E. S. Amis, *Z. physik. Chem.* [N. F.], **8,** 9 (1956); N. Goldenberg and E. S. Amis, *ibid* [N. F.] **22,** 3 (1959); **31,** 10 (1962).

69. J. C. Hindman, private communication.

70. D. Cohen, J. C. Sullivan, and J. C. Hindman, *J. Am. Chem. Soc.* **76,** 352 (1954).

71. D. Cohen, J. C. Sullivan, E. S. Amis, and J. C. Hindman, *J. Am. Chem. Soc.* **78,** 1543 (1956).

72. W. H. Zachariasen, "The Actinide Elements," pp. 784-785. McGraw-Hill, New York, 1954.

73. J. C. Eisenstein and M. H. L. Pryce, *Proc. Roy. Soc. (London)* **A229,** 20 (1955).

74. J. Wyman, Jr. and E. N. Ingalls, *J. Am. Chem. Soc.* **60,** 1182 (1954).

CHAPTER V

SOLVENT EFFECTS COMMON TO REACTIONS INVOLVING VARIOUS CHARGE TYPE REACTANTS

THEORETICAL

Comparison of Rates of Reaction in the Gas and Solution Phases

If the orders of reactions in the gas phase and in solutions are the same, their specific velocities can be compared. If the gaseous reactants are at sufficiently low pressures and if the reactants in solution are dilute enough, all activity coefficients in the equation

$$k' = \nu K^* \frac{f_A f_B}{f_X} \tag{5.1}$$

can be assumed to be unity, and the equation may be written

$$k' = \nu K^*, \tag{5.2}$$

where k' is the specific velocity constant for a particular order of reaction either in the gas phase or in solution, ν is an average vibration frequency, K^* is the equilibrium constant for the formation of the complex, and the f's are the activity coefficients of the respective species. Now if the subscript S refers to the solution phase and the subscript g to the gas phase, then

$$\frac{k_S'}{k_g'} = \frac{\nu_S K_S^*}{\nu_g K_g^*}, \tag{5.3}$$

and if the average vibration frequency is independent of the phase $\nu_S = \nu_g$ and

$$\frac{k_S'}{k_g'} = \frac{K_S^*}{K_g^*}. \tag{5.4}$$

Thus the ratio of the specific velocities is equal to the ratio of the equilibrium

121

constants for formation of the intermediate complex in the two phases. The rate of reaction will be fastest in the phase that favors the more abundant formation of the complex. It might be pointed out that, for two solutions S_1 and S_2,

$$\frac{k_{S_1}}{k_{S_2}} = \frac{K_{S_1}^*}{K_{S_2}^*} \tag{5.5}$$

when the solutions are dilute and all activity coefficients are unity. Remembering that for dilute states

$$K^* = \frac{a_X}{a_A a_B \dots} = \frac{C_X}{C_A C_B \dots} \frac{f_X}{f_A f_B \dots} = \frac{C_X}{C_A C_B \dots}, \tag{5.6}$$

then

$$\frac{k_S'}{k_g'} = \frac{C_{XS}}{C_{AS} C_{BS} \dots} \Big/ \frac{C_{Xg}}{C_{Ag} C_{Bg} \dots} = \frac{C_{XS} C_{Ag} C_{Bg} \dots}{C_{Xg} C_{AS} C_{BS} \dots} \tag{5.7}$$

But, for gases, $C_{ig} = p_{ig}/RT$ where p_{ig} is the partial pressure of the ith component in the gas phase. For dilute solutions the concentration of the component C_{iS} is obtainable from the relationships $p_{iS} = x_{iS} p_i^o$, and $x_{iS} \cong C_{iS} V_S^o$ to be $C_{iS} \cong p_{iS}/V_S^o p_i^o$, where p_i^o is the vapor pressure of the ith component in the pure state and V_S^o is the molar volume of the pure solvent. From Eq. (5.7) we have then an nth-order reaction

$$\frac{k_S'}{k_g'} = \frac{p_A^o p_B^o \dots}{p_X^o} \frac{p_{Ag} p_{Bg} \dots}{p_{Xg}} \frac{p_{XS}}{p_{AS} p_{BS} \dots} \left(\frac{V_S^o}{RT}\right)^{n-1}. \tag{5.8}$$

But for equilibrium $p_{ig} = p_{iS}$, therefore,

$$\frac{k_S'}{k_g'} = \frac{p_A^o p_B^o \dots}{p_X^o} \left(\frac{V_S^o}{RT}\right)^{n-1}. \tag{5.9}$$

This equation was derived by Benson (1) who used the Clausius-Clapeyron equation in the form $\ln (p/p_B) = (\Delta H^o/RT_b)(1 - T_b/T)$ and the Hildebrand-Scott (2) relation

$$\Delta S^o = \frac{V_g^o}{V_{Sb}^o} = RT \ln \frac{RT_b}{V_{Sb}^o P_b} \tag{5.10}$$

to substitute for the values of p_i^o and V_S^o in Eq. (5.9) and obtain

$$\frac{k_S'}{k_g'} = \frac{V_S^o}{V_{SbA}} \frac{V_S^o}{V_{SbB}} \dots \frac{V_{SbX}^o}{V_S^o} \exp \left(\frac{\Delta H_{vap}^o}{RT}\right). \tag{5.11}$$

In these equations ΔH^o_{vap} is the difference in the heats of vaporizations at their boiling points of X and the reactants; n is the number of reactants associating to form the complex; ΔS^o is the entropy of vaporization, $\Delta H^o_{vap}/T_b$; T_b is the boiling temperature; $V_g{}^o$ is the molar volume of the solvent in the gaseous state; V^o_{Sb} is the free volume per molecule of a component at its boiling point, and p_b is one atmosphere.

For nonideal solutions in which the partial pressures have to be corrected for deviations from Raoult's law by including the activity coefficients, $f_i{}^o$, of the components, Eq. (5.11) becomes

$$\frac{k_S{}'}{k_g{}'} = \frac{f_A{}^o f_B{}^o}{f_X{}^o} \frac{V_S{}^o}{V^o_{SbA}} \frac{V_S{}^o}{V^o_{SbB}} ... \frac{V^o_{SbX}}{V_S{}^o} \exp\left(\frac{\Delta H^o_{vap}}{RT}\right) \qquad (5.12)$$

when $p_i = x_i p_i{}^o f_i{}^o$ is used intead of $p_i = x_i p_i{}^o$ in the derivation.

Benson points out that when the energy of activation for a reaction, or the enthalpy of reaction for an equilibrium, is the same for the gas as for the liquid phase, then

$$\Delta H^o_{vap} = \Delta E^o_{vap} - (n-1)RT = -(n-1)RT,$$

since ΔE^o_{vap} is zero. Equation (5.11) becomes

$$\frac{k_S{}'}{k_g{}'} = \frac{V_S{}^o}{V^o_{SbA}} \frac{V_S{}^o}{V_{SbB}} ... \frac{V^o_{SbX}}{V_S{}^o} \exp\left[-(n-1)\right]. \qquad (5.13)$$

Finally by assuming that the molar volumes of the solvent S and solutes A, B... are equal, that the molar volume of X is the sum of the molar volumes of A, B, C..., and remembering that for most liquids the free volumes are about 0.01 of their molar volumes, $V_S{}^o/V^o_{Sbi}$ becomes approximately 100 and $V^o_{Sbx}/V_S{}^o$ approximates $n/100$, therefore, Eq. (5.13) can be written

$$\frac{k_S{}'}{k_g{}'} = (100)^n \frac{n}{100} \exp\left[-(n-1)\right] \qquad (5.14)$$

$$= \frac{(10^2)^{n-1}n}{e^{n-1}} = \frac{(10)^{2n-2}n}{e^{n-1}}. \qquad (5.15)$$

Equation (5.15) predicts that for unimolecular reaction, $n = 1$, the specific velocity constants will be the same in the gas phase and in solution unless the energies of activation are different in the two phases, or perhaps unless the reaction in solution is diffusion controlled. Benson states that, for the few unimolecular reactions studied, the rates are the same in both the gas phase and in solution. He points out, however, that there are uncertainties in both the mechanisms and energies of activation which makes detailed comparisons unfruitful.

For bimolecular reactions, $n = 2$, k_S'/k_g' approximates 75. This approaches two orders of magnitude, and hence, as pointed out by Benson, reactions in solution would be expected to be faster than in the gas phase, provided the mechanisms and the energies of activation are the same in the two phases. However, Benson states that, in the main, both rate and equilibrium constants are approximately the same in the two phases.

For $n = 3$ the ratio k_S'/k_g' would be about 4100. While there are no data available for trimolecular reaction rates in both the gas phase and in solution, the reactions of NO with O_2, Cl_2, and Br_2 have been studied in the gas phase and, according to Benson, would be of interest to study in solution.

For nonideal solution there will probably be greater variations than for ideal solutions in the ratio k_S'/k_g' when in the former case there is great attraction between solute species or between the solvent and one or more of the solute species. Such marked solvent effects will be evidenced by a change in activation energy, which, in general, is partially compensated for by an alteration in the frequency factor. The energy of vaporization of the strongly solvated solute will be increased, but this effect will be balanced by a decrease in its effective free volume. Benson presents data on several reactions to illustrate these points. The data on the addition of cyclopentadiene to benzoquinone (3, 4) in various solvents show that, although the energy of activation E varies by 4 kcal/mole over the range of solvents listed, there is a compensating change in the Arrhenius frequency term A by a a factor of 300, so that the rate changes only by a factor of 5. The similar but less polar dimerization of cyclopentadiene in the gas phase and in a variety of nonpolar and polar solvents evinces only small changes in E and A and very little change in the specific velocity constant. The data used by Benson for this reaction were those compiled by Wasserman (5). A greater effect of changing solvent was found in the case of the first-order rate of decarboxylation of malonic acid. Excluding the water–solvent data of Hall (6) at pH 9, Benson found that the data of Clark (7) and of Hall (6) show a variation in the enthalpy of activation of over 9 kcal/mole and a variation in the specific velocity constant of a factor of about 12. A greater variation in k' was prevented by a compensating change in the entropy of activation with change in enthalpy of activation. On the other hand, if the activated complex is polar and highly solvated, its activity coefficient is small; and if at the same time the less polar reactants are not solvated their activity coefficients are not reduced. In such a case, by Eq. (5.12), the ratio of the rate in a polar, solvating solvent to that in the gas phase or less polar solvent should be large. Laidler (8) found that the reaction of triethylamine with ethyl iodide at 100°C was 39,000 times as fast in nitrobenzene as in hexane. In this reaction, the activated complex probably is polar like the product and hence is extensively solvated by the nitrobenzene but not the benzene. The mechanistic picture would perhaps be somewhat as follows:

$$\begin{array}{c}
CH_3CH_2 \\
\diagdown \\
CH_3CH_2\!-\!N \ + \ CH_3CH_2I \ \longrightarrow \\
\diagup \\
CH_3CH_2
\end{array}$$

$$\begin{array}{c}
CH_3\!-\!CH_2 \qquad\quad I^{-\delta} \\
\diagdown \qquad\qquad\ \vdots \\
+ \ \delta CH_3CH_2\!-\!N^{-\delta}\!\cdots\!CH_2CH_3 \ \longrightarrow \ (CH_3CH_2)_4N^{+\delta}I^{-\delta} \\
\diagup \qquad\quad {}_{+\delta} \\
CH_3CH_2
\end{array}$$

$$(5.16)$$

The Influence of Internal Pressure or Cohesion on Reaction Rates

The internal pressure or cohesion of the solvent influences the rate of reaction. Glasstone *et al.* (9), Laidler (8), and Richardson and Soper (10) discuss this effect. The expression $E\sigma/v^{1/3}$ was used by Harkins *et al.* (11) to evaluate the internal pressure of a medium, where E is the total surface energy in dynes per centimeter and v is the molar volume in cubic centimeters. Stefan (12) used L/v to calculate the cohesion, where L is the latent heat of evaporation in joules per mole.

Empirically, Richardson and Soper (10) stated the rule that solvents of high cohesion accelerate reactions in which the products are substances of higher cohesion than reactants; that solvents of high cohesion retard reaction in which the products are substances of lower cohesion than reactants; and that the solvent has little influence on reaction velocity when products and reactants are substances of like cohesion. Glasstone (13) arrived at a theoretical justification for the rules. According to the first part of the rule, reactions meeting the specifications involved should be accelerated by solvents of high cohesion. That this part of the rule is, in general, obeyed is shown in Table I. That the second part of the rule fairly accurately predicts results is shown by the data of Richardson and Soper in Table II.

In illustrating the third part of the rule, Richardson and Soper indicate that there is relatively little solvent effect for the conversion in the presence

TABLE I

<small>INFLUENCE OF THE COHESION OF THE SOLVENT ON QUATERNARY SALT FORMATION</small>

Solvent	$E_\sigma/v^{1/3}$	L/v	Triethylamine and ethyl iodide	Triethylamine and ethyl bromide
Hexane	9.45	243.3	0.000180	—
p-Xylene	12.12	292.2	0.00287	0.000103
Benzene	15.29	275.0	0.00584	0.000228
Chlorobenzene	14.68	360.8	0.00231	0.000843
Acetone	14.46	408.3	0.0608	0.0024

of ethyl tartrate of anissynaldoxime into antialdoxime. In this reaction the cohesion of the products and of the reactants are probably nearly the same. Laidler attributes deviations from these rules as often being due to solvation effects in solutions which are not regular, i.e., ones in which the molecular distributions are not entirely random.

TABLE II

<small>INFLUENCE OF THE COHESION OF THE SOLVENT UPON THE ESTERIFICATION OF ISOPROPYL AND ISOBUTYL ALCOHOL BY ACETIC ANHYDRIDE</small>

Solvent	$E_\sigma/v^{1/3}$	L/v	Acetic anhydride with isopropyl alcohol	Acetic anhydride with isobutyl alcohol
Hexane	9.45	243.2	0.0855	0.0307
Xylene	12.12	292.5	0.0510	0.0196
Benzene	15.29	275.0	0.0401	0.0148

Laidler writes for the activity coefficient γ of a solute the expression

$$RT \ln \gamma_1 = v_1 \left(\frac{x_2 v_2}{x_1 v_1 + x^2 v^2} \right)^2 \left[\left(\frac{E_1}{v_1} \right)^{1/2} - \left(\frac{E_2}{v^2} \right)^{1/2} \right]^2, \qquad (5.17)$$

where x_1 and x_2 are the mole fractions of solute and solvent, respectively, v_1 and v_2 their respective molar volumes, and E_1 and E_2 the heats of vaporization in the pure states of the solute and solvent, respectively.

For dilute solutions x_1 is small and $x_1v_1 \ll x_2v_2$, therefore,

$$RT \ln \gamma_1 = v_1 \left[\left(\frac{E_1}{v_1} \right)^{1/2} - \left(\frac{E_2}{v_2} \right)^{1/2} \right] \tag{5.18}$$

Now $E/v \cong a/v^2 \cong P$ for any substance, where a is the van der Waals attraction constant for that substance and P is its cohesion or internal pressure. Therefore, Eq. (5.18) becomes

$$RT \ln \gamma_1 = v_1 [P_1^{1/2} - P_2^{1/2}]^2 . \tag{5.19}$$

But

$$\ln k' = \ln k_0' + \ln \frac{\gamma_1 \gamma_2}{\gamma_X} , \tag{5.20}$$

where k_0' is the specific velocity constant for the ideal solution. Substituting from Eq. (5.19) the values of $\ln \gamma$ for the pertinent species into Eq. (5.20), the result is

$$\ln k' = \ln k_0' + \frac{1}{RT} \{ v_A [P_A^{1/2} - P_S^{1/2}]^2 + v_B [P_B^{1/2} - P_S^{1/2}]^2$$

$$- v_X [P_X^{1/2} - P_S^{1/2}]^2 \} \tag{5.21}$$

$$= \ln k_0' + \frac{1}{RT} [v_A \Delta_A + v_B \Delta_B - v_X \Delta_X] .$$

In these equations the subscript S refers to the solvent and the Δ's represent the respective $[P_i^{1/2} - P_S^{1/2}]$ terms, which are always positive.

The Effect of External Pressure on Reaction Rates

A general statement based on the Le Chatelier principle might be made that, if the volume of the critical complex molecules is less than the sum of the volumes of the molecules of reactants, the rate of reaction will be increased by increase of pressure, while if the volume of the critical complex molecules is greater than the sum of the volumes of the reactant molecules the rate of the reaction will be slowed down by an increase of pressure. The opposite would be true for a decrease of pressure. Evans and Polanyi (14) first studied the effect of external variables on reaction rates. Their finding would indicate that most reactions are increased by increase of pressure, and, therefore, that the volume of the critical complex molecules is less than the sum of the volumes of the reactant molecules. Evans and Polanyi qualify the statement that reactions accompanied by contraction should be accelerated by increase of pressure, to limit these reactions to those in which the density of the transition state is intermediate between that of the initial and final states. Thus the *cis–trans* isomerism of fumaric acid, which involves a contraction between initial and final states, is not accelerated by pressure, due perhaps to the extension of the C–C linkage in the transition state, thus causing this state to be less dense than either the initial or final states.

Reactions such as the combination of hydrogen and carbon monoxide to give formaldehyde and polymerization reactions, in which it is reasonable to suppose that the transition state will have a volume density intermediate between that of the initial and final states of the system, will be accelerated by increase of pressure.

Evans and Polanyi write the equation for the change of rate with pressure as

$$\frac{d \ln k'}{dP} = \frac{\Delta V}{RT},\tag{5.22}$$

where $\Delta V = V_1 - V_2$, the difference in the volumes in the initial and transition states. If, however, there are changes in the nature and strength of the forces between solvent molecules and the solute molecules as they go from the initial to the transition state, as well as changes in the solute molecules themselves as they go from the initial to the transition state, ΔV must be modified to include two terms accounting for both resultant volume changes. The solute–solvent interaction term will presumably not be great for neutral particles reacting to give neutral particles but could be large for charged particles giving a neutral intermediate or, in general, when reactants and intermediate complex are considerably different in charge or polarity. The modified equation is

$$\frac{d \ln k'}{dP} = \frac{\Delta_1 V}{RT} + \frac{\Delta_2 V}{RT},\tag{5.23}$$

where $\Delta_1 V$ is determined by the difference in the volumes of the initial and transition states without regard to the solvent, and $\Delta_2 V$ is dependent on the difference in electrostatic forces between the reactant species and the solvent and the electrostatic forces between the transition state and the solvent. Thus for an organic substitution reaction of the type $X^- + RY \rightarrow X^- RY$ (or $XRY^-) \rightarrow XR + Y^-$ the second term would be important, since the electrostatic forces between the transition state and the solvent would be less than those between the reactants and solvent, since in the transition state the molecule RY would screen off the solvent on one side of X^-. Van't Hoff (15) suggested an equation similar to Eq. (5.22). Other internal variables could be treated in a manner analogous to that of pressure.

Benson derives Eq. (5.22) by dividing the transition state rate equation into kinetic and thermodynamic contributions and considers the effect of change of external thermodynamic variables of state, here limited to change in pressure, on the rate. For the rate of appearance of products from the transition state the equation for the rate constant k' is

$$k' = \varkappa_X K_X \bar{v}_X \frac{f_A f_B \, \cdots}{f_X \, \cdots} , \tag{5.24}$$

where K_X is the equilibrium constant between the transition state X and the reactants A, B, ..., $\bar{v}_X$ is the mean frequency with which X passes through the critical configuration on the potential energy diagram, $\varkappa_X$ is the transmission coefficient, and $f_A, f_B, f_X, \ldots$ are the activity coefficients of the respective species.

K_X and the f's are thermodynamic and $\varkappa_X$ qnd v_X are kinetic in nature Taking $\varkappa_X$ as one and setting

$$K_X = K_X{}^* \frac{kT}{h\bar{v}_X} , \tag{5.25}$$

since $\bar{v}_X$ can be assumed as a normal vibration frequency whose partition function can be factored out of K as kT/k gives

$$k' = \frac{kT}{h} K_X{}^* \frac{f_A f_B}{f_X} . \tag{5.26}$$

Although the resolution of K_X by the method of partition functions will necessitate the analysis and factoring of $\bar{v}_X$, as an approximation Eq. (5.26) contains only thermodynamic factors, and only these need be analyzed in the theoretical analysis of specific velocity constants. The only external variables that need to be dealt with with respect to their influence on the reaction rate are the thermodynamic variables of state, and of these pressure is to be dealt with here. Therefore

$$\frac{\partial \ln k'}{\partial P} = \frac{\partial \ln T}{\partial P} + \frac{\partial \ln K_X{}^*}{\partial P} + \left(\frac{\partial \ln f_A}{\partial P} + \frac{\partial \ln f_B}{\partial P} - \frac{\partial \ln f_X}{\partial P} \, \cdots \right). \tag{5.27}$$

If we limit ourselves to dilute solutions at constant temperature

$$\left(\frac{\partial \ln k'}{\partial P}\right)_T = \left(\frac{\partial \ln K_X^*}{\partial P}\right)_T. \tag{5.28}$$

But, from thermodynamics,

$$RT \ln K_X^* = -\Delta \bar{F}_X^* \tag{5.29}$$

and

$$\left(\frac{\partial \bar{F}_X^*}{\partial P}\right)_T = \Delta \bar{V}_X^*. \tag{5.30}$$

Therefore,

$$RT\left(\frac{\partial \ln k'}{\partial P}\right)_T = RT\left(\frac{\partial \ln K_X^*}{\partial P}\right)_T = -\left(\frac{\partial \Delta \bar{F}_X^*}{\partial P}\right)_T \tag{5.31}$$

$$= -\Delta \bar{V}_X^*,$$

where $\Delta \bar{V}_X^* = \bar{V}_X - \bar{V}_A - \bar{V}_B - ...$, and $\bar{V}_i$ is the partial molal volume of the ith component of the solution. For other than strong interactions between A, B..., $\bar{V}_X$, $\bar{V}_A$, and $\bar{V}_B$,... approximately equal V_X, V_A, V_B,..., respectively; $\Delta V_S^* = \Delta V_X$; and Eq. (5.31) becomes identical with Eq. (5.22) when solute-solvent interactions are neglected. If the interactions A, B... are small enough $V_X^* - V_A - V_B = \Delta V_X^* = O$, and the rate of reaction is independent of pressure.

For dilute, non-ideal solutions Benson writes

$$RT\left(\frac{\partial \ln k'}{\partial P}\right)_{T,x_i} = -\Delta V_X^* = RT\left(\frac{\partial \ln k'}{\partial P}\right)_{T,c_i} + (n-1)RT\beta_S \tag{5.32}$$

where n is the order of the reaction and β_S is the coefficient of compressibility of the solution. Equation (5.32) predicts that the change of the logarithms of the rate constant with pressure at constant temperature will depend on the change of the partial molal volume for the transition-state reaction. The change of pressure will be influential in changing the rate constant only at pressures of 100 atm or greater, since $\Delta V_X^*/RT$ is of the order of magnitude of 0.001 atm^{-1}.

By determination of the rate over a sufficiently wide range of pressures, V_X^* and β may be determined since V_A and V_B can be determined independently.

From van Laar's (16) relations Moelwyn-Hughes (17) writes for the chemical potential μ_2 (partial molal free energy) of component 1 in a binary liquid mixture the equation

$$\mu_2 = - kT \ln f_2 + kT \ln x_2 + u_2{}^0 + P v_2 + x_1{}^2 \Delta u^0, \qquad (5.33)$$

where f_2 is the activity coefficient, x_2 the mole fraction, $u_2{}^0$ the average potential energy of one molecule in the pure state, and v_2 the partial molecular volume in the solution of component 2; x_1 is the mole fraction of component 1 in the solution; P the external pressure; and Δu^0 is the interchange energy, or the average increase in energy per molecule of either kind when it exchanges all its neighbors for neighbors of another kind. At the same temperature and pressure the chemical potential of pure component 2 may be written

$$\mu_2{}^0 = - kT \ln f_2 + u_2{}^0 + P v_2{}^0, \qquad (5.34)$$

where v^0 represents the molecular volume of the pure component. From Eqs. (5.33) and (5.34) the equation

$$\mu_2 = \mu_2{}^0 + kT \ln x_2 + P(\bar{v}_2 - v_2{}^0) + x_1{}^2 \Delta u^0 \qquad (5.35)$$

is obtained. For an activated molecule of the same species under identical conditions a similar equation

$$\mu_2{}^* = \mu_2{}^0 + kT \ln x_2{}^* + P(\bar{v}_2{}^* - v_2{}^0) + x_1{}^2 \Delta u^0 + e', \qquad (5.36)$$

where e' is the additional interval energy of an active molecule and $\bar{v}_2{}^*$ is its partial molecular volume at the temperature and pressure of the system.

Now x_1, for dilute solutions, is one and at equilibrium $\mu_2 = \mu_2{}^*$, therefore, from Eqs. (5.35) and (5.36), there results

$$\mu_2{}^0 + kT \ln x_2{}^* + P(\bar{v}_2{}^* - v_2{}^0) + x_1{}^2 \Delta u^0 + e'$$
$$= \mu_2{}^0 + kT \ln x_2 + P(\bar{v}_2 - v_2{}^0) + x_1{}^2 \Delta u^0, \qquad (5.37)$$

from which can be found the relationship

$$\frac{n_2{}^*}{n_2} = \frac{x_2{}^*}{x_2} = \exp\left[- \frac{P(\bar{v}_2{}^* - \bar{v}_2)}{kT} \right] \exp\left(- \frac{e'}{kT} \right), \qquad (5.38)$$

since the ratio of mole fractions of activated and normal molecules of solute is equal to the ratio of their concentrations. If v is the probability per second that an activated molecule will react, the number of molecules decomposing per cubic centimeter per second is n^*v, and the specific first order velocity constant k' is the fraction of the molecules decomposing per cubic centimeter per second, which is n^*v/n. Therefore, multiplying both sides of Eq. (5.38) by v yields

$$k' = v \exp\left[-\frac{P(\bar{v}_2{}^* - \bar{v}_2)}{kT}\right] \exp\left(-\frac{e'}{kT}\right), \qquad (5.39)$$

and in terms of moles per liter, Eq. (5.39) can be written

$$\ln k' = \ln v - \frac{E}{RT} - \frac{P(\bar{V}_2{}^* - \bar{V}_2)}{RT}, \qquad (5.40)$$

which, if v is independent of pressure when subscripts are dropped, becomes

$$\ln k' = \ln k_0' - \frac{P(\bar{V}^* - \bar{V})}{RT}. \qquad (5.41)$$

This is Moesveld's (18) law. A plot of $\ln k'$ versus P should be a straight line of negative slope if $\bar{V}^* > \bar{V}$ and of positive slope if $\bar{V}^* < \bar{V}$. The slope, in any case, should make possible the calculation of $(\bar{V}^* - \bar{V})$, the difference in the partial molar volumes of the activated and normal molecules; and, if the partial molar volume $\bar{V}$ of the normal molecules are determined independently, the partial molar volume $\bar{V}^*$ of the activated molecules should be calculable from the difference. Moelwyn-Hughes (17) gives a table of $(\bar{V}^* - \bar{V})$ values for various reactions in various solvents. These values vary from -21.8 cc/gram-mole for the dimerization of cyclopentadiene in the liquid state to 3.3 cc/gram-mole for the formation of the quaternary ammonium salt $(CH_3)(C_3H_5)(C_6H_5)(C_6H_5CH_2)NBr$ in chloroform.

Perrin (19) classified reactions into three groups with respect to the effect of pressure on their rates. One group is illustrated by the "normal" bimolecular reactions, which take place at a rate which can be calculated from the rate of collision between molecules having the necessary activation energy. Their Arrhenius frequency factors are of the order 10^{10} liters/mole-sec. Examples of this class of reaction are the reactions between sodium ethoxide and ethyl iodide and the hydrolysis of sodium monochloracetate by sodium ethoxide. A second group includes "slow" reactions which progress at a rate several powers of ten less than that calculated from the rate of collision between molecules having the necessary energy of activation.

The esterification of acetic anhydride with ethyl alcohol and the formation of quaternary ammonium salts are examples of this group. The Arrhenius frequency factor for these reactions varies from 1.33×10^7 for the reaction between pyridine and n-butyl bromide in acetone at 60°C to 3.85×10^9 for the esterification of acetic anhydride with ethyl alcohol in excess ethyl alcohol solvent at 20°C. In other words, the Arrhenius frequency factor is smaller often much smaller, in magnitude than 10^{10} liters/mole-sec. In the third group are slow unimolecular decompositions such as that of phenyl-benzyl-methyl-allyl-ammonium bromide.

For the first group mentioned, the "normal" reactions, the reaction rates increase, to a relatively small extent, with an increase in pressure. The increase in reaction rate is roughly linear with pressure, but tends to fall off at higher pressures. In these reactions the Arrhenius frequency factor shows a tendency to decrease slightly with increase of pressure, and the increase in rate apparently arises from a decrease in energy of activation with pressure.

The second group of "slow" reactions shows a much larger increase of rate with increase of pressure than do the "normal" reactions. In this group the Arrhenius frequency factor is increased markedly by an increase of pressure, and this increase of the frequency factor more than compensates for an increase in the energy of activation. At least in acetone both the frequency factor and the energy of activation for the Menschutkin reaction increase with increasing pressure, but in methanol the frequency factor is increased but the energy of activation is little affected (20). In this group of reactions the esterification of acetic anhydride by ethanol in excess ethanol as a solvent show both a decrease in the frequency factor and energy of activation with increase of pressure. Harris and Weale (20) attribute structural variations in the Menschutkin reaction, not to the commonly accepted theory of steric interference between nonbonded atoms and groups in the transition state, but to differences in degree and type of solvation of the transition states imposed by the configurations of the reacting molecules. The acceleration of rate by increase of pressure is greater for more complex reactants. Thus, in the reaction of dimethylaniline with the four iodides, acceleration with pressure increases in the order Me < Et ≈ Bu < Pr. Harris and Weale show that for the reactions of dimethylaniline with the four alkyl iodides in methanol solvents the volume of activation ΔV^* varies inversely as the entropy of activation ΔS^*, so that the largest volume decrease corresponds to the lowest entropy change. See Table III.

A model of the NNN-trimethylanilium iodide transition state shows four segments of the nitrogen atom to which an oxygen of methanol may

TABLE III

VOLUMES AND ENTROPIES OF ACTIVATION IN METHANOL AT 52.5° C

Reaction	$-\Delta V^*$ (cc/mole)	$-\Delta S^*$ (e.u.)
NPhMe$_2$ + MeI	26	30.4
NPhMe$_2$ + EtI	34	29.0
NPhMe$_2$ + n-BuI	34	28.5
NPhMe$_2$ + i-PrI	47	20.4

approach to within about 1 Å and be tightly bound. This will give rise to a relatively high entropy of activation but not an exceptionally large volume of activation. The N-ethyl-NN-dimethyl transition state has only two segments of nitrogen atom as accessible as in the trimethyl compound. Development of a positive charge may produce binding of a greater number of more distant methanol molecules. Since the solvation shell will have greater volume in this case, there will be an increase in the volume change, but a smaller entropy decrease will occur because of the less rigid binding of the solvent molecules. In the case of N-isopropyl-NN-dimethyl transition state there is only one easily accessible segment of the nitrogen atom available to a methanol molecule. This will presumably result in the weak binding of still a larger number of more distant methanol molecules, resulting in the greatest ΔV^* and the lowest ΔS^*.

Laidler (21) proposed that weaker binding will be associated with larger values of the temperature coefficient of ΔV^*, which can be evaluated from the relation $\partial \Delta V^*/\partial T = -\partial \Delta S^*/\partial P$. Harris and Weale give the value of $-\Delta S^*$ to be 6 e.u. for the reaction of dimethyaniline with methyl iodide between 1 and 2875 atm, while for the reactions with ethyl and butyl iodide ΔS^* decreases by about 11 e.u. These data support Laidler's suggestion. For the reaction with isopropyl iodide, however, $\Delta S^*_{2875} - \Delta S_1^* = 4.8$ e.u. Harris and Weale suggest that the correlation is not applicable when there are large variations between the binding of different solvent molecules.

The rate of reaction of the third group, according to Perrin's classification, is decreased with an increase in pressure, though the decrease is not marked in the case of the unimolecular decomposition of pheny-benzyl-methyl-ally-ammonium bromide. In this group both the frequency factor and the energy of activation decrease with increasing pressure for the unimolecular decomposition studied, but the decrease in the frequency factor dominates and accounts for the decrease in rate.

Figure 1, which is after Perrin, shows the difference in behavior of the rates with change of pressure. The values of k_P'/k_0', rather than values of $\ln k_P'/k_0'$ are plotted as a function of the pressure for the hydrolysis of sodium monochloroacetate, the reaction between pyridine and ethyl iodide and for the unimolecular decomposition of phenyl-benzyl-methyl-allyl-ammonium bromide. The "slow" reaction between pyridine and ethyl iodide shows, from density measurements on solutions of reactants and products, a decrease in volume of 54.3 cc/gram-mole. The corresponding value of $(V_{\text{reactants}} - V_{\text{products}})/RT$ is 2.22×10^{-3} atm^{-1}. The value of

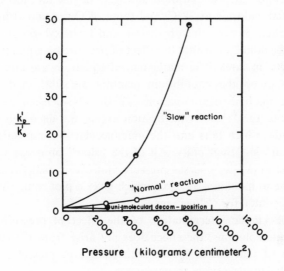

FIG. 1. k_p'/k_0', versus P. ○, Hydrolysis of monochloracetate; ◕, reaction between pyridine and ethyl iodide; ●, decomposition of phenyl-benzyl-methyl-allyl-ammonium bromide.

$d \ln k'/dP$ at 3000 kg/sq cm is 0.69×10^{-3}. The difference in the two quantities is not great, and if, as has been suggested, the volume of the transition complex is intermediate between that of the reactants and products, but more nearly equal to the latter, then the actual value of $(\bar{V} - \bar{V}^*)/RT$ in Eq. (5.41) would be less than the value calculated for $(\bar{V}_{\text{reactants}} - \bar{V}_{\text{products}})/RT$ and would more nearly agree with the observed value of $d \ln k'/dP$.

Plots of the values of $\ln (k_P'/k_0')$ versus pressure for the esterification of acetic anhydride with ethyl alcohol in toluene (19), for the hydrolysis of methyl acetate by normal hydrochloric acid at 14°C (17, 22), and for the reaction of ethyl iodide with ethylate ion in ethyl alcohol at 25°C (1, 23) are linear with positive slopes; and a similar plot for the normal hydro-chloric acid catalyzed hydrolysis of sucrose is linear with a negative slope (*17, 22*). These observations are in accord with the predictions of Eq. (5.41).

Other studies of the effect of pressure on reaction velocity constants and equilibrium constants include those of Newitt and Wasserman (*24*) on the dimerization of cyclopentadiene at 50°C in monomer as a solvent, those of Hamann and Teplitzky (*25*) on the association cited above in methanol at 25°C and on other reactions, those of Nicholson and Norrish (*26*) on the decomposition of benzoyl peroxide in carbon tetrachloride at 70°C. those of Ewald (*27*) on the association of nitrogen dioxide into nitrogen tetroxide in carbon tetrachloride at 22°C, and those of Owen and Brinkley (*28*) on the dissociation of water into hydrogen and hydroxyl ion at 25°C.

It might be pointed out that the effect of pressure on equilibria are similar to the effects on rates. The mathematical equations are alike for the two phenomena, except that equilibrium quantities are substituted for rate quantities in the mathematical equations. Of the above listed processes all of them fall into the "slow" classification except the decomposition of benzoyl peroxide, which falls into the unimolecular decomposition class with a k_P'/k_0' ratio less than unity. All of the "slow" processes can be placed on the same $\ln (k_P'/k_0')$ versus pressure plot except that for the ionization of piperidine in methanol at 25°C, which gives a plot somewhat above that of the other data for "slow" reactions.

As to the kinetic interpretation of the effect of pressure on reaction rates in dilute solutions, the effect does not arise from the change of rate of collision of reactant molecules with change of pressure. The collision rate is relatively insensitive to pressure.

For the "slow" reactions the increase in the temperature-independent frequency factor with pressure is probably due to an increase in the "probability", which in these reactions is much smaller than unity.

For "normal" reactions, an increase of pressure increases the time of contact when collision takes place between reactant molecules. There is a correspondingly greater chance for a correct distribution of the energy in molecules, and thus there is greater chance of reaction because of the increase in the "probability factor."

As to the unimolecular decompositions the decrease in rate with increasing pressure arises perhaps from the decreased rate of separation of the two product molecules after their formation.

Clearly, hydrostatic pressure markedly influence rates of reaction in dilute solutions, being more influential in instances of reactions involving large and complex organic molecules and which occur at atmospheric pressure at rates much slower than would be expected from elementary considerations of energy relationships and collision frequencies.

Norrish (29) states that in abnormal reactions the nascent particles require stabilization by the solvent, and that, for a reaction of the type

$$A + B \; \underset{k_2'}{\overset{k_1'}{\rightleftarrows}} \; AB \; \overset{k_3'}{\longrightarrow} \; C + D, \tag{5.42}$$

the rate is subject to a probability factor $\tau_1/(\tau_1 + \tau_2)$, which will steadily increase with increase of hydrostatic pressure. Here τ_1 is the average life of the activated complex AB and increases with hydrostatic pressure, and τ_2 is the time required for solvation of C or D, and in a given solvent remains practically constant. "Normal" reactions are not subject to such a probability factor and are therefore unaffected by pressure. Norrish's requirement of stabilization of the nascent particles by solvent would make the pressure effect on "abnormal" reactions very solvent dependent.

Angus (30) suggested that reversibility of reactions might play an important role in the influence of solvent with respect to the pressure effects on reaction rates. He cites the work of Edwards (31) on the association of dimethylaniline and methyl iodide in polar and nonpolar solvents as an example of a reaction showing appreciable reversibility, especially in nonpolar solvents.

In summary, it can be said that when a reaction takes place with an over-all contraction in volume, the rate or equilibrium process will be favored by an increase of pressure; and that when a reaction takes place with an over-all increase in volume, the rate or equilibrium process will be inhibited by increase in pressure.

Differentiating Eq. (5.40) with respect to pressure at constant temperature gives

$$\left(\frac{\partial \ln k'}{\partial P} \right)_T = \left(\frac{\partial \ln v}{\partial P} \right)_T - \frac{1}{RT} \left(\frac{\partial E}{\partial P} \right)_T - \frac{(\bar{V}_2{}^* - \bar{V}_2)}{RT}$$
$$- \frac{P}{RT} \left[\frac{\partial(\bar{V}_2{}^* - \bar{V}_2)}{\partial P} \right]_T. \tag{5.43}$$

If the second term on the right is zero, or nearly so, that is, if the energy of activation is independent of pressure or only relatively slightly dependent upon pressure, Eq. (5.43) becomes

$$\left(\frac{\partial \ln k'}{\partial P} \right)_T = \left(\frac{\partial \ln v}{\partial P} \right)_T - \left(\frac{\bar{V}^* - \bar{V}}{RT} \right) - \frac{P}{RT} \left[\frac{\partial(\bar{V}^* - \bar{V})}{\partial P} \right]_T, \tag{5.44}$$

which is the equation given by Moelwyn-Hughes (17) who points out that the last term on the right is negligibly small compared to the second term, but that the first term on the right may be significant.

By defining the apparent energies at constant volume E_V and at constant pressure E_P, respectively, as

$$E_V = RT^2 \left(\frac{\partial \ln k'}{\partial T} \right)_V \tag{5.45}$$

and

$$E_P = RT^2 \left(\frac{\partial \ln k'}{\partial T} \right)_P, \tag{5.46}$$

we can derive from thermodynamic procedures the relation between them to be (17)

$$E_V = E_P + RT^2 \frac{\alpha}{\beta} \left(\frac{\partial \ln k'}{\partial P} \right)_T, \tag{5.47}$$

where α is the coefficient of isobaric expansion and β is the coefficient of isothermal compression. Newitt and Wassermann (24) applied this equation to the dimerization of cyclopentadiene in the liquid phase. They concluded that both E_V and A_V, the parameters of the Arrhenius equation obtained at constant volume, were, within experimental error, unaffected by pressure, while E_P and A_P, the parameters found at constant pressure, increased markedly with increase of pressure. Thus E_P increased from 16.6 ± 0.4 to 18.8 ± 0.3 kcal/mole in going from 1 atm to 3000 atm. The corresponding change in $\log A_P$ (A_P in liters per mole-second) was from 6.1 ± 0.3 to 8.8 ± 0.3. E_V and A_V were appreciably larger than the corresponding values of E_P and A_P. Thus, at 1 atm, E_V was 20.7 ± 0.7 kcal/mole and $\log A_V$ (A_V in liters per mole-second) was 9.2 ± 0.5.

But from Eq. (5.44) if $(\partial \ln v)/\partial P)_T$ and $- (P/RT) [\partial(\bar{V}^* - \bar{V})/\partial P]_T$ are negligible compared to $- [(\bar{V}^* - \bar{V})/RT]$, then

$$- (\bar{V}^* - \bar{V}) = - \Delta \bar{V} = RT \left(\frac{\partial \ln k'}{\partial P} \right)_T. \tag{5.48}$$

Hence, from Eqs. (5.47) and (5.48),

$$E_V = E_P - T \frac{\alpha}{\beta} \Delta \bar{V}, \tag{5.49}$$

which was derived by Moelwyn-Hughes who points out that since $\Delta \bar{V}$

is usually negative E_V is usually greater than E_P, as was noted by Newitt and Wasserman for the dimerization of cyclopentadiene in the liquid state.

Kleinpaul (32) discusses the effect of pressure on energy of activation. He found the alteration of activation energy with pressure to be in the first approximation equal to the volume change of activation at 0°K.

In the case of many-step enzymatic reactions, many studies (33, 34) have been made on the effect of temperature and pressure on parameters which are in no sense to be regarded as rate constants or equilibrium constants. More thorough experimental work is required before the results can be accepted as chemically meaningful (35).

The Effect of the Viscosity of the Solvent on Reaction Rates

There are some types of reactions on which the viscosity of the solvent exerts an influence upon their rates. Thus the quenching of fluorescence and coagulation show viscosity dependence on their rates. Any rate process which is governed by the rate of diffusion of interacting particles to within reacting distance would presumably be influenced by the viscosity of the medium. Such processes are characterized by relatively small energies of activation. The Smoluchowski (36) equation

$$z = \frac{kT}{3\eta} \frac{(r_A + r_B)^2}{r_A r_B} n_A n_B , \qquad (5.50)$$

for the frequency of collision of particles A and B having concentrations n_A and n_B particles per cubic centimeter and radii r_A and r_B in a medium of viscosity η at temperature T, was modified by Moelwyn-Hughes (17) to give an equation for the specific velocity constant of a second-order rate constant by multiplying both sides by $(N/1000) \exp (E/RT)$, giving

$$k' = \frac{N}{1000} \frac{kT}{3\eta} \frac{(r_A + r_B)^2}{r_A r_B} \exp \left(- \frac{E}{RT} \right). \qquad (5.51)$$

This equation, however, does not reproduce chemical reaction rate data for reactions involving appreciable energies of activation as well as does the collision bimolecular rate expression from kinetic theory, namely,

$$k' = \frac{N}{1000} (r_A + r_B)^2 \left[8\pi kT \left(\frac{1}{m_A} + \frac{1}{m_B} \right) \right]^{1/2} \exp (- E/RT). \qquad (5.52)$$

Equation (5.51) would require a plot of ln $(k'\eta/T)$ versus $1/T$ to be a straight line of negative slope from which slope the energy of activation E can be calculated provided r_A and r_B are known. Equation (5.52) would require a plot of ln $(k'/T^{1/2})$ versus $1/T$ to be a straight line of negative slope from which slope E can be calculated provided r_A and r_B are known. Moelwyn-Hughes states that the former plot of the data of Hecht and Conrad (37) on the rate of reaction of methyl iodide and sodium ethoxide in ethyl alcoholic solution gave a continuous curve, the changing slope of which indicated an increase in E with increasing T. A plot of ln $(k'/T^{1/2})$ versus T, however, gave a straight line which resulted in a value of E of 18,970 cal/mole confirmed by later data to within a hundred calories.

Moelwyn-Hughes points out that the kinetic theory of gases does not relate collision frequency and viscosity, but will lead to equations in which the collision frequency is related, in the case of spherical molecules of the same kind, both directly and inversely to the viscosity. Elimination of the viscosity term yields the more familiar binary collision frequency equation.

Bowen and Wokes (38) arrive at the approximate equation for the specific velocity constant of quenching k' as a function of the viscosity of the solution

$$k' = \frac{T}{200\eta}. \tag{5.53}$$

Oester and Adelman (39) assuming spherical species of equal size with encounter radii the same as the diffusional radii, give the number of collisions n per cubic centimeter per mole per second to be

$$n = \frac{8RT}{3\eta}, \tag{5.54}$$

where R is the gas constant in ergs per mole per degree, η the viscosity of the medium in poise, and T the absolute temperature. If each encounter is effective, this would give the maximum velocity constant for the process.

For diffusion-controlled reaction rates in solution, the general equation for the molar rate constant k' in liters per second per gram-mole is, according to Umberger and LaMer (40),

$$k' = \frac{P(1 - \alpha)}{f} \frac{8NkT}{3000\eta} \left(\frac{2 + r_a/r_b + r_b/r_a}{4} \right) \tag{5.55}$$

where P is the probability factor, α is the fraction of the bulk concentration $n_b{}^0$ of reactant B at the surface of reactant A, f is the factor expressing the effect of interionic forces on the rate of diffusion, r_a is the radius of species A, r_b is the radius of species B, and the other terms have been defined previously. If P and f are each taken as unity, and if the concentration αn_b^0 of species B at reaction distance a of species A is kept at zero by the reaction, then assuming $r_a = r_b$ and remembering that $Nk = R$, we have the equation given by Oester and Adelman [Eq. (5.54)], where the volume is in cubic centimeters and it is assumed that every collision produces reaction.

Oester and Adelman studied the photoreduction of eosin Y with allyl thiourea and postulated a long-lived excited state of the dye formed by the transition of the first electronically excited singlet state of the dye. This metastable state of eosin could react with thiourea to give leuco eosin, or the metastable state could revert to the ground state by radiationless trans-mission or by a delayed fluorescence or posphorescence. Still again the metastable state could be quenched by the eosin Y in the ground state. If D′ stands for the dye in the metastable state, and A stands for the allyl thiourea, the three steps can be written

$$D' + A \xrightarrow{k_1'} \text{products}, \tag{5.56}$$

$$D' \xrightarrow{k_2'} D, \tag{5.57}$$

$$D' + D \xrightarrow{k_3'} 2D, \tag{5.58}$$

and if Eq. (5.54) is used to calculate the number of collisions per cubic centimeter per mole per second, the answer for water at room temperature is 6.6×10^9 collisions/sec in a liter of molar solution. If each collision represented by Eq. (5.58) is effective in quenching the metastable state the rate constant is given by the collision frequency. The ratio of k_3'/k_2' was found to be 4.88×10^4, and since k_3' was found to be 6.6×10^9, k_2' is 1.35×10^5. The mean lifetime for the metastable state is the reciprocal of k_2', and is calculated to be 7.4×10^{-6} sec. This mean lifetime was found to be about fifty times that of the first electronically excited singlet state. This metastable state was formed by the transition of the first electronically excited singlet state to the metastable state.

Noyes (41) has treated the solvent as a conventional structureless contin-uum in explaining the effect of the solvent on the quantum yields for dis-sociation. For a molecule consisting of two identical atoms having radii b and a minimum energy $\mathscr{E}$ to cause dissociation, then during photochemical dissociation, the energy $h\nu \cdot \mathscr{E}$ is divided equally as kinetic energy between the two atoms separating in opposite directions. The solvent, owing to viscous drag, will decrease the velocity u of the atom according to the relation

$$-\frac{du}{dt} = \frac{6\pi\eta bu}{m},$$ (5.59)

where η is the viscosity of the medium and m is the mass of the atom. Now r_0, the separation attained by the centers of the atoms before random diffusion begins, is given by the equation

$$r_0 = 2b + \frac{\sqrt{m(h\nu - \mathscr{E})}}{3\pi\eta b}.$$ (5.60)

The probability of subsequent recombination varies inversely as this separation, and therefore, the quantum yield φ can be written

$$\varphi = 1 - \frac{2b\beta'}{r_0},$$ (5.61)

where β' is the probability that a pair of atoms separating with normal thermal energies would recombine. Now if it is assumed that the two atoms are influenced by a frictional drag proportional to the viscosity of the medium, as would be the case for macroscopic bodies moving through the medium, and that for a particular initial velocity the body will be brought to rest at a distance inversely proportional to this viscosity, then each atom will lose its kinetic energy within the distance $(r_0 - 2b)/2$, and this distance should be inversely proportional to η for a constant energy of the absorbed quantum. Equation (5.61) becomes

$$\varphi = 1 - \frac{\beta'}{1 + A/\eta},$$ (5.62)

where A is a constant for a particular wave length of light and is independent of the solvent. If the atom has a diameter ϱ and if the frictional force per unit area is the same as that on a macroscopic sphere, then

$$A = \frac{2\sqrt{(h\nu - \mathscr{E})m}}{3\pi\varrho^2}.$$ (5.63)

If two particles separate at thermal velocities, the center of mass of the system, as well as the separation of the particles, has an average kinetic energy of $\frac{3}{2}kT$. Therefore,

$$hv - \mathscr{E} = \frac{3}{2}kT \tag{5.64}$$

and

$$A = \frac{\sqrt{6kTm}}{3\pi\varrho^2}. \tag{5.65}$$

Hence

$$\varphi = 1 - \frac{\beta'}{1 + (\sqrt{6kTm}/3\pi\varrho^2\eta)}.$$

The quantity β' is found to be given by the equation

$$\beta' = \frac{1}{1 + A/\eta} \tag{5.66}$$

by arguments based on the assumption that two atoms produced by thermal dissociation separate with an average kinetic energy component of $\frac{3}{2}kT$ along the line of centers.

Noyes and co-workers (42, 43) have applied this theory to the dissociation of iodine in inert solvents varying by a factor of 10^3 in viscosity. They found that, although there were quantitative discrepancies, the theory predicted the quantum yield well.

The quantitative discrepancies suggested the magnitude of proximity effects. At long wave lengths the quantum yields were less than predicted, and at short wave lengths in hexane the quantum yields are greater than predicted. The long wave length deviations arise because surrounding solvent molecules tend to force back together iodine atoms whose perimeters have not been separated by half a molecular diameter. The short wave length discrepancies originate because the most probable distribution of solvent molecules will tend to force the atoms apart rather than together.

Another shortcoming of the theory is its prediction of a very small quantum yield for viscosities greater than 0.1 poise, while experiments up to viscosities of 3.8 poise suggest that the quantum yield falls with increasing viscosity down to a limiting value that is not affected by further viscosity increase. It seems that increase of molecular complexity beyond a certain value has little effect on the behavior of solute particles, although there may be considerable influence on macroscopic viscosity.

Table IV contains calculated and observed quantum yields for the dissociation of iodine at different wave lengths in hexane and hexachlorobutadiene. The viscosity of the first solvent is 0.0029 poise and that of the second solvent is 0.030 poise. In hexane the agreement between theory and experiment is good at the wave lengths 4358, 5461, and 5790 Å. In hexachlorobutadiene the agreement is good only at 4358 Å. At the short wave length 4047 Å, the observed value of the quantum yield is greater than the calculated value, while at long wave lengths the observed value is less than the calculated value.

TABLE IV

COMPARISON OF THEORY AND EXPERIMENT

λ (Å)	Hexane		Hexachlorobutadiene	
	Obs	Calc	Obs	Calc
4047	0.83	0.54	—	—
4358	0.66	0.52	0.075	0.087
5461	0.46	0.46	0.036	0.070
5790	0.36	0.44	0.018	0.065
6430	0.14	0.40	0.023	0.055
7350	0.11	0.31	0.020	0.040

When the energy received is high enough and the atoms separate energetically enough and far enough, the solvent seems to aid in the separation of the atoms; while if the energy and, therefore, perhaps distance of separation are less than a critical amount, the solvent hinders the separation. Noyes and co-workers point out that these results are in agreement with the original "cage" picture of Rabinowitch and Wood (44).

TABLE V

HEAT OF DIFFUSION H AND ENERGY OF ACTIVATION E FOR THE I_2–AgCN REACTION
IN iso-BuOH-CCl$_4$ SOLUTIONS

Volume percent CCl$_4$ in liquid	H (kcal/mole)	E (kcal/mole)
0	—	3.10
67	3.00	2.80
91	2.91	2.67
100	2.50	2.65

Kataev and Gosteva (45) studied the reaction of iodine with silver cyanide in mixtures of isobutyl alcohol and carbon tetrachloride at various temperatures. The rate depended on the viscosity of the medium. The similarity between the heat of diffusion and the activation energy of the reaction indicated that diffusion limited the rate. Table V shows the comparison between the heat of diffusion and the energy of activation.

REFERENCES

1. S. W. Benson, "The Foundations of Chemical Kinetics," Chapter XV. McGraw-Hill, New York, 1960.

2. J. H. Hildebrand and R. S. Scott, "Solubility of Non-electrolytes." Reinhold, New York, 1950.

3. A. Wasserman, *Trans. Faraday Soc.* **34**, 128 (1938).

4. R. A. Fairclough and C. N. Hinshelwood, *J. Chem. Soc.* **1928**, 236.

5. A. Wasserman, *Monatsh. Chem.* **83**, 543 (1952).

6. G. A. Hall, Jr., *J. Am. Chem. Soc.* **72**, 1906 (1950).

7. L. W. Clark, *J. Phys. Chem.* **62**, 79 (1958).

8. K. J. Laidler, "Chemical Kinetics," Chapter 5. McGraw-Hill, New York, 1950.

9. S. Glasstone, K. J. Laidler, and H. Eyring, "The Theory of Rate Processes." McGraw-Hill, New York, 1941.

10. M. Richardson and R. G. Soper, *J. Chem. Soc.* **1929**, 1873.

11. W. D. Harkins, D. S. Davis, and G. Clark, *J. Am. Chem. Soc.* **39**, 555 (1917).

12. J. Stefan, *Ann. Physik* [3] **29**, 655 (1886).

13. S. Glasstone, *J. Chem. Soc.* **1936**, 723.

14. M. G. Evans and M. Polanyi, *Trans. Faraday Soc.* **31**, 875 (1935); **32**, 1333 (1936).

15. J. H. van't Hoff, "Vorlesungen über theoretische und physikalische Chemie." vol. I, Braunschweig, 1901.

16. J. J. van Laar, *Z. physik. Chem.* **15**, 457 (1894), and succeeding papers up to 1929.

17. E. A. Moelwyn-Hughes, "Physical Chemistry," Chapter XXIV. Pergamon Press, New York, 1957.

18. A. L. Th. Moesveld, *Z. physik. Chem.* **105**, 442, 450, 455 (1923).

19. M. W. Perrin, *Trans. Faraday Soc.* **34**, 144 (1938).

20. A. P. Harris and K. E. Weale, *J. Chem. Soc.* **1961**, 146.

21. K. J. Laidler, *Discussions Faraday Soc.* **22**, 88 (1956).

22. V. Rothmund, See E. A. Moelwyn-Hughes, Reference 17, p. 1230.

23. R. O. Gibson, E. W. Fawcett, and M. W. Perrin, *Proc. Roy. Soc.* (*London*) **A150**, 223 (1935).

24. D. M. Newitt and A. Wasserman, *Trans. Chem. Soc.* **1940**, 735.

25. S. D. Hamann and D. R. Teplitzky, *Discussions Faraday Soc.* **22**, 114 (1956).

26. A. E. Nicholson and R. G. W. Norrish, *Discussions Faraday Soc.* **22**, 97 (1956).

27. A. H. Ewald, *Discussions Faraday Soc.* **22**, 138 (1956).

28. B. B. Owen and S. R. Brinkley, *Chem. Rev.* **29**, 461 (1941).

29. R. G. W. Norrish, *Trans. Faraday Soc.* **34**, 154 (1938).

30. W. R. Angus, *Trans. Faraday Soc.* **34**, 153 (1938).

31. G. E. Edwards, *Trans. Faraday Soc.* **33**, 1294 (1937).

32. W. Kleinpaul, *Z. physik. Chem. (Frankfurt)* **26**, 313 (1960).

33. M. Dixon and E. C. Webb, "Enzymes," Chapter 4. Academic Press, New York.

34. K. J. Laidler, "The Chemistry of Enzyme Action." Oxford Univ. Press, London, and New York, 1958.

35. L. Peller and R. A. Alberty, Physical Chemical Aspects of Enzyme Kinetics, in "Progress in Reaction Kinetics" (G. Porter, ed.), Chapter IX. Pergamon Press, New York, 1961.

36. M. V. Smoluchowski, *Z. physik Chem.* **92**, 129 (1917); A. Ölander, *ibid.* **144**, 118 (1929).

37. W. Hecht and M. Conrad, *Z. physik Chem.* **3**, 450 (1889).

38. E. J. Bowen and F. Wokes, "Fluorescence of the Solutions." Longmans, Green, New York, 1953.

39. G. Oester and A. H. Adelman, *J. Am. Chem. Soc.* **78**, 913 (1956).

40. J. Umberger and V. K. LaMer, *J. Am. Chem. Soc.* **67**, 1099 (1945).

41. R. M. Noyes, *J. Am. Chem. Soc.* **78**, 5486 (1956); **82**, 1868 (1960); Z. Electrochem. **64**, 55 (1960); "Progress in Reaction Kinetics" Chapter 5. (G. Porter, ed.), Pergamon Press, New York, 1961.

42. D. Booth and R. M. Noyes, *J. Am. Chem. Soc.* **82**, 1868 (1960).

43. L. F. Meadows and R. M. Noyes, *J. Am. Chem. Soc.* **82**, 1872 (1960).

44. E. Rabinowitch and W. C. Wood, *Trans. Faraday Soc.* **32**, 1381 (1936).

45. G. A. Kataev and G. S. Gosteva, Uch. Zap. Tomsk. Gos. Univ. 1959, No. 29, 32.

CHAPTER VI

CORRELATIONS INVOLVING SOLVENT EFFECTS

CORRELATIVE

Introduction

Correlations involving the influence of solvents on rates of reactions are often of the linear free energy type. In such a correlation the left-hand side of the equation consists of the difference in the logarithm of the rate constant for a reaction in a selected solvent and that of the logarithm of the rate constant for the same reaction in a standard solvent, say 80% by volume ethanol in water. The left-hand sides of such equations are proportional to the difference in the free energies of reactions in the selected and standard solvents (if the k's are equilibrium constants) or to the difference in energies of activation (if the k's are rate constants). Another way of explaining the term "linear free energy relationship" is found in the fundamental assumption made in their formulation. It is assumed in these correlations that, when a selected molecule will undergo two similar equilibrium or rate processes, the respective relative free energy or relative energy of activation changes in the two processes will be influenced similarly by changes in structure. Some of these linear free energy correlations were not formulated primarily to relate equilibrium and rate constants to the effects of solvent on these processes. However, even those which are not primarily aimed at solvent effects, in many cases, contain quantities whose values are different for different solvents, and thus measure, in a sense, the effect of solvent on the chemical process. The Brønsted and Hammett equations fall in this latter class of correlations and should therefore be discussed in a treatment of solvent effects.

Brønsted's Equation for Acid- or Base-Catalyzed Reactions

Snethlage (1) noted the correlation between rates of acid-catalyzed reactions and the strengths of the catalyzing acids after it had become clear that acid catalysis could be separated into a catalysis by hydrogen ion and one by undissociated acid.

Taylor (2) also studied the relative rates of catalysis by undissociated acid molecules and by hydrogen ions. He expressed the relationship quantitatively by the equation

$$\frac{k_m'}{k_H'} = \frac{C_H}{C_m^{1/2}} \tag{6.1}$$

where k_m' and k_H' are the catalytic rate constants for the undissociated acid molecules and for the hydrogen ions, respectively, and C_m and C_H are the respective concentrations of these two catalysts.

Taylor's equation has been largely superceded by the more adaptable equation of Brønsted (3), to a consideration of which we will now turn.

The rate constant, k_{a0}', for an acid-catalyzed reaction involving an acid catalyst possessing one ionizable hydrogen atom is given by Brønsted as

$$k_{a0}' = G_1 K_{a0}^a , \tag{6.2}$$

where G_1 is a constant for structurally similar acid, α is a constant positive fraction for a given reaction and solvent, and K_{a0} is the acid dissociation constant, which, for a typical acid–base reaction such as

$$AH + H_2O \; \rightleftarrows \; A^- + H_3O^+ , \tag{6.3}$$

can be defined as

$$K_{a0} = \frac{k_{dissoc}'}{k_{assoc}'} \tag{6.4}$$

if the reaction from left to right be (arbitrarily) called a dissociation and from right to left an association, and if k_{dissoc}' and k_{assoc}' represent the respective velocities of the dissociation and association processes.

For a basic catalysis by a base having one point at which a proton may attach itself the relationship is

$$k_{b0}' = G_2 K_{b0}^\beta , \tag{6.5}$$

where G_2 is a constant for structurally similar bases, β is a constant positive fraction for a given reaction and solvent, and

$$K_{b0} = \frac{1}{K_{a0}} \tag{6.6}$$

correcting Eqs. (6.2) and (6.3) for the probability factor that arises from the possibility of more than one acid or basic center, and letting p represent the number of acid centers and q the number of basic centers, the equations

$$k_a' = G_1 K_a^\alpha p^{1-\alpha} q^\alpha \tag{6.7}$$

and

$$k_b' = G_2 K_b^\beta p^\beta q^{1-\beta} \tag{6.8}$$

can be written for the acid- and base-catalyzed reactions, respectively.

Brønsted and Guggenheim (4) give plots of log k_a' and log k_b' versus log K_a for the mutarotation of glucose in the presence of various acids and bases. In these plots the influence of electric types to be expected theoretically and also statistical effects have been omitted. The graph is evidence that water acts as both an acidic and as a basic catalyst. Except for the base having a double positive charge, the points corresponding to basic catalysis lie approximately on a straight line. The curve for acid catalysis is not so well established as the one for basic catalysis, but the increasing catalytic effect with increasing acid strength is obvious. The relationship between the catalytic effect of an acid or a base and its strength is striking, when it is remembered that the plot covers a range of 10^{18} in K_a and 10^8 in K_b.

In Pfluger's (5) plot of the catalytic constants for the base-catalyzed glucose mutarotation against those of the nitramide reaction, the data on a variety of catalysts are given by a single straight line. The same data expressed by the Brønsted law [Eqs. (6.7) and (6.8)] would require several lines.

Many other confirmations of these equations have been made. Benson (6) used two parallel lines to represent the base-catalyzed decomposition of nitramide at 18°C in aqueous solution involving four different types of bases. The bases of the types B^0, B^{-1}, and B^{-2} fall on one line and the base of the type B^{2+} fall on another line. The reaction presumably takes place through the tautomeric azo acid with the rate-determining step being the removal of the proton from this acid. Thus, according to Benson,

$$\tag{6.9}$$

$$\xrightarrow{\text{fast}} NNO + OH^-. \tag{6.10}$$

Hine (7) plots the logarithms of the catalytic rate constants for various bases in the decomposition of nitramide versus the logarithms of the basicity constants for these bases using the data of Brønsted and Pedersen (8). Again all the data except that for the positively double-charged base $[M(NH_3)_m(H_2O)_n(OH)]^{2+}$ fell on a single line.

Bell and co-workers (9) have made extensive application of the Brønsted equation to data. Bell has discussed reasons for the deviation of data from the requirements of the Brønsted equation. In aqueous solutions he believes that the anomalies may be present in the aqueous dissociation constants rather than in the kinetic data or in the structure of the amines themselves. Thus for the amine-catalyzed decomposition of nitramide in anisole solution, plots of the logarithms of the catalytic constants versus the logarithms of the dissociation constants in water resulted in two straight lines for primary and tertiary amines. The difference between the primary and tertiary amines in this case, however, is much greater than when plots are made of the logarithms of the catalytic constants in aqueous solutions versus the logarithms of the dissociation constants also in aqueous solutions. In the latter case the values of β are the same in the two cases, but G_b for the tertiary amines is about four times as great as for the primary amines. For the plots using the decomposition data in anisole solvent for the tertiary amine, G_b is about ten times as great for the primary amine. Bell attributes this difference in the two solvents to a much reduced solvation in anisole solution.

Bell also observed that when the catalytic constants were compared with basic strengths measured in anisole or phenyl chloride solutions instead of dissociation constants in water, a single equation served to represent the data for primary, secondary, and tertiary amines. This supports Bell's reasoning that the apparent kinetic anomalies observed with amines have their origin in the variable contribution of hydration to the dissociation constants in aqueous solutions.

The above discussed causes of deviation from the Brønsted relation arise purely from solvent effects.

Another cause of both positive and negative deviations arises from a considerable charge distribution difference in the transition state and in the base of the conjugate acid (or vice versa). Benson (6) however points out that if we are to compare the free energy changes for a series of similar reactions in terms of molecular structure we must first eliminate all contributions to ΔF^o which are accidental in nature such as symmetry change in reaction, and properties such as molecular weight, moment of inertia, and vibration frequency. The symmetry corrections turn out to be very similar, with occasional differences resulting from ambiguities in the use of the p and q corrections, to the p and q corrections of Brønsted. Benson (10) has given a general discussion of the use of p and q corrections.

Bell also discussed deviations which will occur if some stabilizing effect is present in the transition state but cannot occur either in the initial or final state. These deviations are found in the bromination of various ketones and esters catalyzed by carboxylate anions (11). It was found that when both the substrate and the catalyst contain large groups, such as halogens or hydrocarbon radicals, near the seat of reaction, positive deviations up to 300% occurred. The proximity of these two large groups in the transition state was accepted as the cause of these deviations. The van der Waals forces between the groups may be partly responsible for the stabilization of the transition state, but more responsible no doubt is the reduction in interaction between the water molecules brought about by the necessity of making a smaller cavity in the solvent to accommodate the transition state with its closely spaced large groups than was required by the two reactants with their large groups farther apart. Thus there is separation of fewer water molecules by the transition state and a consequent stabilization of the state. Bell suggests that energies and entropies of activation might throw further light on the problem. This effect arises then from the properties of the solvent. In solvents with smaller interaction forces among its molecules such deviations should be reduced.

In Fig. 1, the data of Bell and Wilson (12) for the kinetics of the amine-catalyzed decomposition of nitramide are shown. In the figure, the decadic logarithms of the catalytic rate constants are plotted against the pK_a values for the amine cations. The primary and tertiary amines fall on two lines, the separation of which is not so great as when the reaction velocities in anisole are used. The tertiary bases are more effective catalysts than primary bases of the same basic strength. This arises from the different effect on the dissociation constants of the primary and tertiary amines of hydrogen bonding between the amine cation and the water molecule. Many anomalies in the effect of substituents on the basic strength of amines in water are clarified by the same concept (13). The greater solvation of amine ions than free amines makes the amine ions appear more stable when their dissociation constants are measured in water than when measured in an inert solvent. The stabilization of amine ions is in the order primary > secondary > tertiary.

Hammett's Equation

Hammett (14) proposed an equation that takes into account the effect of a substituent upon the rate or equilibrium constant of a general reaction for a member of a class of aromatic compounds. A general reaction is one that a compound and its derivatives will commonly undergo. This equation is

$$\log k' = \log k^{0'} + \sigma\varrho \qquad (6.11)$$

where $k^{0'}$ is the rate or equilibrium constant or the general reaction for an unsubstituted member of a class; σ is a substituent constant, depending upon the substituent, and ϱ is a reaction constant, depending upon the reaction, the medium, and the temperature.

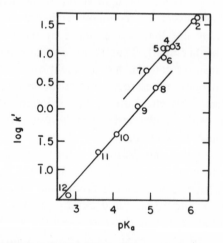

FIG. 1. A plot of the decadic logarithm of the catalytic rate constant versus pK_a for the amine-catalyzed decomposition of nitramide. (1) α-picoline, (2) γ-picoline, (3) dimethyl-o-toluidine, (4) isoquinoline, (5) pyridine, (6) dimethyl-m-toluidine, (7) quinoline, (8) p-toludine, (9) aniline, (10) p-chloroaniline, (11) m-chloroaniline, (12) o-chloroaniline.

Now the only available data give the $\sigma\varrho$ product; therefore it is necessary to assign some arbitrary value to some σ or ϱ. Hammett chooses $\varrho = 1$ for the ionization equilibrium of substituted benzoic acids. Thus the difference in logarithms of the ionization constants of substituted and unsubstituted benzoic acids gives the value of sigma for that substituent. Taking these σ values, ϱ values for other reactions can be obtained, and from these, in turn, other σ values can be calculated for substituents whose effects upon the ionization of benzoic acid have never been determined. He found that thirty-eight reactions, involving derivatives of benzoic acid, of phenol, of aniline, of benzenesulfonic acid, of phenylboric acid, and of phenyphosphine and including both equilibrium and rate constants, gave a mean value of the probable errors of the values of log k' calculated from Eq. (6.11), as compared to the corresponding observed values to be only 0.067. Figure 2

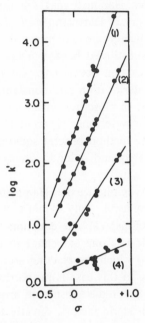

FIG. 2. Relationship between log k' and σ for various chemical processes. Curve (1), acidity constants of substituted anilinium ions in water at 25°C; curve (2), ionization of substituted phenylboric acids in 25% ethyl alcohol at 25°C; curve (3), reaction of substituted benzoyl chlorides with methyl alcohol at 0°C; curve (4), base-catalyzed bromination of substituted acetophenones in acetic acid–water medium with sodium acetate as catalyst at 35°C.

gives Hammett's plot of log k' versus σ for the ionization and kinetic processes listed. Thus, for the rate constant for the reaction

$$p - BrC_6H_5CH_2Cl + KI \longrightarrow p - BrC_6H_5CH_2I + KCl, \qquad (6.12)$$

we have, from Hammett's data, log $k^{0'}$ of Eq. (6.11) is 0.167, ϱ is 0.785, and σ is 0.232. The $k^{0'}$ is the rate constant for the reaction of the unsubstituted benzyl chloride with potassium iodide

$$C_6H_5CH_2Cl + KI \longrightarrow C_6H_5CH_2I + KCl, \qquad (6.13)$$

ϱ is the constant characteristic of the general reaction of the substituted phenylchloride with potassium iodide

$$XC_6H_4CH_2Cl + KI \longrightarrow XC_6H_4CH_2I + KCl \qquad (6.14)$$

and σ is a constant for *para* substitution of chlorine.

Hammett gives an extensive list of σ values. The ϱ values must be evaluated for each general reaction. Hammett makes a list of ϱ values for several general reactions. The work of Hartmann *et al.* (*15*) should be mentioned in support of the theory. The authors determined the hydrion-catalyzed esterification of twenty-two substituted benzoic acids with cyclohexanol. They took into account the reaction between the catalyst and the solvent. The experimental values of esterification rate constants for *meta* and *para* substituents are found to be in close agreement with those using Hammett's equation. The values of σ were those reported by Hammett, and the value of ϱ was obtained by the method of least squares. The average difference between $-\log k'$ calculated and $-\log k'$ observed by these authors was 0.082 or about 2%.

Branch and Calvin (*16*) emphasize that Hammett's equation is an approximate empirical one, applicable only within certain limits. These authors offer the following criticisms: (1) The equation is a good approximation when the substituent effect is one of change in the electric charge of an aromatic carbon atom to which the reacting group is attached directly by a chain. (2) The σ value in this case is proportional to the difference between the polar constants for the substituted aromatic group and for the phenyl group. (3) When steric factors, directly transmitted polar effects, and important resonances other than those between the substituent group and the aromatic resonances are caused by the substitution, considerable error is introduced by using either polar constants for complex groups or values determined only by substitution. (4) The values assigned to groups are subject to criticism since the normal effect of the substituent is difficult to obtain, because there may be more or less perturbation in any chosen general reaction.

Branch and Calvin worked out the following relationship between Brønsted's and Hammett's equations. According to Brønsted's theory, the rate constants for the aromatic base and one of its *meta* or *para* derivatives are given by the equations

$$\log k_u' = \log G - x \log K_u \qquad (6.15)$$

and

$$\log k_\sigma' = \log G - x \log K_\sigma. \qquad (6.16)$$

In these equations K_u and K_σ are the dissociation constants of the conjugate acids of the aromatic base and its derivatives. But according to Hammett's equation

$$\log K_\sigma = \log K_u + \sigma \varrho' , \tag{6.17}$$

combining Eqs. (6.15), (6.16), and (6.17) yields

$$\log k_\sigma' = \log k_u' - \sigma \varrho' x . \tag{6.18}$$

Here x is determined by the substrate and ϱ' is determined by the aromatic acids, hence $\varrho' x$ is determined by the general reaction of a class of aromatic acids, that is, $\varrho' x$ is ϱ for this general reaction. Knowing ϱ and ϱ', x can be calculated. Thus ϱ' for aniline is 2.73, and ϱ for the reaction of substituted anilines with dinitrochloronaphthalene in ethyl alcohol medium is 3.69; therefore x is 1.3.

Hammett (17) has suggested that the free energy change is made up of the sum of the entropy change, the kinetic energy change, and the potential energy change. The validity of the expression is limited to substituents in the *meta* and *para* positions of the benzene ring; and in many side-chain reactions of benzene derivatives, the entropy is not appreciably affected by substituents in these positions. Hammett believes that the constancy of the entropy term implies a similar constancy in the kinetic energy changes, and hence changes in the potential energy only determine the differences in free energy.

Jaffé (18) believes that the validity of the Hammett equation should be in terms of the substituents upon the electron density at the reaction site when the reaction site is insulated from the benzene ring, e.g., by one or more methylene groups. The effect of substituents upon the difference in resonance energy between ground and transition states must also be considered if the reaction site is not insulated from the benzene ring.

The Hammett equation remains fundamentally an empirical relation, even though some attempts have been made to derive the equation theoretically. Price (19), for example, has dealt with this problem.

Hammett (20) has suggested that the reaction constant ϱ is of the form

$$\varrho = \left[\frac{B_1}{D + B_2} \right] \bigg/ RTd^2 . \tag{6.19}$$

In this equation the constant B_1 is assumed to depend only on electrostatic interaction between the reacting benzene derivative and the medium, the constant B_2 presumably measures the susceptibility of the reaction to change in charge density at the reaction site; R is the gas constant, T the absolute temperature, D the dielectric constant of the solvent, and d the distance of the substituent from the reaction site.

Since D enters the equation for ϱ, this constant should show a dependence of ϱ on the solvent. Hine (21) points out that ϱ commonly increases with decrease in ion-solvating power and dielectric constant of the medium. Equation (6.19) would predict an increase of ϱ with decreasing dielectric constant. This is illustrated by ϱ-values for the ionization of benzoic acid in different solvents, the data for which is given in Table I.

TABLE I

VALUES OF ϱ FOR THE IONIZATION OF BENZOIC ACID AT 25°C IN DIFFERENT SOLVENTS

Solvent	Water	Methanol	Ethanol
ϱ	1.000	1.537	1.957
D	78.55	31.5	24.33

The increase of ϱ with decreasing dielectric constant is explained in different ways, all of which contribute to the effect, and the over-all effect probably is the resultant of the different contributions. Thus as the dielectric constant decreases the electrostatic interactions between the substituents and the reaction center increase. Then too the nature of the equilibrium or rate process may be different in the different solvents due to the difference in the chemical nature of the active center solvated by different solvents. For example, the water-solvated and the ethanol-solvated carboxylate anion group will probably differ markedly.

While the σ-factor is assumed to depend on the electrical effects of substituents, yet Hine points out that the σ's for some reaction-center groupings must be dependent on the solvent, since if they were not, all ϱ's would change to the same extent with changing solvent, a phenomenon which is not supported by experiment.

Jaffé (18) gives an extensive summary of the applications of Hammett's equation to data. A special substituent constant, designated by Jaffé as σ^*, has been assigned to phenols and anilines since the substituent constant ϱ for the p-nitro group, applicable to most reactions, does not give good results when applied to reactions of phenols and anilines. Groups such as CN, COOH, and CHO may also require special constants. In the alkaline hydrolysis of phthalide and its derivatives, the reacting side chain is attached to two places in the ring bearing the rate-affecting substituent; therefore, the substituent effect must be transmitted to the reaction site by two paths. If the effects through both parts of the side chain are included, separate reaction constants must be used for each path, and we must write

$$\log \frac{k'}{k^{0'}} = \sigma_m \varrho_1 + \sigma_0 \varrho_2 \qquad (6.20)$$

The "English School" mentions four separate effects as making up the total effects of the substituents. Remick (22) discusses these effects. The *inductive effect* deals with the alteration of the partial moment of a given bond by substitution in other parts of the same molecule. The *mesomeric effect* is a resonance effect. The magnitudes of dipole moments give evidence of the existence of mesomerism. For aliphatic amines, only the inductive effect is operative. For aromatic amines mesomeric displacement will oppose inductive displacement and perhaps outweigh inductive displacement, with the result that the moment associated with the linkage in question be reversed from that of the aliphatic amine, especially since nitrogen has a low electronegativity:

$$\underset{\text{—}\overset{\cdots}{N}R_2}{\xleftarrow{\quad}} \qquad H_3C\text{—}NR_2 \xrightarrow{\quad} .$$

The mesomeric structure in keto–enol tautomerism prototropy is a structure which is more or less midway between keto and enol forms. Remick writes the mesomeric structure as

$$R_2\overline{C} \overset{\text{H}}{\underset{\overset{\shortparallel}{O}}{\text{—}C}} \qquad \text{or} \qquad R_2C\text{—}CH\text{—}O \underset{\ominus}{} .$$

The *inductometric effect* is an induced polarity effect, which is the only kind admissible in saturated compounds in general. Thus an attacking nitro group may induce polarity in the methyl groups of isobutane causing the tertiary hydrogen atom to be most readily attacked. The *electrometric effect* is the displacement of electron pairs by the tautomeric mechanism under the influence of the electric field external to the molecule under consideration. Such a movement of electrons must frequently take place in compounds containing double bonds under the influence of an attacking reagent.

Jaffé lumps the inductometric and electrometric effects into a single polarizability effect. These effects are usually neglected. Polarizability effects operate only to favor a reaction, that is, they are unidirectional. They may be responsible for some of the deviations from the Hammett equation. If σ' expresses the polarizability effects, the Hammett equation can be written

$$\log \frac{k'}{k_0'} = [\sigma + \sigma'f(p)]\varrho . \tag{6.21}$$

The polarizability effects vanish in many instances.

Jaffé lists several causes, other than polarizability effects, for deviations from Hammett's equation. One of these is the approximation involved in the assumption that the substituent constants are independent of the nature of the side chain, the reaction, and the solvent. There seems to be general recognition that both ϱ and σ are solvent dependent. At present, data are too scarce to permit the determination of a number of σ's in a variety of solvents.

Jaffé calculated the correlation coefficient r for 371 reaction series and the standard deviation s for the best straight line (regression line) for these reaction series. According to the criteria set up for the values of r and s in relation to values of ϱ Jaffé found that only 26 of 371 reaction series violated the Hammett equation, although many others were expressed only to a rough approximation.

According to Jaffé, the Hammett equation applies to substituents that carry an integral charge. The equation applies to electrophilic as well as to nucleophilic attack on the substituted benzene derivative. However, Pearson et al. (23) believe that the substituent constants tabulated by Hammett are valid only for reactions involving nucleophilic attack on the substituted benzene derivative.

Certain physical constants, such as the polarographic reduction potentials and the effect of substituents upon the infrared frequencies associated with side chains, have been correlated with the Hammett equation. Jaffé believes that a correlation might be expected between substituent constants and nuclear magnetic resonance phenomena which is believed to depend on the electron density at the side-chain atom. He also believes that the Hammett equation might be used to correlate optical rotatory power in aromatic compounds in general, since the optical rotatory power of certain Schiff bases can be correlated with the pK's of benzoic acids corresponding to the benzaldehydes used.

At variance with the view that $\Delta\sigma$, the difference in the substituent constant in the *meta* and *para* positions, depends primarily upon electrostatic effects; it is believed that $\Delta\sigma$ measures the resonance effect of the substituents, since, as pointed out by Jaffé, the shift which a substituent causes in the primary absorption band of benzene is proportional to the difference in the substituent constants in the *meta* and *para* positions. A satisfactory quantitative explanation has not been found as yet for the relationship between the substituent constants and the resonance and inductive effects. Taft (*24, 25*) and his co-workers made major efforts with respect to this problem. These workers have also greatly extended linear free energy relationships by dealing with steric and polar effects as independent of each other.

Taft's Equation

Taft assumed that steric effects and also resonance effects were the same for both acid and basic hydrolysis of esters.

The specific velocity constant k_B' for the basic hydrolysis of the ester XCO_2R in which X is varied and R is held constant is related to the specific velocity constant $k_{0,B}'$ for the ester defined as the standard reactant by the equation

$$\log \frac{k_B'}{k_{0,B}'} = \varrho_B\sigma_X + S + R, \tag{6.22}$$

where S is a factor that measures the steric contribution to the rate and R is a factor that measures the resonance contribution to the rate. For the acid hydrolysis a similar relation holds, so that

$$\log \frac{k_A'}{k_{0,A}'} = \varrho_A\sigma_X + S + R, \tag{6.23}$$

where the stearic (*26*) and resonance factors are the same for both basic and acid catalysis. Subtracting Eq. (6.23) from Eq. (6.22) gives

$$\log \frac{k_B'}{k_{0,B}'} = \log \frac{k_A'}{k_{0,A}'} + \sigma_X(\varrho_B - \varrho_A). \tag{6.24}$$

Taft (*27*) assumed that the nonpolar $\log (k_A'/k_{0,A}')$ measures nearly quantitatively the net potential and kinetic energy steric effects, i.e.,

$$\log \frac{k_A'}{k_{0,A}'} = E_S. \tag{6.25}$$

Exceptions occur when there are unsaturated substituents conjugated with the carbonyl group, or when there are substituent groups which cause changes in attractive interaction between reactant and transition states, for example, internal hydrogen bonding. Also if $(\varrho_B - \varrho_A)$ is defined as 2.48 for hydrolysis in aqueous alcoholic and aqueous acetone solutions then, from Eqs. (6.24) and (6.25),

$$\log \frac{k_B{}'}{k_{0,\,B}'} = E_S + 2.48\sigma^*, \qquad (6.26)$$

where σ^* is a substituent constant dependent only upon the net polar effect of the substituent relative to that of the standard of comparison, corresponding to the rate constant k' compared to the rate constant $k_0{}'$ chosen for the standard of comparison ($k_0{}'$, $R = CH_3$).

The substituent constant σ^* is analogous to, but different in nature and origin from, the Hammett substituent constant σ, that is, $\sigma^* \neq \log (K/K_0)$ for the ionization of carboxylic acids. The factor 2.48 is introduced in order to place the polar effects dealt with by Taft on a similar scale to the Hammett σ values. σ^*, according to Taft is identical to σ_X, and from Eq. (6.24), if $(\varrho_B - \varrho_A)$ is taken as 2.48, it is given by the equation

$$\sigma^* = \frac{1}{2.48} \left[\log \frac{k_B{}'}{k_{0,\,B}'} - \log \frac{k_A{}'}{k_{0,\,A}'} \right]. \qquad (6.27)$$

Taft has correlated the rates of a number of reactions, in which stearic and resonance effects are negligibly small, using the simple equation

$$\log \frac{k'}{k_0{}'} = \sigma^*\varrho^*, \qquad (6.28)$$

where again σ^* is the polar substituent constant for the R group relative to the standard CH_3 group and ϱ^* is a constant representing the susceptibility of a given reaction series to polar substituents. Its values varies with the nature of the reaction center, the attacking reagent, etc. This equation applies to certain reactions of *ortho*-substituted benzene derivatives, as o-X-C_6H_4-Y, where X is the *ortho* substituent and Y is the reaction center. In these *ortho* substituents, k' and σ^* values are generally related to the unsubstituted (X = H) derivative.

Equation (6.28) is not a linear free energy relationship but rather a free energy–polar energy relationship since σ^*-values are measures of polar energies which have been separated from free energies.

Equation (6.16) is more limited than the Hammett equation, since the former does not apply to those reaction series in which there are marked steric and resonance effects upon the rate constants. These conditions frequently prevail for reaction series involving bulky and unsaturated substituents at the reaction center.

Acetate esters were used in the determination of σ^*-values except for *ortho* substitution, in which case benzoates were used as the standard esters. A linear relation between $\log (k'/k_0')$ requires that the polar effect be the only variable operating.

Taft states that a nonlinear relation between $\log (k'/k_0')$ values for any reaction and the corresponding σ^*-values does not invalidate the proportional nature of polar effects, but shows that these rates or equilibria are determined by influences other than polar effects.

Taft (*24*) tabulates many substituent constants both for the R-Y aliphatic series and for the *ortho*-substituted benzene derivatives. Table II (*28*) contains σ^* constants for the group R relative to a CH_3 group calculated from Eq. (6.27), using k_B' and k_A' as the rate constants for the normal basic and acidic hydrolyses, respectively, of an ethyl ester of the type $RCO_2C_2H_5$, and $k_{0,B}'$ and $k_{0,A}'$ for the rate constants for the basic and acidic hydrolyses, respectively, of ethyl acetate. These σ^* values correlate quantitatively the effects of structure on rates and equilibria for a variety of reactions, which it is believed are nearly free from steric and resonance contributions to the free energy term according to Eq. (6.28). The σ^* values for some groups were calculated by multiplying σ^* for the RCH_2 group by 2.8.

TABLE II

TYPICAL σ^* VALUES

R	σ^*	R	σ^*
CH_3N^+	+5.04	CH_3CO	+1.65
CH_3SO_2	+3.70	HO	+1.55
$N \equiv C$	+3.64	$ClCH_2$	+1.05
F	+3.08	$C_6H_5OCH_2$	+0.85
HO_2C	+2.94	C_6H_5	+0.60
Cl	+2.94	$HOCH_2$	+0.56
Br	+2.80	H	+0.49
Cl_3I	+2.65	CH_3	0.00
I	+2.38	$(CH_3)_3C$	—0.30
Cl_2HC	+1.94	$(CH_3)_3Si$	—0.72

Qualitative relationship of σ^* values to substituent constitution have been established, and these relationships relate quantitative correlations involving the effects of structure on reactivity to the concepts of bonding and electronegativity.

If the Hammett equation or Taft equation is not obeyed with usual precision, and there are variable entropies of activation with changing substituents, then there may be substantial variation in steric hindrances of internal motions or steric hindrances of solvation within a reaction series. The principal cause of variable entropy terms in the reaction of *meta-* and *para-* substituted benzene derivatives are kinetic energy terms which result from solvent interaction with poles and dipoles.

Correlation of Solvolysis Rates

Grunwald and Winstein (*29*) found that for reactions occurring by the S_N1 mechanism, when values of $\log k'$ for the solvolysis of any one compound in a number of solvents were plotted against values of $\log k'$ for the solvolysis of some other compound (tertiary butyl chloride was chosen as this compound) in the same solvents, a straight line was obtained. They chose therefore $\log k'^{BuCl}$ (BuCl = tertiary butyl chloride) as satisfying the requirements for a useful quantitative measure of the ionizing power of a given solvent. They defined a function Y by the equation

$$\log k'^{BuCl} = Y + \log k_0'^{BuCl}, \qquad (6.29)$$

where k'^{BuCl} is the specific velocity constant for the solvolysis of tertiary butyl chloride at 25.0°C in a given solvent and $k_0'^{BuCl}$ is the specific velocity constant for the solvolysis of tertiary butyl chloride at 25°C in an aqueous ethanol solvent which is 80% by ethanol volume. The definition of a S_N1 type reaction should be given since the work of Grunwald and Winstein involves this type. In the S_N1 type of mechanism the rate-controlling step is the loss of a nucleophilic atom or group with the formation of a carbonium ion. The carbonium ion then combines rapidly with another nucleophilic reagent. Thus the RX molecule or group may lose the X^- ion leaving R^+ a carbonium ion. The R^+ then combines rapidly with a nucleophilic Y^- to give RY. The reaction rate is thus, under certain conditions, independent of the concentration of the nucleophilic substance Y^-. The mechanism is thus termed a substitution nucleophilic reaction of the first order or S_N1 reaction.

Bateman *et al.* (*30*) found the nucleophilic displacement of chloride in benzhydryl chloride by the nucleophilic reagent fluoride ion, pyridine, or triethylamine had the same initial rate of reaction, independent of the nucleophilic reagent used in liquid sulfur dioxide solution. The rates were the same and independent of concentration of the nucleophilic reagent, since the rate-controlling step in each case was the loss of chloride ion by the benzhydryl chloride to form carbonium ion. The reaction using fluoride as the nucleophilic reagent can be written

$$(6.30)$$

$$(6.31)$$

Returning to the solvolysis reactions, if the specific velocity constants for solvolysis in solvents 1 and 2 are, respectively k_1' and k_2', then the general equations relating the rate constants to the "ionizing power" Y_1 and Y_2 of the respective solvents can be written

$$\log k_1' = m(Y_1 - Y_2) + \log k_2' . \qquad (6.32)$$

The factor m is a characteristic constant at a given temperature of the compound being studied.

But from Eq. (6.29) m is unity for the solvolysis of tertiary butyl chloride at 25.0°C, and Y_2 is zero for the 80% by volume ethanol in water solvent. The straight line relationships, found when the values of $\log k'$ for the solvolysis of different compounds in different solvents were plotted against $\log k'$ for the solvolysis of tertiary butyl chloride in 80% by volume of alcohol in water, indicate that the slope m of Eq. (6.32) is not in general unity. Therefore, from these considerations, Y_2 in Eq. (6.32) is zero, and designating k_1' as k' and k_2' as k_0' and Y_1 as Y, the general equation becomes

$$\log k' = mY + \log k_0' \qquad (6.33)$$

or

$$\log \frac{k'}{k_0'} = mY , \qquad (6.34)$$

where k' is the specific velocity constant for the solvolysis of a given compound in a given solvent at a given temperature and k_0' is the specific velocity constant for the solvolysis of the compound in "80% aqueous ethanol" at the same temperature. In Table III are listed values for k', ΔE, and Y at 25.0°C for solvolysis of the compounds listed in the specified solvents. In water–acetone solvents the values for Y were found to depend on the compound which was solvolyzing by the S_N1 mechanism. The quantity

TABLE III

SOLVOLYSIS RATES FOR SOME HALIDES AND Y VALUES FOR SOLVENTS AT 25.0°C

Compound	Solvents	ΔE (kcal)	k' (sec⁻¹)	Y
$(CH_3)_3CCl$	H_2O	—	3.3×10^{-2}	3.56
	40% EtOH	—	1.29×10^{-3}	2.151
	50% EtOH	22.92	3.67×10^{-4}	1.604
	60% EtOH	—	1.27×10^{-4}	1.139
	70% EtOH	—	4.07×10^{-5}	0.644
	80% EtOH	23.06	9.24×10^{-6}	0.000
	90% EtOH	—	1.73×10^{-6}	—0.727
	absolute EtOH	25.97	9.70×10^{-8}	—1.974
	5% $(CH_3)_2CO$	—	2.57×10^{-2}	3.449
$(CH_3)_3CBr$	70% $(CH_3)_2CO$	—	5.15×10^{-4}	0.205
	80% $(CH_3)_2CO$	20.8	1.10×10^{-4}	—0.527
	90% $(CH_3)_2CO$	20.8	1.27×10^{-5}	—1.549
	95% $(CH_3)_2CO$	22.7	2.10×10^{-6}	—2.402
$(C_6H_5)_2CHCl$	70% $(CH_3)_2CO$	19.6	4.60×10^{-6}	—3.377
	80% $(CH_3)_2CO$	21.0	7.24×10^{-5}	—1.797
	90% $(CH_3)_2CO$	—	3.20×10^{-4}	—0.945

Y in general is specific for each solvent and is a quantitative measure of its ionizing power, as evidenced by its influence on the rate of solvolysis reactions by the S_N1 mechanism, and m is supposed to be characteristic of the compound being solvolyzed and of the temperature but actually has also been found to be a function of the solvent pair in which the solvolysis is being studied.

Winstein and Fainberg (31–33) found that for certain solvolysis reaction, for example, the solvolysis of α-phenylethyl chloride, when log k' was plotted versus Y there was a strong tendency toward dispersion into rather satisfactory separate lines for each binary solvent set. Especially, there tended to be a separation into two separate lines when the data for non-carboxylic acid-containing solvents, and carboxylic acid-containing solvents are compared. Hine (34) suggests modifying Eq. (6.34) to make it clear that k_0' and m are solvent dependent, while k' and Y retain their previous definitions. Thus

$$\log \frac{k'}{k_0^{s'}} = m_S Y. \tag{6.35}$$

Figure 3 contains a plot of selected data of Winstein et al. (35) for the solvolysis of benzhydryl chloride at 25.0°C in various solvent pairs. The lines must curve in higher water regions, since they must all intersect in pure water. The plot does show that using Eq. (6.34) each solvent pair should be considered separately. There are thus definite structural limitations to the linear free energy relationship represented by the equation.

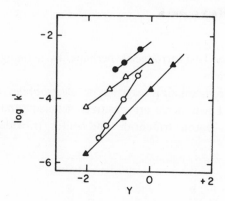

FIG. 3. Plot of log k' versus Y for solvolysis of benzhydryl chloride at 25°C in H$_2$O–MeOH, ●; H$_2$O–EtOH, △; H$_2$O–AcOH, ○; H$_2$O–Dioxane, ▲.

The dispersion evidenced by the mY relation is believed to be due in part to the specificity of the leaving group, which is described in terms of the importance of hydrogen bonding to the leaving group in the solvolysis transition state. Data on solvolysis suggest that the importance of hydrogen bonding falls off in the order $F \gg$ OTs > Cl > Br, where OTs represents toluene sulfonate. It is suggested that a measure of the effect of the leaving group specificity is the vertical gap between the aqueous alcohol line and the acetic acid point in a plot of log k_X' for the substituent X versus log k_{Cl}' for chloride.

A second cause of the mY dispersion is the disturbance caused by ion pair return, or the return to the covalent condition of a varying fraction of the ion pair intermediates. Young *et al.* (*36*) first demonstrated this phenomenon in the acetolysis and rearrangement of α, α-dimethyl allyl chloride. Where there is such return present, the titrimetric rate constant k_t' for solvolysis is only a fraction F of the true ionization rate constant k_I', the difference in these rate constants depending on the substituent and on the solvent. For the solvolysis of exo-norbonyl bromide k_I exceeds k_t by a factor of 24 in acetic acid but by only 5 in 80% ethanol. In the acetolysis of benzhydryl chloride and bromide, k_I in acetic acid is at least 10 times k_t for the chloride and 15–20 times for the bromide (*35*). Ion pair return would account in part for the slowness of bromide as compared to chloride solvolysis in acetic acid. It also accounts in part for slowness of these compounds in acetic acid as compared to other solvents in which ion pair return is less important.

Swain's Equation

This section could be best introduced, perhaps, by defining electrophilic and nucleophilic reagents.

Electrophilic (electron-seeking) reagents or electrophiles are electron-deficient substances. Lewis acids are electron deficient and are electrophiles. For example, boron trifluoride and sulfur trioxide are electrophiles, thus

$$
\begin{array}{ccc}
& :\overset{\cdot\cdot}{F}: & & :\overset{\cdot\cdot}{O}: \\
:\overset{\cdot\cdot}{F}:\overset{\cdot\cdot}{B} & & \text{and} & :\overset{\cdot\cdot}{O}:\overset{\cdot\cdot}{S} \quad . \\
& :\overset{\cdot\cdot}{F}: & & :\overset{\cdot\cdot}{O}:
\end{array}
$$

Electrophilic reagents include positive ions. These include metal ions such as Ag^+ and nonmetallic ions such as carbonium R_3C^+, bromonium Br^+, and nitrnoim NO_2^+. Double bonds such as $C = O$, $C = N$, and $N = O$, and triple bonds such as $C \equiv N$ are electrophiles. Halides such as $SnCl_4$ and $TiCl_4$, in which the central atom may hold more than an octet of electrons and in which there are vacant d orbitals of relatively low energy, may be acidic in nature and show electrophilic properties. A $C = C$ double bond may act as an electrophile if its electron density has been reduced by an electron-attracting group such as $-CH_3$, $-CH_2R$, $-CHR_2$, and $-C-O^-$.

Nucleophilic (electron-donating) reagents or nucleophiles are substances belonging to one of the following classes: (a) negative ions (halide, hydroxide, carbanions, and others); (b) molecules involving fifth- or sixth-group elements having unshared electron pairs (ammonia, phosphine, amines, alcohols, ethers, and mercaptans); olefins and aromatic hydrocarbons. These reagents generally react with another molecule at a position where the atomic nucleus is poorly shielded by outer electrons. Nucleophiles are Lewis bases and in most cases behave also as Brønsted bases. As Lewis bases they donate a pair of electrons to form a compound with a Lewis acid. As a Brønsted base they accept a proton from a Brønsted acid. Thus

$$
\begin{array}{ccc}
\text{H} & \ddot{:}\text{F}\ddot{:} & \text{H} \;\; \ddot{:}\text{F}\ddot{:} \\
\text{H} : \ddot{\text{N}} : \; + \; \ddot{\text{B}} : \ddot{\text{F}} : & \longrightarrow & \text{H} : \ddot{\text{N}} : \ddot{\text{B}} : \ddot{\text{F}} : \quad (6.36) \\
\text{H} & \ddot{:}\text{F}\ddot{:} & \text{H} \;\; \ddot{:}\text{F}\ddot{:}
\end{array}
$$

$$
\begin{array}{cc}
\text{Lewis base} \quad \text{Lewis acid} & \text{Adduct}
\end{array}
$$

$$
\begin{array}{cc}
\text{H} & \left[\text{H} \right]^{+} \\
\text{H} : \ddot{\text{N}} : \; + \; \text{H} : \ddot{\text{Cl}} : \; \longrightarrow & \left[\text{H} : \ddot{\text{N}} : \text{H} \right]^{+} \; + \; \ddot{:}\text{Cl}\ddot{:}^{-} \quad (6.37) \\
\text{H} & \left[\text{H} \right]
\end{array}
$$

Brønsted base   Brønsted acid   Conjugate acid   Conjugate base
                                of base NH_3 of acid HCl

Electrophilic and nucleophilic reactions include acid–base reactions as illustrated above, substitution reactions, and addition reactions.

The reaction

$$ S_2O_3^{2-} + BrCH_2CO_2^{-} \longrightarrow S_2O_3^{-}CH_2CO_2^{-} + Br^{-} \quad (6.38) $$

is a nucleophilic substitution reaction. Both the attacking reagent and the leaving group are Lewis bases. The reaction

$$:\ddot{Br}^{+} + C_6H_6 \longrightarrow C_6H_5Br + H^{+} \quad (6.39) $$

is an electrophilic substitution in which an electron-deficient bromonium ion attacks the benzene substrate giving the products bromobenzene and the electron-deficient hydrogen ion. In this type of substitution both the attacking agent and the departing group are electron-deficient Lewis acids.

Turning to Swain's equation we may note that it is another linear free energy relationship. Basically, one should be able to correlate all solvolysis rates with one equation, including those in which the solvent participates as a nucleophile. Swain and co-workers (37, 38) represented the polar displacement reaction of an uncharged organic molecule S with a nucleophilic reagent N and an electrophilic reagent E by the equation

$$N + S + E \; \rightleftharpoons \; [\text{Transition state}] \; \longrightarrow \; \text{Products} \qquad (6.40)$$

They relate the over-all rate constant k' of this reaction to the corresponding rate constant k_0', where water acts as both N and E in the same medium at the same temperature, thus:

$$H_2O + S + H_2O \; \rightleftharpoons \; [\text{Different transition state}] \; \longrightarrow \; \text{Products} \qquad (6.41)$$

The linear free energy equation correlating these rates would be

$$\log \frac{k'}{k_0'} = sn + s'e. \qquad (6.42)$$

In this equation n, the nucleophilic constant, is a quantitative measure of the nucleophilic reactivity of N ($n = 0$ for water); e, the electrophilic constant, is a measure of the electrophilic reactivity of E ($e = 0$ for water); and s and s', substrate constants, are measures of the discrimination of S among different N and E reagents.

From Eq. (6.42) three corollaries are derivable which are well known and useful quantitative relationships between structure and reactivity.

For a fixed N and S when e is proportional to the acid ionization constant of E, that is $e = (-\alpha/s')\,(pK_a)$, the Brønsted catalytic law for acids results. Thus

$$\log \frac{k_a'}{k_a''} = \alpha \log \frac{K_a}{K_a'}, \qquad (6.43)$$

therefore

$$\log k_a' = \alpha \log K_a + C \qquad (6.44)$$

and

$$k_a' = G_a K_a{}^{\alpha}. \qquad (6.45)$$

For the case where S and E are fixed and n is proportional to the basic ionization constant of N, that is $n = (-\beta/S)\,(pK_b)$, the Brønsted catalytic law for bases is obtained. Thus

$$\log \frac{k_b'}{k_b''} = \beta \log \frac{K_b}{K_b'}, \qquad (6.46)$$

hence

$$\log k_b' = \beta \log K_b + C \qquad (6.47)$$

and

$$k_b' = G_b K_b^\beta .$$ (6.48)

For $sn \ll s'e$, Eq. (6.42) reduces to

$$\log \frac{k'}{k_0'} = s'e .$$ (6.49)

If E is the solvent and one substitutes $s' = m$ ($m = 1$ for tertiary butyl chloride) and $e = Y$ (Y is the "solvent ionizing" power), the Grunwald-Winstein correlation of solvolysis rates results, that is,

$$\log \frac{k'}{k_0'} = mY .$$ (6.50)

A fourth corollary of equal importance arises when $s'e \ll sn$, then

$$\log \frac{k'}{k_0'} = sn .$$ (6.51)

This would be the case when the electrophilic reagent is held constant, for example, when water is the solvent and acts as the only important E. In this case we set $e = 0.00$ for E $= E^0$.

The standard substance for use of Eq. (6.51) was chosen as methyl bromide with $s = 1.00$. Values of n for azide, hydroxide, iodide, and thiosulfate ions and aniline were then defined by the equation

$$\log \frac{k'}{k_0'} = n ,$$ (6.52)

where k_0' was the rate of hydrolysis of the methyl bromide and k' was the rate of reaction of methyl bromide with a given one of the substances listed. Thus n could be calculated for each of the substances.

Then the rate of the reaction of epichlorohydrin (2,3 epoxypropyl chloride) with water and each of the substances mentioned above was measured, and the values of $\log (k'/k_0')$ obtained were plotted versus the values of n found for the substances (azide, hydroxide, iodide, and thiosulfate ions and aniline) using methyl bromide. The slope of this line gave the substrate constant s for the epoxychlorohydrin, as indicated by Eq. (6.51). In a similar manner the nucleophilic constant n was found for each nucleophilic reagent, and the substrate constant s was found for each substrate. The straight lines used to determine the s constant were determined from the data, using the method of least squares. For forty-seven reactions studied, this equation correlated variations of more than 10^5 for the average substrate with a probable error factor of 1.5 in the calculation of k'/k_0' and therefore in the calculation of k'.

For the constants in the more general equation (6.53) given below, 146 values of $\log (k'/k_0')$ involving 25 values of c_1, 25 values of C_2, 17 values of d_1, and 17 values of d_2 were determined by Swain et al. (38):

$$\log \frac{k'}{k_0'} = c_1d_1 + c_2d_2 \,. \tag{6.53}$$

In this equation k' is the first-order rate constant for the solvolysis of any compound in any solvent, k_0' is the corresponding rate constant in a standard solvent (here chosen as 80% ethanol–20% water by volume) at the same temperature, c_1 and c_2 are constants depending only on the compound undergoing solvolysis, and d_1 and d_2 are constants depending only on the solvent. By their mode of definition, d_1 and d_2, while not exactly equivalent to n and s, are intended to measure nucleophilicity and electrophilicity.

The data represent a wide range of structural variation. The compounds range from paranitrobenzoyl and methyl to triphenylmethyl and from fluorides to arylsulfonates. Some have strong neighboring group participation, and the pinacolyl compound even rearranges. The solvents range from anhydrous alcohols and water to glacial acetic acid and anhydrous formic acid. An 80% by volume ethanol in water solvent was chosen as the standard solvent because more data were available for this than for any other solvent.

In order that all 146 $\log (k'/k_0')$ values be weighted alike, the condition

$$\Sigma \left[\log (k'/k_0')_{\text{obs}} - (c_1d_1 + c_2d_2)\right]^2 \tag{6.54}$$

was chosen to define the best fit. Eighty-four simultaneous equations resulted from setting the partial derivatives with respect to each of the $25c_1$, $25c_2$, $17d_1$, and $17d_2$ unknown parameters equal to zero. These 84 equations were solved by a iterative procedure using a digital computer. To make the solution unique it was necessary to impose the following conditions:

$$d_1 = d_2 = 0.00 \qquad \text{for 80\% ethanol}, \tag{6.55}$$

$$c_1 \quad\;\; = 3.00c_2 \qquad \text{for MeBr}, \tag{6.56}$$

$$c_1 = c_2 = 1.00 \qquad \text{for } t\text{-BuCl}, \tag{6.57}$$

$$3.00c_1 \;\; = c_2 \qquad \text{for } (\text{Ph})_3\text{CF}. \tag{6.58}$$

Swain *et al.* point out that the Eqs. (6.56), (6.57), and (6.58) are qualitatively in the right order since sensitivity to nucleophilic reagents increases in the order methyl bromide, tertiary butyl chloride, trityl fluoride. The reactivity of a compound is characterized by a convenient single number, the ratio c_1/c_2. High values of this ratio indicate that the compound discriminates relatively more highly among nucleophilic than among electrophilic reagents. This ratio decreases from paranitro to paramethyl in the order methyl, ethyl, isopropyl, tertiary butyl, benzhydryl, trityl. These are the expected results.

The solvent is characterized by a convenient single number, the difference $(d_1 - d_2)$. The difference is greatest for the most nucleophilic solvents, and the difference decreases in the order anhydrous alcohols, acetone–water and alcohol–water mixtures, water, glacial acetic acid, anhydrous formic acid.

In Table IV are listed some of the constants obtained. The values in parentheses are the ones arbitrarily assigned. Using these constants, logarithms of the rate constants were calculated for the various reactions in the various solvents. The mean error for all 25 compounds in 17 solvents was 0.124 in $\log k'_{calc}$. This is a factor of 1.33 error in k'.

Hine (*39*) says that, to the extent that Eq. (6.42) holds, the solvating power for cations (carbonium ions, at least) is proportional to nucleophilicity and that there is a smooth and continuous transition from S_N1 to S_N2 mechanism. According to Hine, n and e indicate some correlation of n with basicity and e with acidity, and the values of s and s' show compounds reacting with the S_N1 mechanism to be relatively more susceptible to electrophilic attack and compounds reacting by the S_N2 mechanism are more susceptible to nucleophilic attack.

Swain and Scott (*37*) and Swain *et al.* (*40*) have related rates of solvolysis with a special two-parameter equation.

The former authors used Eq. (6.51), as already discussed. The latter authors used the equation for the solvolysis of organic chlorides and bromides:

$$\log\left(\frac{k'}{k_0'}\right)_A - \log\left(\frac{k'}{k_0'}\right)_{A^0} = ab \qquad (6.59)$$

Methyl bromide is chosen as the standard and is represented by the subscript A^0. Other chlorides and bromides are represented by the subscript A. k' is then the first-order rate constant for the solvolysis of any organic chloride or bromide (A) or of the standard compound (A^0) in any solvent, k_0' is the corresponding rate constant in a standard solvent, here chosen as 80%

TABLE IV

VALUES OF COMPOUND AND SOLVENT CONSTANTS

Compound	c_1	c_2	c_1/c_2
$NO_2PhCOCl$	1.09	0.21	5.2
NO_2PhCOF	1.67	0.49	3.4
$PhCOCl$	0.81	0.52	1.6
$PhCOF$	1.36	0.66	2.1
$MePhCoCl$	0.82	0.65	1.3
$MePhCOF$	1.29	0.80	1.6
$MeBr$	0.82	0.27	3.0
$EtBr$	0.80	0.36	2.2
$EtOTs$	0.65	0.24	2.7
$n\text{-}BuBr$	0.77	0.34	2.2
$PhCH_2Cl$	0.74	0.44	1.7
$PhCH_2OTs$	0.69	0.39	1.8
$i\text{-}PrBr$	0.90	0.58	1.5
$i\text{-}PrOBs$	0.63	0.48	1.33
$MeOCxOBs$	0.57	0.57	1.00
$BrCxOBs$	0.80	0.87	0.92
$PinOBs$	0.76	0.87	0.86
$PhCHClMe$	1.47	1.75	0.84
$(Ph)_2CHCl$	1.24	1.25	0.99
$(Ph)_2CHF$	0.32	1.17	0.27
$t\text{-}BuCl$	(1.00)	(1.00)	(1.00)
$(Ph)_3CSCN$	0.19	0.28	0.69
$(PH)_3COPhNO_2$	0.18	0.59	0.31
$(PH)_3CF$	0.37	1.02	(0.33)

Solvent	n	e	$n - e$
MeOH 100	—0.05	—0.73	+0.7
MeOH 96.7	—0.11	—0.05	—0.1
MeOh 69.5	—0.06	+1.32	—1.4
EtOH 100	—0.53	—1.03	+0.5
EtOH 90	—0.01	—0.54	+0.5
EtOH 80	(0.00)	(0.00)	(0.00)
EtOH 60	—0.22	+1.34	—1.6
EtOH 50	+0.12	+1.33	—1.2
EtOH 40	—0.26	+2.13	—2.4
Me_2CO 90	—0.53	—1.52	+1.0
Me_2CO 80	—0.45	—0.68	+0.2
Me_2CO 70	—0.09	—0.75	+0.7
Me_2CO 50	—0.25	+0.97	—1.2
H_2O 100	—0.44	+4.01	—4.5
AcOH 100	—4.82	+3.12	—7.9
Ac_2O 97.5	—8.77	+5.34	—14.1
HCOOH 83.3	—4.44	+6.26	—10.7
HCOOH 100	—4.40	+6.53	—10.9

ethanol–20% water by volume at the same temperature, a is a constant depending on only the chloride or bromide, and b is a constant depending on only the solvent.

The best values of a and b were obtained by the method of least squares using an interative procedure. Three conditions were imposed to make the solution unique:

$$b = 0.00 \quad \text{for } 80\% \text{ EtOH}, \tag{6.60}$$

$$a = 0.00 \quad \text{for MeBr}, \tag{6.61}$$

$$a = 1.00 \quad \text{for } t\text{-BuCl}. \tag{6.62}$$

From Eqs. (6.59) and (6.62), if tertiary butyl chloride is chosen for comparison of its solvolysis rate with the standard compound methyl bromide, $a = 1$ and

$$\log \left(\frac{k'}{k_0'} \right)_{t-\text{BuCl}} - \log \left(\frac{k'}{k_0'} \right)_{\text{MeBr}} = b. \tag{6.63}$$

Using this relation, crude values of b were obtained for solvents in which the rates with both tertiary butyl chloride and methyl bromide were measured. Crude a values were then determined using Eq. (6.59) and any solvent for which b values had been found. Crude b values for other solvents in which methyl bromide had not been studied were obtained using the equation

$$\log \left(\frac{k'}{k_0'} \right)_{\text{A}} - \log \left(\frac{k'}{k_0'} \right)_{t-\text{BuCl}} = (a - 1.00)b \tag{6.64}$$

and any compound A for which a had been determined.

To minimize experimental error and to obtain better values of b for all solvents from the crude values of a, the method of least squares with equal weighting of all the usable $\log (k'/k_0')$ values was used. From the better b values, better values of a were obtained in the same way.

Table V compares the value of a for different substituents on the carbon atom at which reaction occurs, and Table VI compares the value of b with dielectric constant of the solvent.

In general, as the electron-supplying ability of the substituents increases a increases. Nitro groups are electron attracting, and hence substituents with nitro groups have the smallest a values. In the stepwise replacement of hydrogen atoms by methyl groups a increases. A resonance rather than an inductive effect might cause the increase in a when a phenyl group replaces a methyl group. This resonance effect might involve the shift of electron

TABLE V

VALUES OF a FOR COMPOUNDS $R_1R_2R_3CX$

Compound	No. of reactions	a	R_1-, R_2-, R_3-, X
PicCl	8	−0.42	$-C(NO_2)=CH-C(NO_2)=CH-(NO_2)=$, Cl
NO_2PhCOCl	7	−0.37	$4\text{-}NO_2C_6H_4-$, O=, Cl
$PhCOCH_2Br$	8	−0.04	C_6H_5CO-, H—, H—, Br
MeBr	10	(0.00)	H—, H—, H—, Br
PhCOCl	12	+0.06	C_6H_5, O=, Cl
EtBr	5	+0.15	CH_3-, H—, H—, Br
i-BuBr	4	+0.16	$(CH_3)_2CH-$, H—, H—, Br
n-BuBr	12	+0.18	$CH_3CH_2CH_2CH_2-$, H—, H—, Br
$PhCH_2Cl$	8	+0.19	C_6H_5-, H—, H—, Cl
MePhCOCl	5	+0.41	$4\text{-}CH_3C_6H_4-$, O=, Cl
i-PrBr	5	+0.42	CH_3-, CH_3-, H—, Br
PhCHClMe	5	+0.64	C_6H_5-, CH_3-, H—, Cl
$(PH)_2CHCl$	13	+0.78	C_6H_5-, C_6H_5-, H—, Cl
t-BuBr	7	+0.93	CH_3-, CH_3-, CH_3-, Br
t-BuCl	15	(+1.00)	CH_3-, CH_3-, CH_3-, Cl

distribution from the phenyl ring toward the reaction site. The value of a, however, is a function of the temperature and of the nature of the leaving group as well as of the polar effects of the substituents on the reaction.

Of more interest to the present discussion is the relation of b to the properties of the solvent. Electrostatic interactions between reactant particles and between reactant and solvent particles are strongly influenced by the dielectric constant of the medium. From Table VI it can be seen that, in general, there is an increase in b with increasing dielectric constant of the medium. There are several exceptions, the most prominent of which is acetic acid. The values of b also tend to increase as the basicity of the solvents decreases and their acidity increases. Aniline, acetic acid, and formic acid are comparatively more acidic than their neighbors. This might explain their having greater b-values than their dielectric constants would indicate. Also as solvents become richer in water the b values increase. This may be due in part to increase in water content of the solvent. Since b is a function of the logarithms of the k' and k_0' values and since, in electrostatic ef-

TABLE VI

VALUES OF b FOR SOLVENTS

Solvent	No. of reactions	b	Dielectric constant at or near 20°C
Et$_3$N	3	−17.27	3.2
n-BuNH$_2$	5	−10.15	5.3
C$_5$H$_5$N	5	−9.66	12.4
PhNH$_2$	5	−4.78	7.3
MeOH	6	−0.94	33.7
EtOH	14	−0.74	23.2
Me$_2$CO 90	7	−0.72	24.6
EtOH 90	4	−0.52	28.0
MeOH 96.7	6	−0.51	34.2
EtOH 80	15	(0.00)	33.9
Me$_2$CO 80	7	+0.04	30.9
Me$_2$CO 70	7	+0.42	36.5
AcOH	5	+0.57	9.7
MeOH 69.5	5	+0.61	47.3
EtOH 60	4	+0.88	44.7
Me$_2$CO 50	8	+1.02	49.5
EtOH 50	8	+1.14	51.3
H$_2$O	4	+2.95	79.2
HCOOH	6	+4.00	58.5

fects, log k' is a function of the reciprocal of the dielectric constant, it was thought that a plot of b versus $100/D$ might show some consistency in the data. Such a plot is given in Fig. 4 for water–ethanol, water–acetone, and water–methanol solvents. The data show some consistency. The data for the water–ethanol and water–acetone solvents follow approximately the same shape of curve, but the curve for the water–methanol solvents is different in position and curvature from the plots of the data in the other solvents.

The correlation was restricted to chloride and bromide as leaving groups to ensure that the vibrational energies and partition functions will always change in the same way from one solvent to another. The correlation might be extended to organic sulfonates, fluorides, or thiosulfates by changing the standard compound to a sulfonate ester, a fluoride, or a thiocyanate.

Because of excess solvent interaction with the polar nitro groups or because of excessive resonance in the transition state for reaction with more nucleophilic solvents, picryl chloride is less well correlated than the other compounds.

The fit of the data to the equation was measured by φ, where φ is given by the equation

$$\varphi = \left(1 - \frac{\varepsilon}{\theta}\right) 100\%. \tag{6.65}$$

In this equation

$$\varepsilon = \frac{1}{n} \sum_n \left(\left|\log q_{\text{obs}} - \log q_{\text{calc}}\right|\right) \tag{6.66}$$

and

$$\theta = \frac{1}{n} \sum_n \left(\left|\log q_{\text{obs}} - \frac{1}{n} \sum_n \log q_{\text{obs}}\right|\right), \tag{6.67}$$

where n is the number of q values observed and q is a rate or equilibrium constant or a ratio of rate or equilibrium constants. The vertical bars denote absolute values, irrespective of sign. For perfect correlation ε is zero and φ is 100%. For poor correlation φ may be small or even negative. The calculated typical fits varied from poor to excellent.

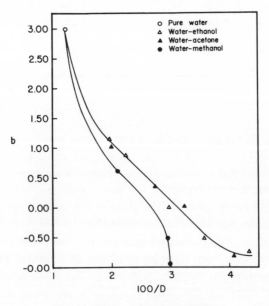

FIG. 4. Plots of b versus $100/D$.

The Equation of Edwards

Edwards (*41, 42*) has proposed an equation of the type

$$\log \frac{k'}{k_0'} = \alpha E_n + \beta H, \tag{6.68}$$

which is a combination of the nucleophilic scale and basicity scale to correlate the reactions of electron donors. The ratio (k'/k_0') represents rate or equilibrium constants for reaction with the given nucleophilic reagent and with water. E_n is a characteristic constant of the electron donors, depending on both their polarizabilities (measured as molar refraction R_∞) and on their basicities ($H = pK_a + 1.74$). Thus

$$E_n = aP + bH, \tag{6.69}$$

where P is a measure of the polarizability of the nucleophilic reagent and is given by

$$P = \log \frac{R_\infty}{R_{H_2O}}. \tag{6.70}$$

From Eqs. (6.68) and (6.69)

$$\log \frac{k'}{k_0'} = \alpha aP + \alpha bH + \beta H$$

$$= \alpha aP + (\alpha b + \beta)H \tag{6.71}$$

$$= AP + BH,$$

where the product of constants αa is set equal to the constant A and the combination of constants $(\alpha b + \beta)$ is set equal to the constant B. The constants A and B, therefore, measure the sensitivity of the nucleophilic reagent to a change of polarizability and basicity, respectively.

The equation correlates rather well rates of displacement reactions of carbon, oxygen, hydrogen, and sulfur and equilibrium constants for complex ion associations, solubility products, and iodine and sulfur displacements.

In all cases except for the displacement of hydrogen, the αE_n term in Eq. (6.68) dominates, and the βH term contributes only slightly to $\log (k'/k_0)$.

E_n, it might be pointed out, for a reagent (N^-) was taken as the electrode potential E_0 for the oxidative dimerization (to N : N) corrected for the potential, -2.60, of the couple

$$2H_2O \rightleftharpoons H_4O_2^{2+} + 2e, \qquad (6.72)$$

that is,

$$E_n = E_0 + 2.60. \qquad (6.73)$$

The H values ($H = pK_a + 1.74$) were determined by the acidities in water of the conjugate acids of the reagents relative to water.

It is seen that, since E_n and H values are determined relative to water, the correlations are solvent dependent.

Equation (6.68) or Eq. (6.71), compared to other free energy relations, provides independent means of parameter determination and seems to correlate a wider range of reaction types than, for example, the linear free energy relations of Swain and co-workers. A major achievement of the Edwards equation for displacements on carbon is the correlation of the hydroxyl ion. The Edwards equation does fail to correlate either the E_n value or the P value of cyanide ion with its nucleophilicity. This failure may arise from either resonance interactions within the cyanogen molecule (43), or the necessity of consideration of the polarizability of more than one atom, a feature which is general to polyatomic nucleophiles. In the case of displacement on hydrogen there is low steric requirement. So far this feature has not been assessed by any reagent correlation.

Wells (44) points out that Eq. (6.71) might be a powerful tool for mechanistic investigation, where a large number of accurately known electrode potentials or reliable values of polarizability of many nucleophiles are available. Nevertheless, he is not convinced that the oxidative dimerization process is a valid model for acceptor–donor interactions in displacement reactions. He believes that the separation of total reactivity into the two factors, nucleophilicity and acidity, is not too fruitful, especially since basicity is only one special case of nucleophilicity.

The Hansson Equation

An expression correlating variations in two substituents has been proposed by Hansson (45). The equation, surprising in its simplicity and accounting for structural variations in both reactants in the reaction of amines with epoxides, is

$$\log \frac{k'}{k_0'} = \varrho(\sigma_a + \sigma_0) + \tau, \qquad (6.74)$$

where k' is the specific velocity constant for the reaction of any amine with any epoxide; k_0' is the specific velocity constant for the reaction of ammonia with ethylene oxide; ϱ depends only on the reaction and is unity for the reactions of amines with epoxides in water at 20°C; σ_a depends only on the structure of the amine and is zero for ammonia; σ_0 depends only on the structure of the epoxide and is zero for ethylene oxide; and τ, which is zero for ethylene oxide, is a correction factor, apparently arising from steric effects, and necessitated by failure, in almost all cases, in the correlation of trimethylamine and triethylamine.

Now since both σ_0 and τ are zero for ethylene oxide, values of σ_a were found by averaging the values of log (k'/k_0') for each of the amines reacting with ethylene oxide, propylene oxide, glycidol, and epichlorohydrin.

Since, also, σ_a was zero for ammonia, values of σ_0 were found by averaging the values of log (k'/k_0') for each epoxide reacting with eight amines including ammonia.

Table VII contains the structural parameters for the Hansson equation (44).

TABLE VII

STRUCTURAL PARAMETERS FOR THE HANSSON EQUATION

Amine	σ_a	Amine	σ_a	$\overset{\displaystyle O}{\overset{\diagup\diagdown}{R—CH—CH_2}}$ R	σ_0
NH_3	0.00	$C_2H_5NH_2$	0.74	H	0.00
CH_3NH_2	0.94	$(C_2H_5)_2NH$	0.63	CH_3	—0.02
$(CH_3)_2NH$	1.47	$(C_2H_5)_3N$	0.36	CH_2OH	—0.04
$(CH_3)_3N$	2.23	Pyridine	0.14	CH_2Cl	0.51

The Hansson equation assumes that the attack occurs exclusively at the same site in all four epoxides. Parker and Isaacs (46) mention the fact that the direction of ring opening must be accounted for in examining structural effects on epoxide reactions.

It would be expected that steric factors would be involved in all epoxide–amine reactions. In the case of a reaction of a series of methylpyridines with propylene oxide the τ-term was found not to correct for steric factors (47), and calculated σ-terms showed no additive relationships.

The Hansson equation correlates the basicity of amines well, whereas the Taft equation cannot correlate these by a single line (48). On the other hand, the Hansson equation correlates the solvolysis of tertiary alkyl iodides poorly, using σ_a-values for $R_1R_2R_3N$ in aqueous alcohol and a τ-value of -0.34; while with the Taft equation the tertiary alkyl chlorides are correlated with a single line (49). Using some additional rate data for the reactions of propylene oxide Hansson (50) derived σ_a-values for some other tertiary amines and found excellent correlation for their dissociation constants using Eq. (6.74) taking $\varrho = -0.54$, $\sigma_0 = 0$, and $\tau = 0$.

Conclusions

The equations discussed in this chapter assume the free energy changes are linearly dependent on the variables involved. The purpose in formulating the equation has been to correlate variations in substrate, reagent, and reaction medium. In the author's opinion as was stated with regard to the Hammett equation, the equations are basically empirical though Wells (44) points out that, when the variables can be identified, the derived parameters are useful theoretical quantities.

The Hammett equation has been extended, modified, and increased in applicability by intensive studies, especially those of Jaffé. It remains the most satisfactory relationship under the most rigid limitation of variations. The more limited Taft equation may be applicable with less stringent limitations.

Wells suggests that an analysis of the Hammett parameters in terms of solvent and reaction parameters might be one of worthwhile new correlations which will undoubtedly be made. Parameters correlating variations in solvent and reagents have been proposed for a few reactions. These equations have been extended by inserting more adjustible parameters, which in some cases have doubtful significance. Most data can be correlated by the use of sufficient parameters, but definiteness of the origin and significance of parameters, in general, becomes less as their numbers increase.

REFERENCES

1. H. C. S. Snethlage, *Z. Elektrochem.* **18**, 539 (1912); *Z. physik. Chem.* **85**, 211 (1913).

2. H. S. Taylor, *Medd. Vetenskapsakad. Nobelinst.* **2**, No. 37, 1 (1913); *Z. Elektrochem.* **20**, 201 (1914).

3. J. N. Brønsted, *Trans. Faraday Soc.* **24**, 630 (1928); *Chem. Rev.* **5**, 231 (1928).

4. J. N. Brønsted and E. A. Guggenheim, *J. Am. Chem. Soc.* **49**, 2554 (1927).

5. M. L. Pfluger, *J. Am. Chem. Soc.* **60**, 1513 (1938).

6. S. W. Benson, "The Foundations of Chemical Kinetics," Chapter XVI. McGraw-Hill, New York, 1960.

7. J. Hine, "Physical Organic Chemistry," Chapter V. McGraw-Hill, New York, 1962.

8. J. N. Brønsted and K. Pedersen, *Z. physik. Chem.* **108**, 2554 (1924).

9. R. P. Bell and W. C. E. Higginson, *Proc. Roy. Soc.* (*London*) **A197**, 141 (1949); R. P. Bell, *J. Phys. & Colloid Chem.* **55** 885 (1951); R. P. Bell and J. C. Clunie, *Proc. Roy. Soc.* (*London*), **A212**, 33 (1952).

10. S. W. Benson, *J. Am. Chem. Soc.* **80**, 5151 (1958).

11. R. P. Bell, E. Gelles, and E. Moller, *Proc. Roy. Soc.* (*London*) **A198**, 308 (1949).

12. R. P. Bell and G. L. Wilson, *Trans. Faraday Soc.* **46**, 407 (1950).

13. A. F. Trotman-Dickenson, *J. Chem. Soc.* **1949**, 1293.

14. L. P. Hammett, *J. Am. Chem. Soc.* **59**, 96 (1937).

15. R. J. Hartmann, H. M. Hoogsteen, and J. A. Moede, *J. Am. Chem. Soc.* **66**, 1714 (1944).

16. G. E. Branch and M. Calvin, "The Theory of Organic Chemistry." Prentice-Hall, Englewood Cliffs, New Jersey, 1941.

17. L. P. Hammett, "Physical Organic Chemistry." McGraw-Hill, New York, 1940.

18. H. H. Jaffé, *Chem. Rev.* **53**, 191 (1953).

19. C. C. Price, *Chem. Rev.* **29**, 60 (1941).

20. L. P. Hammett, *J. Am. Chem. Soc.* **59**, 96 (1937).

21. J. Hine, "Physical Organic Chemistry," Chapter 4. McGraw-Hill, New York, 1962.

22. A. E. Remick, "Electron Interpretations of Organic Chemistry," 2nd ed. Chapter V. Wiley, New York, 1949.

23. D. E. Pearson, J. F. Baxter, and J. C. Martin, *J. Org. Chem.* **17**, 1511 (1952).

24. R. W. Taft, Jr., in "Steric Effects in Organic Chemistry" (M. S. Newman, ed.), Chapter 13. Wiley, New York, 1956.

25. R. W. Taft, Jr., *J. Phys. Chem.* **64**, 1805 (1960), and earlier references listed therein.

26. C. K. Ingold, *J. Chem. Soc.* **1930**, 1032.

27. R. W. Taft, Jr., *J. Am. Chem. Soc.* **74**, 3120 (1952).

28. R. W. Taft, Jr., *J. Chem. Phys.* **26**, 93 (1957).

29. E. Grunwald and S. Winstein, *J. Am. Chem. Soc.* **70**, 846 (1948).

30. L. C. Bateman, E. D. Hughes, and C. K. Ingold, *J. Chem. Soc.* **1940**, 1011.

31. A. H. Fainberg and S. Winstein, *J. Am. Chem. Soc.* **79**, 1597 (1957).

32. A. H. Fainberg and S. Winstein, *J. Am. Chem. Soc.* **79**, 1602 (1957).

33. A. H. Fainberg and S. Winstein, *J. Am. Chem. Soc.* **79**, 1608 (1957).

34. J. Hine, "Physical Organic Chemistry," Chapter VII. McGraw-Hill, New York, 1962.

35. S. Winstein, A. H. Fainberg, and E. Grunwald, *J. Am. Chem. Soc.* **79**, 4146 (1957).

36. W. G. Young, S. Winstein, and H. Goering, *J. Am. Chem. Soc.* **73**, 1958 (1951).

37. C. G. Swain and C. G. Scott, *J. Am. Chem. Soc.* **75**, 141 (1953).

38. C. G. Swain, R. B. Mosley, and D. E. Brown, *J. Am. Chem. Soc.* **77**, 3731 (1955).

39. J. Hine, "Physical Organic Chemistry," p. 137. McGraw-Hill, New York, (1956).

40. C. G. Swain, D. C. Dittmar, and L. E. Kaiser, *J. Am. Chem. Soc.* **77**, 3737 (1955).

41. J. O. Edwards, *J. Am. Chem. Soc.* **76**, 1540 (1954).

42. J. O. Edwards, *J. Am. Chem. Soc.* **78**, 1819 (1956).

43. M. F. Hawthorne and D. J. Cram, *J. Am. Chem. Soc.* **77**, 486 (1955).

44. P. R. Wells, *Chem. Rev.* **63**, 171 (1963).

45. J. Hansson, *Svensk Kem. Tidskr.* **66**, 351 (1954).

46. R. E. Parker and N. S. Isaacs, *Chem. Rev.* **59**, 766 (1959).

47. J. Hansson, *Svensk Kem. Tidskr.* **67**, 246 (1955).

48. H. K. Hall, Jr., *J. Am. Chem. Soc.* **79**, 5441 (1957).

49. A. Streitwieser, Jr., *J. Am. Chem. Soc.* **78**, 4935 (1956).

50. J. Hansson, *Acta Chem. Scand.* **8**, 365 (1954).

CHAPTER VII

VARIOUS TYPES OF SOLVENT EFFECTS ON DIFFERENT REACTIONS

EXPERIMENTAL, HYPOTHETICAL, AND THEORETICAL

Rates, Mechanisms, and the Solvent

The solvent can influence both the rates and mechanisms of reactions. Sometimes the solvent alters the rate without influencing the mechanism, but it would be a coincidence if the solvent changed the mechanism without changing the rate. A solvent can change a rate without changing the mechanism by changing the force between reacting particles and hence altering the readiness with which they approach each other. Such a phenomenon is illustrated by the effect of dielectric constant on electrostatic forces among reacting particles. The solvent may change the rate of diffusion of particles by its viscosity effect and hence alter the frequency of collision between reactant types, and in this way alter the rate of diffusion-controlled reactions. In some cases solvation or selective solvation of reactants may influence the rate and leave the mechanism essentially unaltered. For example it is though that, in electrostatically influenced rates, the trend of the rates at lower dielectric constants toward the rates at the higher dielectric media is due to selective solvation by the higher dielectric constant component of the mixed solvent with the results that when the lower dielectric constant of the solvent is added its effect will be less than if the solvent molecules were randomly distributed. Such results are seen quite frequently (*1–6*). In other instances solvation or selective solvation of a reactant or of reactants will change both the rate and mechanism of a reaction (*7, 8*). Many other properties of the solvents, such as solvolyzing reactants, nucleophilicity, electrophicity, cohesion, influence on hydrogen bonding, etc., influence rates and mechanisms. We now turn to a more detailed examination of some of these effects, not discussed in detail in earlier chapters.

Selective Solvation and Rates and Mechanisms

In the case of the bromoacetate ion–thiosulfate ion, tetraiodophenolsul-fonphthalein ion–hydroxyl ion, and ammonium ion–cyanate ion reactions, as the dielectric constant of mixed solvent decreases to low values, say 50 or less, the rates tend to vary in the direction of the rates in the solvent of higher dielectric constant (1, 9). These deviations have been attributed (2), as mentioned above, to selective solvation of the reactant ions by the higher dielectric component of the solvent, in these cases water. Presumably only the rates and not the mechanisms of the reactions in all ranges of the

$$(7.1)$$

solvent compositions were influenced by the solvent. In the higher dielec-tric constant regions, the influence of the solvent on the rates was predom-inantly that of modifying the electrostatic forces among reactant particles. In the lower dielectric regions of the solvent the electrostatic effects were modified strongly by selective solvation of the reactant species, causing the rates at low dielectric constants to be more like those at high dielectric constants than would be the case were the solvent molecules distributed at random.

The mechanisms of these reactions seem to be the same throughout the solvent range. Thus for the tetraiodophenolsulfonphthalein ion–hydroxyl ion reaction, the mechanism is presumably as presented above, with step one being the rate-governing step (10).

In the case of the ammonium cyanate reaction it might be pointed out that, while the mechanism does not seem to alter due to change in composition of solvent, electrostatic effects cannot be used to distinguish whether the mechanism is ionic or molecular. It might be instructive to examine the point in more detail.

Chattaway (11), points out that, due to the mobile equilibrium

$$NH_4^+ + CNO^- \rightleftarrows NH_3 + HCNO$$

$$NH_4CNO,$$

$$(7.2)$$

the product of the concentration of the ions is proportional to the concentrations of ammonia and the cyanic acid in solution; and that, therefore, the view that the rate of formation of carbamide urea by the interaction of ammonia and undissociated cyanic acid does not appear to be inconsistent with the experimental findings of Walker (12) and his co-workers that the rate of formation of carbamide is proportional to the concentrations of the two ions. Chattaway (11) gave the mechanism of the reaction as the formation of the amide from the ammonia and isocyanic acid, which amide then suffers a hydrogen atom shift to give carbamide. Thus,

$$NH_4 \cdot N:C:O \rightleftarrows H \cdot N:C:O + NH_3 \rightleftarrows H \cdot N:C \overset{\displaystyle OH}{\underset{\displaystyle NH_2}{\diagup}} \rightleftarrows H_2N \cdot CO \cdot NH_2.$$

$$(7.3)$$

Lowry (13) supports this mechanism. Frost and Pearson (14) write the mechanism as

$$H-N=C=O \longrightarrow H-N=C-O^- \longrightarrow HN-C=O.$$

$$\underset{\displaystyle H}{\underset{\displaystyle HNH}{|}} \qquad \underset{\displaystyle H}{\underset{\displaystyle HNH^+}{|}} \qquad \overset{\displaystyle H}{\underset{\displaystyle H}{\underset{\displaystyle NH}{|}}}$$

$$(7.4)$$

From Chattaway we have

$$K_b = \frac{C_{NH_4^+} \cdot C_{OH^-}}{C_{NH_3}}, \tag{7.5}$$

$$K_a = \frac{C_{H^+} \cdot C_{NCO^-}}{C_{HNCO}}, \tag{7.6}$$

and

$$K_a K_b = \frac{C_{\mathrm{NH_4^+}} \cdot C_{\mathrm{NCO^-}}}{C_{\mathrm{NH_3}} \cdot C_{\mathrm{HNCO}}} \cdot C_{\mathrm{H^+}} \cdot C_{\mathrm{OH^-}} . \tag{7.7}$$

But

$$C_{\mathrm{H^+}} \cdot C_{\mathrm{OH^-}} = K_w , \tag{7.8}$$

and therefore

$$C_{\mathrm{NH_3}} \cdot C_{\mathrm{HNCO}} = \frac{K_w}{K_a K_b} \cdot C_{\mathrm{NH_4^+}} \cdot C_{\mathrm{NCO^-}} \tag{7.9}$$

The rate r of the reaction between neutral reactants then is given by the equation

$$r = k_n' \, C_{\mathrm{NH_3}} \cdot C_{\mathrm{HNCO}} = \frac{k_n' K_w}{K_a K_b} \cdot C_{\mathrm{NH_4^+}} \cdot C_{\mathrm{NCO^-}} , \tag{7.10}$$

and including activity coefficients f the equation becomes

$$r = k_n' \, C_{\mathrm{NH_3}} \cdot C_{\mathrm{HNCO}} \frac{f_{\mathrm{NH_3}} \cdot f_{\mathrm{HNCO}}}{f_{\mathrm{X}}}$$

$$= \frac{k_n' K_w}{K_a K_b} C_{\mathrm{NH_4^+}} \cdot C_{\mathrm{NCO^-}} \frac{f_{\mathrm{NH_4^+}} \cdot f_{\mathrm{NCO^-}}}{f_{\mathrm{X}}} \tag{7.11}$$

For ionic reactants the expression for the rate is

$$r = k_i' \, C_{\mathrm{NH_4^+}} \cdot C_{\mathrm{NCO^-}} \frac{f_{\mathrm{NH_4^+}} f_{\mathrm{NCO^-}}}{f_{\mathrm{X}}} . \tag{7.12}$$

In these equations k_n' and k_i' are the specific velocity constants for the rates of reaction between the electrically neutral ammonia and isocyanic acid molecules and between the ammonium and cyanate ions, respectively. Thus

$$k_i' = \frac{k_n' K_w}{K_a K_b} . \tag{7.13}$$

Now the ionic strength effect upon reaction rates can be expressed in terms of the activity coefficients (1), namely,

$$\frac{f_A f_B}{f_X} = \exp\left(\frac{2 Z_A Z_B \varepsilon^2}{1 + \beta a_i \sqrt{\mu}} \right), \tag{7.14}$$

and the dielectric constant effect upon rates can be expressed in terms of activity coefficients as written in Eq. (1.10), that is,

$$\frac{f_A f_B}{f_X} = \exp\left(\frac{Z_A Z_B \varepsilon^2}{kTr} \right)\left(\frac{1}{D_0} - \frac{1}{D} \right). \tag{7.15}$$

Thus from Eqs. (7.11)–(7.15), the ionic and molecular mechanisms are indistinguishable as far as the effects of ionic strength and dielectric constant on activity coefficients are concerned.

This discussion has been given to illustrate the fact that, in general, mechanisms are not unique. Mechanisms are really hypothetical explanations of the rate and of steriochemical, collisional, isotopic, and other forms of data. However new ideas or a different experimental approach may lead to a different mechanism from the one commonly accepted. A few mechanisms are so simple and straightforward that they are generally accepted as correct.

In some reactions not only the rate but also the mechanism is changed by the solvent. Certain nucleophilic reactions fall into this class. In some solvents reactions occur with retention of configuration; for certain nucleophilic reactions in other solvents there is no retention. Cowdrey *et al.* (7) used the $S_N i$ mechanism, to explain those nucleophilic reactions that occur with retention of configuration. These authors suggest that the reactions involving thionyl chloride occurred by the internal decomposition of an intermediate chlorosulfite.

Lewis and Boozer (15) found that the thermal decomposition of secondary alkyl (2-butyl, 2-pentyl, and 2-octyl) chlorosulfites in dilute solutions in dioxane were first-order reactions yielding olefine and alkyl chloride as the principal products. The chloride had the same configuration as the alcohol from which it was derived and was only slightly racemized. The decomposition rate was slower in "isooctane" solution or in the absence of solvent, and in these cases the chloride had the opposite configuration from the alcohol. In both dioxane and isooctane solvents, the reaction was first order. The $S_N i$ mechanism would involve a first-order rate with a retention of configuration. Lewis and Boozer point out that if reaction with retention of configuration occurs in isooctane, then it is much slower in this solvent than in dioxane. The reduction in rate in changing to the less polar solvent (unless due to specific solvent effect) implies that the transition state has a much higher dipole moment than the normal chlorosulfite and that the mechanism would involve the four structures for the transition state represented below:

$$
\begin{array}{cccc}
\overset{\displaystyle O}{\underset{\displaystyle \overset{|}{Cl}}{\underset{\|}{R-O-S}}} & \overset{\displaystyle O}{\underset{\displaystyle Cl^-}{R^+\ O=S}} & \overset{\displaystyle O}{\underset{\displaystyle Cl^-}{\underset{\|}{R-O-\overset{\cdot}{S}{}^+}}} & \overset{\displaystyle O}{\underset{\displaystyle \diagdown Cl}{\underset{\|}{RO=S}}} .
\end{array}
\quad (7.16)
$$

Boozer and Lewis (16) give the specific velocity constant k' for the decomposition of butyl chlorosulfite in dioxane at 99°C to be 56×10^{-4} sec^{-1} and the specific velocity constant for the same decomposition in isooctane at 96°C to be 0.167×10^{-4} sec. These authors give the mechanism for the reaction to give retention of configuration in dioxane to be a three-step process: (1) Ionization of the chlorine–sulfur bond, and consequent weakening of the carbon–oxygen bond. (2) Solvation of the resulting carbonium ion by dioxane and the loss of the now unnecessary sulfur dioxide of solvation. These solvated ions do not racemize due to the fact that the solvation introduces an asymmetry which is present even if the carbonium ion itself is planar. (3) The carbonium ion pair collapses to give a neutral molecule. In this displacement of the dioxane from the oxonium ion or the reaction of chloride ion with solvated carbonium ion, the chlorine becomes attached to the carbon on the same side of the plane of R, CH$_3$, and H from which SO$_2$ left. Thus configuration is retained. The three steps of the mechanism are represented below. In step (1) the cation is written as a resonance hybrid to show identity with a carbonium ion solvated with sulfur dioxide. The steps can be written:

$$(7.17)$$

In toluene which is neither nucleophilic nor a good solvating agent for carbonium ions for any other reason, reversible reaction (1) can, but step (2) cannot take place. The carbonium ion does not ordinarily lose sulfur dioxide, and attack on that side of the carbonium ion is difficult. Hence a slower reaction consisting of the relative motion of the ions within the ion pair can occur, after which a rapid reaction analogous to (3) can now take place to give the inverted chloride. The steps would be

$$(7.18)$$

In acetonitrile, the ketones, and other solvents studied, the stereochemical result is nearer that in toluene than in dioxane. Nearly all these solvents have higher dielectric constants than either dioxane or toluene, and the dissociation of ion pairs is no longer prohibitively difficult. The ionic chain reaction

$$Cl^- + ROSOCl \longrightarrow ClR + OSO + Cl^- \qquad (7.19)$$

is possible and would give inversion.

Cram (17) generalizes the circumstances under which $S_N i$ reactions have been observed to occur. These generalizations are: (1) The $S_N i$ reaction has the best chance of predominating over competing reactions when electron-releasing groups (for example, phenyl) are attached to the carbon undergoing substitution, that is, when a relatively stable carbonium ion is formed. (2) When phenyl alkyl carbinols are treated with the usual halogen-

substituting reagents in liquid sulfur dioxide, the presence of added chloride
ion appears to suppress the competing S_N1 process and results in S_Ni reaction. (3) In the decomposition of chlorosulfites of ordinary dialkyl carbinols,
the S_Ni reaction predominates in better ion-solvating media such as dioxane
but is not detectable in poorer ion-solvating media such as isooctane.
(4) Strongly basic media such as pyridine promote S_N1 type processes and
give inversion of configuration in the reactions of secondary alcohols with
halogen-substituting reagents.

Thus, specific or selective solvation can change either the rate or the
rate and mechanisms of chemical reactions.

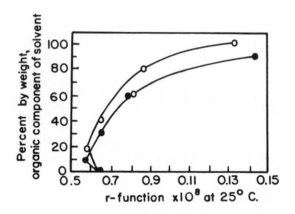

FIG. 1. The r-function in various percentages by weight of the organic component of
mixed solvents: $\bigcirc$ = water–methanol; $\bullet$ = water–acetone.

Winstein and Fainberg (18) have found that the effects of solvent changes
on free energies, enthalpies, and entropies of activation were quite complex
and often depend to the same extent on changes in solvation of reactants
as on solvation of the transition state.

That selective solvation does occur is evidenced by many phenomena
other than kinetic measurements. Using the Walden equation for univalent
electrolytes, Amis (19) wrote the equation,

$$\Lambda_0 \eta = \frac{\varepsilon F}{1800 \, \pi} \left(\frac{r_0^+ + r_0^-}{r_0^+ \, r_0^-} \right). \tag{7.20}$$

Using Eq. (7.20) to calculate the term in parenthesis and taking the reciprocal of the term, namely, $(r_0^+ r_0^-)/(r_0^+ + r_0^-)$, some interesting conclusions can be drawn from the results (19). In Fig. 1 is a plot of r-function versus the percent organic component of the solvent for potassium chloride in water–methanol and water–acetone solvents at 25°C. The reciprocal is termed the r-function since it has the dimensions and magnitude of an ionic radius. The minimum at low percentages of the organic component of the solvent may be significant. Landskroener and Laidler (20) discuss kinetic anomalies which occur in the region of 10% by weight of the organic component of the solvent.

Apparently the relatively slow change of the r-function with the first additions of the solvent can be interpreted as meaning that the potassium and chloride ions cling rather exclusively to the more polar water up to 30 or 40% by weight of the organic component. The ions seemingly cling relatively tightly to water until it is entirely replaced, as is indicated by the sudden large increase of the r-function as the larger molar volume component replaces the last few percent of water.

It might pay to pause and define the terms in Eq. (7.20). Λ_0 is the equivalent conductance at infinite dilution of an electrolyte possessing positive ions of radius r_0^+ and negative ions of radius r_0^-, ε is the electronic charge (the charge on each of the ions) π has its usual meaning, F is the faraday, and η is the viscosity of the medium.

Another confirmation of the selective solvation of ions is the data on the equivalent conductance of perchloric acid in methanol, ethanol, and methanol–ethanol mixtures containing 0.3% by weight of water (21). It was found that if 0.3% by weight of water were added to either pure methanol or pure ethanol, the equivalent conductance Λ_0 dropped precipitously, in the case of ethanol from about 96 to about 67 and in the case of methanol from about 215 to about 171, the temperature being 25°C. This is explained by a combination of two effects. These are the predominance of a proton jump mechanism along the hydrogen bonded chains in the pure alcohols and the more prominent role of the movement of $H_9O_4^+$ ion as a unit in the conductance mechanism of the 0.3 wt % water solvents. The 0.3 wt % of water selectively solvates the proton forming the $H_9O_4^+$ complex.

It is to be hoped that other methods of investigation will give a clearer insight into the real role that specific or selective solvation plays in the determination of the solvent sheath around reactant particles and will give insight as to the true nature of the dielectric constant in the vicinity of the solute particle.

Cage Effects

When two particles become near neighbors in a liquid they will retain this relationship for a time that is long compared to the frequency of molecular vibration. Franck and Rabinowitch (22) and Rabinowitch and Wood (23) deal with this prolonged time of nearest neighbor residence of newly created fragments of dissociation in liquid media. Einstein and also Smoluchowski (24) propose the equation

$$\bar{\Delta}^2 = 2Dt \tag{7.21}$$

for the mean displacement $\bar{\Delta}$ of a particle during the time t if its diffusion coefficient is D. Rabinowitch and Wood take D, for uncharged molecules in solutions of low viscosity, for example, iodine in carbon tetrachloride, to be of the order of 2×10^{-5}. They point out that in a liquid displacement distance over a single molecular diameter, taken to be 3×10^{-8} cm, is composed of a number of segments over a zigzag path, and hence Eq. (7.21) applies and

$$t = \frac{(3 \times 10^{-8})^2}{2 \times 2 \times 10^{-5}} = 2.3 \times 10^{-11} \text{ sec.} \tag{7.22}$$

A gas molecule would travel a distance of the order of a thousand times as far in a tenth of the time. Thus the collision partners, due to being caged by the solvent particles, would be agitated in this time interval but not separated, so that they would collide more than once but probably not more than four or five times. Collision sets consisting of 1, 2, or 3 collisions exist in solution. Thus, in addition to the "normal" probability of recombination governed by the law of mass action, there will be a probability of "primary recombination" of dissociation fragments.

Franck and Rabinowitch (22) point out that this effect exists in the gaseous state but can be neglected there. They calculate the probability of primary recombination in the liquid state in the following manner. The mechanism would be somewhat as follows:

$$RA + h\gamma \longrightarrow R + A^*, \tag{7.23}$$

$$A^* + S \longrightarrow A + S^*, \tag{7.24}$$

$$R + A \longrightarrow RA, \tag{7.24a}$$

where S is a solvent molecule or some other third body.

The mean free path, in which distance from R there is a loss of excess energy by A, is called λ. After the third body collision suffered by A it starts moving again with equal probability in all directions. If d, the molecular diameter, is the same for R and A and if the difference between $1/\bar{\lambda}^2$ and $(\overline{1/\lambda^2})$ is neglected, then the probability p of A again meeting the radical R is given by the equation

$$p = \frac{\pi d^2}{4\pi\lambda^2} = \left(\frac{d}{2\lambda}\right)^2. \qquad (7.25)$$

For a gas at one atmosphere pressure, putting $d = 3 \times 10^{-8}$ cm and taking λ as of the order 1×10^{-5} cm, one obtains a value of 2.5×10^{-6} for p. Since only one out of 10^3 triple collisions at one atmopshere leads to recombination, the probability of primary recombination is, under these conditions, only 2.5×10^{-9}. However, this probability increases as the third power of the pressure and would be of the order of 1 at 1000 atm if ordinary gas kinetics could be applied at densities at which the mean free path is of the same order as the diameter of molecules. Franck and Rabinowitch point out that this calculation indicates that the probability of primary recombination in a liquid system cannot be neglected and could reasonably be expected to be of the order of 0.1.

In addition to the recombination effect there is a dissipation effect in which the electronically excited atom or molecule may interact with the solvent to convert the activation energy of the atom or molecule into vibrational energy of the solvent molecule or to produce chemical change in the colliding molecule. There is a definite probability of both kinds of dissipation interactions with the solvent when an atom or molecule immersed in a liquid is excited through absorption of light. This is to be expected when an excited particle is held in a solvent cage. This is represented by step (7.24) in the mechanism written above.

Both the primary recombination and dissipation effects will have quantitative efficiencies which will depend on the nature of the dissociation products and of the solvent, for example, its viscosity, molecular mass, strength of its hydrogen bonding, etc., as well as on the frequency of the light absorbed. The greater the frequency of the absorbed radiation, the less the recombination effect will be, since the greater excess energy will enable the dissociation particles to escape through the walls of the solvent cage and thus prevent multiple collisions with their former partners.

Ogg et al. (25) noted a change in the quantum efficiency with wavelength of light in the formation of amides in alkali metal solutions in liquid ammonia. Warburg and Rump (26) noted a wavelength effect in the dissociation of HI and H_2S and the reduction of NO_3^- in solution.

If the dissociation products are heavier than the molecules of the solvent, their excess kinetic energy allows them to break through the walls of the solvent cage and thus escape recombination. This is according to the law of the conservation of momentum. For dissociation products lighter than the molecules of solvent, whatever their kinetic energy, they may be stopped or reflected back by the first collision with a molecule of the solvent. Especially with solvents with very light molecules, a dependence of recombination upon the amount of excess energy, that is, upon the wavelength of the absorbed light, may be expected. In aqueous solutions a dependence of quantum yield on wave length is often observed. There would be no dependence of quantum yield with wavelength expected in the case of iodine in carbon tetrachloride, since the dissociation product iodine atom has a mass (127) less than that of the solvent (156). That the excess of energy of the excited electronic state $^2P_{1/2}$ of one of the iodine atoms formed photochemically is not the cause of the failure of primary recombination to take place was excluded as a possibility by proving that no decrease in quantum yield occurred at wavelengths greater than 5000 Å. In this region of wavelength the absorbed energy is too small to produce an excited iodine atom.

Noyes (27–30) has treated rate processes and diffusion in liquids. These theories were applied to the quenching of fluorescence among other rate processes, and it was shown that the individual parameters of diffusive motion can be estimated from the effect of scavenger concentration on the quantum yield of photochemical decomposition. These papers indicate that diffusion in liquids involves practically continuous motion and small individual displacements. "Jumps" of the order of a molecular diameter that are opposed by significant potential barriers are not involved in diffusion.

If we assume that there is an above-average probability that two separated particles or groups will recollide before final separation to the average distance of pairs in the liquid, we can write a possible mechanism for a photochemical activated process as follows:

(1) separation into near neighbors in a solvent cage and geminal recombination by reincounter,

$$AB + h\nu \underset{k_2'}{\overset{k_1'}{\rightleftharpoons}} A \ldots B^* ; \qquad (7.26)$$

(2) removal of excess energy of pair by solvent and recombination of

the pair by three-body collision while in the cage,

$$\text{A} \ldots \text{B*} + \text{S} \xrightarrow{k_3} \text{AB} + \text{S*} \; ; \qquad (7.27)$$

(3) diffusion of pair from cage and separation to the average distance of pairs in the liquid and their secondary recombination,

$$\text{A} \ldots \text{B*} \underset{k_5'}{\overset{k_4'}{\rightleftarrows}} \text{A} + \text{B} . \qquad (7.28)$$

The activation might be thermal but the steps would be similar. The kinetic rate expression for the disappearance of AB can be obtained in the following manner:

$$- \frac{d[\text{AB}]}{dt} = k_1'[\text{AB}] - k_2'[\text{A} \ldots \text{B*}] - k_3[\text{A} \ldots \text{B*}] \, [\text{S}] . \qquad (7.29)$$

But the solvent concentration can be considered constant, and $k_3[\text{S}]$ can be taken as a constant k_3'. Therefore

$$- \frac{d[\text{AB}]}{dt} = k_1'[\text{AB}] - k_2'[\text{A} \ldots \text{B*}] - k_3'[\text{A} \ldots \text{B*}] , \qquad [7.30]$$

$$- \frac{d[\text{A} \ldots \text{B*}]}{dt} = - k_1'[\text{AB}] + k_2'[\text{A} \ldots \text{B*}] + k_3'[\text{A} \ldots \text{B*}] \, [\text{S}]$$
$$+ k_4'[\text{A} \ldots \text{B*}] - k_5'[\text{A} \ldots \text{B*}] \qquad (7.31)$$
$$= - k_1'[\text{AB}] + k_2'[\text{A} \ldots \text{B*}] + k_3'[\text{A} \ldots \text{B*}]$$
$$+ k_4'[\text{A} \ldots \text{B*}] - k_5'[\text{A} \ldots \text{B*}] .$$

At the steady state

$$\frac{d[\text{A} \ldots \text{B*}]}{dt} = 0 \qquad (7.32)$$

and

$$[\text{A} \ldots \text{B*}] = \frac{k_1'}{k_2' + k_3' + k_4' - k_5'} [\text{AB}] . \qquad (7.33)$$

Therefore,

$$- \frac{d[\text{AB}]}{dt} = \frac{k_1'(k_4' - k_5')}{k_2' + k_3' + k_4' - k_5'} [\text{AB}] . \qquad (7.34)$$

Benson (31) neglects the steps involving k_3' and k_5' and obtains

$$-\frac{d[AB]}{dt} = \frac{k_1'k_4'}{k_2' + k_4'}\,[AB]\,, \qquad (7.35)$$

which, by assuming $k_4' \ll k_2'$ due to a great probability for the caged radicals to recombine, reduces to

$$-\frac{d[AB]}{dt} = \frac{k_1'k_4'}{k_2'}\,[AB]\,. \qquad (7.36)$$

However when there is a high steric or energy barrier to recombination, $k_2' \ll k_4'$ and the cage effect becomes vanishingly small.

DeTar and Weis (32) found that, in the decomposition of β-phenyliso-butyl peroxide in carbon tetrachloride solution, the 1-phenyl-2-propanol obtained from the ester produced retained 75% of the original configuration of the peroxide. They explained this by the production in the solvent cage of ester-forming radicals facing each other and reacting faster than the 1-phenyl-2-propyl radical can reorientate to present its other face to the carboxyl radical. Hine (33) writes the reaction

$$(7.37)$$

Upon the induced decomposition of the peroxide the ester is not formed, and when 1-phenyl-2-propyl radical abstracts a chlorine atom from carbon tetrachloride, the 1-phenyl-2-proyplchloride is racemic. These data support the cage theory.

Harkness and Halpern (36) studied the kinetics of the oxidation of uranium (IV) with thallium (III) in aqueous perchloric acid solutions. It was postulated that the oxidation of U(IV) by Tl(III) occurred through a two-equivalent mechanism. They felt that the kinetics of the reaction and the insensitiveness of the rate to such ions as Cu^{2+} ruled out contributions to the kinetics of a chain mechanism similar to that observed in the oxidation of U(IV) by O_2 (37). These authors offered an alternative interpretation in which the reaction,

$$U(IV) + Tl(III) \longrightarrow U(V) + Tl(II), \qquad (7.38)$$

occurred initially, but following this step the U(V) and Tl(II) ions reacted with each other to form U(VI) and Tl(I) in a shorter time than that required for them to diffuse out of their solvent cage and react with other species.

That the oxidation of U(IV) by Tl(III) occurs through a two-equivalent mechanism is in line with the generalizations of Higginson and Marshall (38), and is apparently applicable to reactions of Tl(III) with other two-equivalent reductants (39).

Jones and Amis (40) studied the U(IV)–Tl(III) reaction in water–methanol media. The effect of various cations on the rate was investigated in 25 wt % water–75 wt % methanol. In this media, in contrast to water media where there was no effect, Cu^{2+} and Hg^{2+} accelerated the reaction as they did the rate of oxidation of U(IV) by oxygen (37). In this last reaction a chain mechanism involving intermediates U(V) and HO_2 were postulated. Likewise Jones and Amis postulated a step-wise process for the effect of Cu^{2+} and Hg^{2+} on the rate of the U(IV)–Tl(III) reaction. In the case of Cu^{2+} the steps were written

$$U(IV) + Tl(III) \longrightarrow U(V) + Tl(II), \qquad (7.39)$$

$$U(V) + Cu(II) \longrightarrow U(VI) + Cu(I), \qquad (7.40)$$

$$Tl(II) + Cu(I) \longrightarrow Tl(I) + Cu(II). \qquad (7.41)$$

These cation effects in 75% methanol indicate that the reaction is proceeding in one-electron steps with the formation of the unstable intermediates U(V) and Tl(II). If a solvent cage effect is present, in the case of methanol-containing solvent, hydrogen bonding of the alcohol would be weaker than that of water, and reactants could diffuse out of the cage more readily for reaction with other species. The fact that the reaction does proceed by one-electron steps might be due to difference in solvation in water–methanol solutions as compared with water.

Rembaum and Szwarc (34) and Szwarc (35) have studied the thermal decomposition of acetyl peroxide, $CH_3CO_3COCH_3$, and found that initially acetate radicals CH_3CO_2 were formed. The acetate radicals subsequently yielded methyl radicals CH_3 and CO_2. The methyl radicals combined to form C_2H_6 or accepted an hydrogen from a donor to form CH_4. In the presence of I_2 as a free-radical scavenger, no C_2H_6 was formed in the gas-phase decomposition of acetyl peroxide, but in solution the formation of C_2H_6 could not be eliminated. The formation of C_2H_6 in solution, even in the presence of a scavenger, is attributed to the great tendency for CH_3 radicals formed in the same solvent cage to recombine.

Suppose a scavenger, Sc, is present that can tie up like particles coming from a photochemically activated decomposition; the mechanism must be written to include the deactivation of the activated state by its colliding with molecules of solvent in the walls of the solvent cage. Thus

$$A_2 + h\nu \underset{k_2'}{\overset{k_1'}{\rightleftharpoons}} A \ldots A^*, \tag{7.38'}$$

$$A \ldots A^* + S \xrightarrow{k_3'} A_2 + S^*, \tag{7.39'}$$

$$A \ldots A^* \underset{k_5'}{\overset{k_4'}{\rightleftharpoons}} 2A, \tag{7.40'}$$

$$A + Sc \xrightarrow{k_6'} ASc, \tag{7.41'}$$

$$A \ldots A^* + Sc \xrightarrow{k'_7} ASc + A. \tag{7.42}$$

From this mechanism the following differential expressions for the rates of change in concentration of the indicated species can be found:

$$\frac{d[A \ldots A^*]}{dt} = k_1'A_2I_a - k_2'[A \ldots A^*] - k_3'[A - A^*][S] - k_4'[A \ldots A^*]$$
$$+ k_5'[A]^2 - k_7'[A \ldots A^*][Sc], \tag{7.43}$$

$$\frac{d[A]}{dt} = k_4'[A \ldots A^*] - k_6'[A][Sc] + k_7'[A \ldots A^*][Sc]$$
$$- k_5'[A]^2, \tag{7.44}$$

$$\frac{d[A_2]}{dt} = - k_1'[A_2]I_a + k_2'[A \ldots A^*] + k_3'[A \ldots A^*][S]. \tag{7.45}$$

If it is assumed that the concentrations of A and A...A* reach a steady state, Eqs. (7.43) and (7.44) can be set equal to zero. Then if Eq. (7.44) is solved for A...A*, and if this value is inserted in Eq. (7.43), an equation for the concentration of A can be found and is

$$[A]^2 k_5'(k_4' + k_7'[Sc] - k_2' - k_3'[S] - 1) - [A][Sc]k_6'(k_2' + k_3'[S] + 1)$$
$$+ k_1'[A_2]I_a(k_4' + k_7'[Sc]) = 0. \tag{7.46}$$

The fact that the concentration of solvent [S] and the intensity of incident radiation I_a are essentially constant might be incorporated in the constants of the terms in which they appear.

The assumption that [A] and [A...A*] are constant implies an infinite reservoir of A_2, which insures a constancy of $[A_2]$. That the assumption of a steady state for [A...A*] and [A] leads to the conclusion that $d[A_2]/dt$ is zero can be proved by substituting the value of [A...A*] at $d[A]/dt = 0$ from Eq. (7.44) into Eq. (7.45). If we set Eq. (7.45) equal to zero and solve for [A...A*] and if we insert this value into Eqs. (7.43) and (7.44), the remarkably simple equation

$$k_6'[A][Sc] = 0 \tag{7.47}$$

can be derived. This equation probably implies that [A] is reduced to zero by the scavenger under the conditions of an infinite reservoir of A_2 at constant concentration $[A_2]$ and a true steady state.

In the above mechanism, constancy of [ASc] cannot be assumed since ASc is a final product produced by irreversible processes and does not appear as a reactant. Benson (31) in treating the photolysis of I_2 in liquid solutions in the presence of a scavenger, used the steady states [I...I*], [I], and [ISc], however ISc was assumed to react further with iodine radicals to produce I_2Sc. This author writes the average relative rate of recombination of [I...I*] $\bar{k}_r$ to yield I_2 to the rate of its disapperance $k_7'[Sc]$ due to scavenging to form ASc as

$$\frac{\bar{k}_r}{k_7'[Sc]} = \frac{2k_2'r\,\Delta r}{D}\left[1 - r\left(\frac{k_7'\,Sc}{\pi D}\right)^{1/2}\right], \tag{7.48}$$

where r is the radius of the cage, Δr is the thickness of the cage walls, and D is the diffusion constant of the two I atoms relative to each other, that is $D = 2D_I$. Benson points out that r is the near neighbor distance and $\Delta r \cong r/2$. This same equation would hold for the mechanism written above.

Hydrogen Bonding, Solvents, and Rates

Pimentel and McClellan (*41*) have presented a comprehensive treatment of hydrogen bonding, including its effect on rates and mechanisms of reactions.

The boiling points of water and other hydroxy compounds have been explained on the basis of hydrogen bonding among the molecules. The hydroxylic hydrogen of one molecule forms a bond with the oxygen atom of another molecule. This type of bond can be repeated so that the hydroxy compound can form long chains or agglomerations of molecules. Thus the conductance of hydrogen ion in aqueous solution is explained on the Grotthus mechanism of proton jumps in which a proton attaches itself to one end of a chain of water molecules and another proton detaches itself from the other end almost simultaneously, so that effectively the hydrogen ion jumps the length of the molecular water chain in the time it takes a bond to form or break. This time is much shorter than the time required for a solvated hydrogen ion to diffuse the distance of the length of the chain under the influence of the field created by an applied electromotive force.

In hydrogen bonding, one group must serve as an acidic or proton donating group and must include carboxyl, hydroxyl, amine, and amide groups. The other groups must be basic or proton-accepting groups, and must include oxygen in hydroxyls, carbonyls and ethers, halogen atoms in certain molecular environments, and nitrogen in *N*-heterocyclic compounds.

Hydrogen bonding may be intermolecular, involving the bonding by hydrogen of two molecules of the same or different substances, or it may be intramolecular, involving the bonding by hydrogen of two groups within the same molecule. The intermolecular bonding occurs in water, alcohols, carboxylic acids, amides, proteins, polypeptides, and polyhydroxyorganic and inorganic substances; and may produce chains, cyclic configuration, or three-dimensional structures. Intramolecular hydrogen bonding is termed chelation (after the Greek *chele*, claw), since the closing of the gap between two rings by hydrogen bonding resembles a claw-like action. The hydrogen bond is much weaker than ordinary covalent bonds but is much stronger than the van der Waals attractive forces between molecules. The enthalpy per hydrogen bond per mole is usually from 2 to 7 kcal/mole but can be higher than 7 or less than 2 kcal per hydrogen bond per mole.

The hydrogen bond has been explained in terms of both resonance (*42–46*) and dipole interactions (*47–50*). In the case of the formic acid dimer, the hydrogen is usually not halfway between the bonded atoms, and according to some theorists cannot be due to resonance; however, the proponents of the resonance theory argue that the explanation is that the two resonance structures contribute unequally. Hine (*33*) reasons that, although electrostatic interactions contribute to hydrogen bonding, they cannot be the sole explanation, since if they were there should be a correlation between hydrogen-bonding properties and electronegativity. The ability to accept hydrogen bonds should be in the order alkyl fluorides > alcohols > amines, if electrostatic interaction predominates in determining this property. The reverse of this appears to be the case, with the stronger hydrogen bonds being formed by the more basic atoms. Thus HF forms a strong hydrogen bond with the strongly basic fluoride anion.

To be a bit whimsical, one can see the importance of hydrogen bonding when it is remembered that liquid water, if composed of unassociated molecules with a molecular weight of 18, would boil near the boiling point (− 161°C), of methane with a molecular weight of 16, and thus, at ordinary temperatures, man would have liquid water to neither drink and bathe in, nor catch fish out of. Then what would the world do for the tall tales which fishermen tell, not to mention the problems arising from the offensive odors of unbathed humans, and thirst which whisky could not satisfy! Actually, dipole–dipole interactions, as well as hydrogen bonding, no doubt play an important part in the association of water molecules.

A more pertinent case of the effect on the rate of a reaction of hydrogen bonding between reactant and solvent is the inhibition of phenol alkylation by ethers, studied by Hart *et al.* (*51*) The decrease in reaction rate of the alkylation of phenol by *t*-butyl chloride, brought about by the addition of dioxane, was found to be proportional to the concentration of dioxane. The decrease in rate was quantitatively accounted for by assuming the formation of a 2:1 phenol: dioxane hydrogen-bonded complex, and that phenol so bound could not participate in the alkylation reaction. In the case of tetrahydropyran it was assumed that a 1:1 complex was formed in order to explain the data. The proposed hydrogen-bonded complexes were written for dioxane and tetrahydropyran, respectively, as follows:

$$
\begin{array}{c}
\mathrm{CH_2{-}CH_2} \qquad\qquad \mathrm{C_6H_5} \\
\nearrow \qquad\qquad \searrow \qquad\quad \nearrow \\
\mathrm{O{-}H \ldots O} \qquad\qquad\quad \mathrm{O \ldots H{-}O} \\
\nearrow \qquad\qquad \searrow \qquad\quad \\
\mathrm{C_6H_5} \qquad\quad \mathrm{CH_2{-}CH_2}
\end{array}
\qquad (7.49)
$$

$$
\begin{array}{c}
\mathrm{CH_2{-}CH_2} \qquad\qquad \mathrm{C_6H_5} \\
\nearrow \qquad\qquad \searrow \qquad\quad \nearrow \\
\mathrm{CH_2} \qquad\qquad \mathrm{O \ldots H{-}O} \\
\searrow \qquad\qquad \nearrow \\
\mathrm{CH_2{-}CH_2}
\end{array}
\qquad (7.50)
$$

Calculations of the half-life for the alkylation of phenol using the equation

$$\frac{t_{1/2}}{t_{1/2}{}^0} = \left(\frac{M^0}{M}\right)^n, \tag{7.51}$$

where $t_{1/2}$ refers to half-times of reaction, M is the molarity of the phenol, the superscripts refer to the reaction in pure phenol, the terms without superscripts refer to the reaction with dioxane present, and n is the order of the reaction with respect to phenol. The value of n was found to be 6. The calculations were made assuming (1) dioxane served only as a diluent for the reactants, (2) dioxane forms a 1:1 complex with phenol, and (3) dioxane forms 2-phenol: 1-dioxane complex with phenol. Table I contains the results of the calculations.

TABLE I

Comparison of Experimental Half-Times With Those Calculated for Three Assumptions

Expt. no.	$t_{1/2}$ (obs.) (min.)	$t_{1/2}$ assumption (1) (min.)	$t_{1/2}$ assumption (2) (min.)	$t_{1/2}$ assumption (3) (min.)	Ether / phenol (mole)
1	55.2	—	—	—	—
2	53.1	—	—	—	—
3	96.7	65.5	78.7	95.4	0.0287
4	178.6	84.7	118.2	174.3	0.0621
5	457.8	103.1	217.6	497.7	0.113
6	1497.3	134.0	461.3	1654.7	0.168
7	119.0	85.9	118.4	—	0.0563
8	268.8	111.3	247.6	—	0.116
9	423.7	137.2	400.2	—	0.162
10	146.1	144.5	—	—	0.152

The data for experiments 3–6 are for dioxane as a component of the solvent, while experiments 7–9 are for tetrahydropyran as a component of the solvent. By comparing columns 2 and 5 for experiments 3–6 it is evident that observation and theory are in agreement for a 2-phenol: 1-dioxane complex. By comparing columns 2 and 4 for experiments 7–9, it can be seen that observation and theory more nearly coincide when it is assumed a 1-phenol: 1-tetrahydropyran complex is formed.

These data indicate that phenol bound in a complex with ether does not undergo alkylation. The hydroxyl group apparently plays an important role in alkylation. This function is perhaps the solvation of the halogen atom of the alkyl halide.

Svirbely and Roth (52) studied the kinetics of cyanohydrin formation in aqueous solution. They found that acetaldehyde and propionaldehyde react quantitatively with hydrocyanic acid but that acetone does not. The reactions were second order and were not generally acid catalyzed, except for propionaldehyde where a small general acid catalysis was detected. The mechanism involved hydrogen bonding, specifically of water with the carbonyl group, but generally of any molecular acid with the carbonyl group. Thus,

$$\begin{array}{c}\diagdown \\ \diagup\end{array} C{=}O + H_2O \underset{\longleftarrow}{\overset{fast}{\longrightarrow}} \begin{array}{c}\diagdown \\ \diagup\end{array} C{=}O \ldots HOH \text{ (hydrogen bond)}, \qquad (7.52)$$

$$\begin{array}{c}\diagdown \\ \diagup\end{array} C{=}O \ldots HOH + CN^- \overset{slow}{\longrightarrow} \begin{array}{c} \quad CN \\ \diagup \\ C \\ \diagdown \\ \quad OH \end{array} + OH^- \qquad (7.53)$$

or

$$\begin{array}{c}\diagdown \\ \diagup\end{array} C{=}O + HA \underset{\longleftarrow}{\overset{fast}{\longrightarrow}} \begin{array}{c}\diagdown \\ \diagup\end{array} C{=}O \ldots HA , \qquad (7.54)$$

$$\begin{array}{c}\diagdown \\ \diagup\end{array} C{=}O \ldots HA + B \overset{slow}{\longrightarrow} \begin{array}{c} \quad B \\ \diagdown \diagup \\ C \\ \diagup \diagdown \\ \quad OH \end{array} + A^- \qquad (7.55)$$

Szwarc and Smid (53) believe that the radical PhCOO from benzoyl peroxide forms hydrogen bonds with carboxylic acids which facilitate the hydrogen transfer between the carboxylic acid and the polar radical. This hydrogen transfer reaction is most pronounced in trifluoroacetic acid, which is the strongest acid investigated and, therefore, has the greatest tendency to form hydrogen bonds. The acetyl and propionyl peroxides, on the other hand, form CH_3 and C_2H_5 radicals, since the CH_3COO and C_2H_5COO radicals are unstable and decarboxylate before any other reaction occurs. In the case of benzoyl peroxide–trifluoroacetic acid reaction, CF_3 is formed from the decarboxylation of the resulting CF_3COO radical. The CF_3 either dimerizes or attacks the peroxide, causing further decomposition. However, if isooctane is a component of the solvent, fluoroform is formed from the reaction

$$CF_3 + HR \longrightarrow CF_3H + R. \qquad (7.56)$$

Palit (54) formulates what he terms a basic principle on the effect of the solvent through hydrogen bonding. If the active centers that take part in the reaction are blocked by hydrogen bonding or other interaction with a particular solvent, the speed of the reaction will be diminished by that solvent.

If a solvent, by hydrogen bonding or otherwise, promotes electron shift required in the reaction, the speed of the reaction will be increased by that solvent. Palit accepted the idea that hydrogen bonding can take place between any positive hydrogen and any negative element present in the system.

The low energy of activation for the self-condensation of glycylglycyl-glycine methyl ester in methyl alcohol is attributed by Rees *et al.* (55) to a hydrogen-bonded head-to-tail position complex of the reacting molecules in which the ester and amino groups of the adjoining molecules are held in close proximity, thus establishing conditions favorable for reaction to occur by elimination of methanol. The configuration of the complex is given as

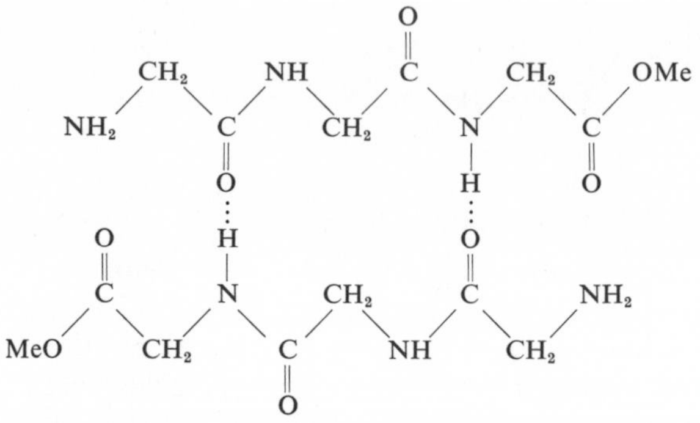

$$
\begin{array}{c}
\text{(structure 7.57)}
\end{array}
$$

$$\tag{7.57}$$

Apparently the methanol solvent did not block the carbonyl groups by forming hydrogen bonds with them. The energy of activation of the above reaction would be less than the energy of activation of a similar reaction not involving such an intermediate by an amount equal to the heat of formation of the hydrogen bonds. It would be of interest to know what a solvent like dioxane, with the strong tendency of its ether oxygens to form hydrogen bonds, would do to this reaction rate.

Swain (56) and Swain and Eddy (57) have postulated hydrogen bond formation in the termolecular mechanism of triphenyl-methyl halide displacements and in the termolecular displacement reactions of methyl halides, respectively.

In the case of triphenyl-methyl halide displacements, the kinetics are third order. Methanol reacts with triphenylmethyl chloride or bromide in benzene solution containing excess pyridine at 25° to give the methyl ether.

When phenol is used instead of methanol, phenyl ether is formed at a slower rate. When both phenol and methanol are present, methyl ether is formed at a rate which is seven times as fast as the combined rates with methanol or phenol separately. In this case the rate is proportional to the concentrations of methanol, phenol, and halide, although phenol is not consumed in the rapid phase of the reaction.

This high rate of reaction in the presence of both phenol and methanol is explained on the basis of a concerted termolecular attack, in which a molecule of phenol hydrogen bonds with the halogen atom weakening its bond with carbon, and a molecule of methanol simultaneously solvates the carbon which is becoming a carbonium ion. In this mechanism the phenol and methanol each plays the part to which it is best suited, and thus the reaction is fast.

In the case of the termolecular displacement reactions of methyl halides, added methanol, phenols, or mercuric bromide facilitate the reaction. Thus in the case of displacement of bromide in methyl bromide in the presence of phenol, the concerted push-pull third-order mechanism involving the formation of a hydrogen bond is written by Swain and Eddy as

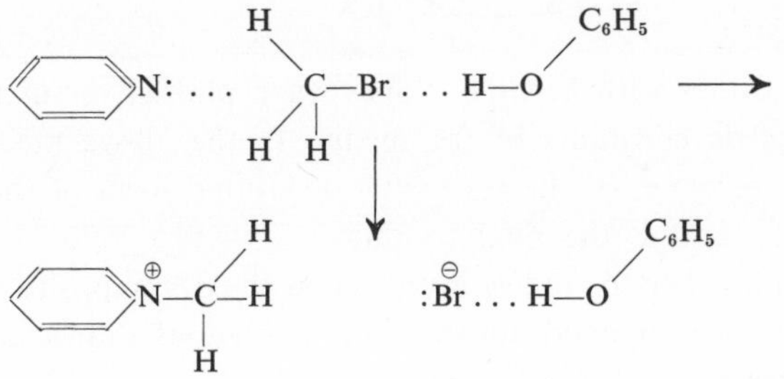

$$(7.58)$$

which is superimposed on the ordinary, slower second-order mechanism that operates when there is present no solvating agents more effective than the solvent, benzene. In the absence of pyridine the reaction of 2.6 M methyl bromide with 24.7 M methanol is five powers of ten less than the rate constant with pyridine present.

Aksnes (58) postulates that the acid-catalyzed reactions of trialkylphosphites, diisopropyl fluorophosphate, and acetamide involve hydrogen bond equilibrium according to the following reaction scheme:

$$S + H_3O^+ \ldots nH_2O \; \underset{}{\overset{K_{assoc}}{\rightleftharpoons}} \; S \ldots H_3O^+ + nH_2O , \qquad (7.59)$$

$$S \ldots H_3O^+ + H_2O \; \overset{k'}{\longrightarrow} \; \text{Product (slow)} . \qquad (7.60)$$

The kinetics of the reaction will be the same as the predictions of Euler's theory (59). However, K_{assoc} in Eq. (7.59) is the hydrogen bond association constant between the substrate and the hydronium ions relative to that of water and is not the base constant of the substrate.

Ono et al. (60) assume a complex involving hydrogen bonding in the α-amylase-catalyzed hydrolysis of amylose. There is hydrogen bonding by the hydrogens of both the imidazolium groups and of solvent water molecules. Addition of methanol affected the rate, but it was concluded that the dielectric constant effect was predominantly effective. Using the theory of Hiromi (61, 62) for the effect of the dielectric constant of the medium on the rates of reactions, the data for the α-amylase-catalyzed hydrolysis of amylose was accounted for by using reasonable values of the parameters involved. Log k_3' and log K_m' for the mechanistic equations

$$ E + S \; \underset{k_2'}{\overset{k_1'}{\rightleftharpoons}} \; ES \; \overset{k_3'}{\longrightarrow} \; E + D , \qquad (7.61) $$

$$ E \; \overset{K}{\rightleftharpoons} \; E' , \qquad (7.62) $$

$$ E' + S \; \overset{K_m'}{\rightleftharpoons} \; E'S \qquad (7.63) $$

gave straight lines with negative slopes when plotted versus the reciprocals of the dielectric constants of the media. In the above mechanisms E represents the enzyme, E′ the reversibly denatured form of the enzyme, S is the substrate, and P the final product.

Shaw and Walker (63) have investigated the decomposition of thioureas. They propose a H-bonded, ionized intermediate as shown in the following reaction scheme:

$$
(CH_3)_2\,N{-}CS{-}NH{-}CH_3 \; \underset{k_2'}{\overset{k_1'}{\rightleftharpoons}} \;
\left[
\begin{array}{c}
\text{S} \\
\parallel \\
H_3C \quad C \\
\diagdown \quad \diagup \\
N \quad N{-}CH_3 \\
\diagup \quad \diagup \\
H_3C \quad\quad H
\end{array}
\right]^{2+}
\overset{k_3'}{\longrightarrow}
$$

$$ CH_3NCS + (CH_3)_2NH , \qquad (7.64) $$

$$ CH_3NCS + H_2O \; \overset{k_4'}{\longrightarrow} \; CH_3NH_2 + COS \qquad (7.65) $$

for the decomposition of the trimethyl thiourea. In acidified aqueous solution, reaction (7.65) is fast, and there should be equal amounts of methyl amine and dimethyl amine produced. They give $k_3' = k_4'$ and therefore argue that the stoichiometry is in accord with the above reasoning. As the solutions become more basic the amines are no longer predominately in the protonated form. The reverse reaction rate k_2' will be magnified by increase of pH, since the forward reaction rate depends on the unionized thiourea. Thus, the acidity of the media is an important factor in determining the rate of this reaction involving hydrogen bonding.

A study was made of tetramethyl thiourea in various media at elevated temperature. It did not show extensive reaction in neutral solution. Hence it was reasoned that a hydrogen atom bonded to nitrogen was essential to the decomposition of thioureas.

So far we have discussed intermolecular hydrogen bonding and its effect on reaction rates and mechanisms. Let us consider a few examples of intramolecular hydrogen bondings.

Rabinowitz and Wagner (64) compared 5(6)-nitrobenzimidazole and 4(7)-nitrobenzimidazole and found the latter to be more rapidly reduced polarographically, more easily reduced catylytically, and more weakly acidic. These observations implied an intramolecular influence that made the nitro group more susceptible to reduction and made the removal of the essential proton more difficult. The 4(7)-isomer was found to be less associated in solution and hence much more volatile. The hydrogens required for meso-hydric linkages were otherwise involved. All these observations led these investigators to postulate intramolecular hydrogen bonding in the 4(7)-isomer. The two isomers would have the structures represented below, where (I) represents the 4(7)-isomer in its unchelated and chelated forms and (II) represents the 5(6)-isomer which does not chelate but can be associated by intermolecular hydrogen bonding.

$$(7.66)$$

(I)   (II)

Wagner and collaborators (65–67) have published papers that further develop the importance of intramolecular hydrogen bonding in other isomeric compounds.

Vavon and Montheard (68) studied the rates of phenyl-hydrazone and oxime formation of aldehydes. In those aldehydes where intramolecular hydrogen bonding could occur, the rates of reaction were more rapid than in the case of their associated isomers. The rates of reaction of the unsubstituted aldehydes fell between those of the other two classes. These investigators did not develop any relation between the physical or chemical properties of the solvent and the rate of reaction. In general the rates were greater in alcohol than in chloroform.

Astle and McConnell (69) have studied the polarography of nitrophenols and Astle and Cropper (70), Astle and Stephenson (71), and Astle (72) have extended the study to nitrocresols and nitrodihydroxybenzenes. The data show that, when intramolecular hydrogen bonds are sterically possible, the reduction of the nitro groups of the compounds in carefully buffered solutions at a dropping mercury cathode are more easily reduced than the nitro groups where no such bonds are possible. Thus the nitro groups in 3-nitrocatechol, 2-nitrohydroquinone and 2-nitroresorcinol are more easily reduced than the nitro groups of 4-nitrocatechol. There seems to be only one intramolecular hydrogen bond in 2-nitroresorcinol even though it is capable of having both nitro oxygens tied up in hydrogen bonding with the ortho OH groups. The medium does influence, to some extent, the intramolecular hydrogen bonding and therefore the rate of reduction, as is evidenced by the findings that between a pH of 5.5 and 8.5 the same type of nitro group seems to be present in both 4-nitroresorcinol and 4-nitrocatechol. Therefore, in this pH range there is probably no intramolecular hydrogen bonding in 4-nitroresorcinol. At lower pH values, there is probably a weak intramolecular hydrogen bond in 4-nitroresorcinol.

Pekkarinen (73) studied, at 0–80° C, the kinetics of the hydrolysis of esters in various solvents. The reaction rate was slower in the intramolecular monoethyl maleate ion and a higher energy of activation was required to overcome this hindrance to hydrolysis. A salt effect is noted for this ion in the decreasing order Li, Na, K. This harmonizes with the formation of a chelate. Deviation from the predictions of absolute rate theory concerning the effect of solvent on the Arrhenius frequency factor and the energy of activation was attributed to preferential absorption of one of the solvent components. The hydrolysis rates of monoethylate malonate and citraconate in water were generally independent of hydrogen ion concentration. However, these rates were increased by increased hydrogen ion concentration in water–acetone solvents at high acetone concentrations. The rate of hydrolysis of the monoethyl maleate ion did not conform to the theory with respect to variation of dielectric constant. These observations on rates of hydrolysis of monoethyl malonate and citraconate were explained on the basis of intramolecular hydrogen bonding.

Another study of the influence of intramolecular hydrogen bonding on the rates of chemical reaction is that of Knorre and Emanuel (74) who reviewed the contributions of the hydrogen bond, chelation, and solvent effects in chemical kinetics. These authors made several other contributions (75–77) to the part played by hydrogen bonding in reaction rates. These authors attempt to systematize the information on chemical manifestations of hydrogen bond. They try to show its presence in many very different reactions, and how it can become discernable in the different reaction types. The intermolecular hydrogen bonding of the reactants with the solvent may exert an important effect in reactions in solution. It was thought that such bonds may determine the effect of the solvent on reaction velocities.

In the biochemical field, Grossman et al. (78) found that the structure of heat-denaturates and ultraviolet-irradiated deoxyribonucleic acid (DNA) was consistent with the hypothesis that the denaturated DNA, due to intra-molecular hydrogen bonds involving the amino group of the purine and pyrimidine bases, exists as randomly coiled structures. Various denaturation, reactivation, and ultraviolet irradiation reactions were studied in elucidating the structure of DNA. They found that rapid cooling after heat denaturation promotes the formation of intramolecular hydrogen bonds. By reheating to 45°C, these bonds may again be broken. The prevention of intramolecular hydrogen bond formation during rapid cooling may be prevented by for-maldehyde which reacts with the amino groups of the bases.

Augenstine et al. (79) studied the inactivation by ultraviolet light of deuterated and protonated trypsin. They postulate the involvement of the intramolecular hydrogen bond in the disruption.

Solvation and Ionization Effects of the Solvent on Rates

Ionization and solvation effects of the solvent as presented here are largely synonomous, since a solvent that promotes ionization will likewise be strongly inclined toward solvating solutes. In some cases the solvation effect only may be pronounced since the reactants exist as ions over a range of solvent compositions and only change in solvation between the complex, and reactant states influences the rate. On the other hand, ionization to produce ionic reactants may take place in solvents of only certain ranges of polarity or of acid or base strength; and therefore, the solvent may influence the rate by influencing the nature and extent of ionization as well as by solvation effects.

A qualitative theory of solvent effects has been proposed by Hughes and Ingold (80). Their theory may be summarized by stating that the creation and concentration of charges will be accelerated and the destruction and diffusion of charges inhibited by an increase in the ion solvating power of the medium. Table II contains a summary of the expected effect of ionizing media on rates of reaction as given by Hughes and Ingold.

Hughes (81) found that the velocity constant for the hydrolysis of t-butyl chloride was increased as the percentage by volume of water in a water–ethanol solvent increased. The data are given in Table III.

This increase of rate with increasing percentages of water in the water–ethanol solvent is in agreement with prediction (a, 1) in Table II. In spite of the change in the specific velocity constant with changing solvent, the values of the activation energies varied only slightly in the examples studied. Thus the mechanism of the reaction has as its rate-controlling step the ionization of the t-butyl chloride:

$$ t\text{—BuCl} \xrightarrow{\text{slow}} t\text{—Bu}^+ + \text{Cl}^- , \tag{7.67} $$

$$ t\text{—Bu}^+ + \text{Cl}^- + \text{H}_2\text{O} \xrightarrow{\text{fast}} t\text{—BuOH} + \text{H}^+, \text{Cl}^- . \tag{7.68} $$

The speeds of hydrolysis of methyl and ethyl iodides were found by de Bruyn and Steger (82) to decrease on increasing the proportion of water in water–alcohol solvents in harmony with prediction (a, 2). The mechanistic picture is then

$$ \text{HO}^- + \text{RI} \xrightarrow{\text{slow}} \overset{\delta}{\text{HO}} \ldots \text{R} \ldots \overset{\delta}{\text{I}} \xrightarrow{\text{fast}} \text{HOR} + \text{I}^- . \tag{7.69} $$

Also, Bergmann et al. (83) found, in agreement with prediction (a, 2), that the rates of racemization of secondary iodides in acetone decreased on the addition of a small proportion of water.

As would be expected from prediction (b, 4), Menschutkin (84) found that ethyl iodide combined with triethyl amine more rapidly in alcohols than in hydrocarbons, the order being $\text{MeOH} > \text{EtOH} > \text{Me}_2\text{CO} > \text{C}_6\text{H}_6 > \text{C}_6\text{H}_{14}$. The mechanism would involve the step

$$ (\text{C}_2\text{H}_5)_3\text{N} + \text{C}_2\text{H}_5\text{I} \xrightarrow{\text{slow}} (\text{C}_2\text{H}_5)_3{}^+\text{N} \ldots \text{C}_2\text{H}_5 \ldots \overset{-}{\text{I}} \xrightarrow{\text{fast}} (\text{C}_2\text{H}_5)_4\text{N}^+ + \text{I}^- . \tag{7.70} $$

Subsequent investigators (85–92) have established the same general sequence for other primary alkyl or methyl halides and other amines or sulfides.

In agreement with (c, 5), von Halban (93) proved that the decomposition of triethyl sulphonium bromide took place at a slower rate in alcohols than in acetone, and Hughes and Ingold (94) showed that the rate of hydrolysis of dimethyl-t-butyl sulphonium cation was decreased by increasing the

TABLE II

EXPECTED EFFECT OF IONIZING MEDIA ON RATES OF REACTION

Type and mechanism		Charges concerned in rate-determining stage of reaction			Effect on charges of forming complex		Expected effect of ionizing media
		Factor	Complex	Products	Magnitude	Distribution	
1	S_N^1 (a)	RX	$\overset{\delta^+}{R}\dots\overset{\delta^-}{X}$	$\overset{+}{R} + \bar{X}$	Increase	—	Accelerate
2	S_N^2 (a)	$\bar{Y} + RX$	$\overset{\delta^+}{Y}\dots R\dots\overset{\delta^-}{X}$	$YR + \bar{X}$	No change	Dispersed	Retard
3	S_N^1 (b)	RX	$\overset{\delta^+}{R}\dots\overset{\delta^-}{X}$	$\overset{+}{R} + \bar{X}$	Increase	—	Accelerate
4	S_N^2 (b)	$Y + RX$	$\overset{\delta^+}{Y}\dots R\dots\overset{\delta^-}{X}$	$\overset{+}{Y}R + \bar{X}$	Increase	—	Accelerate
5	S_N^1 (c)	$\overset{+}{R}X$	$\overset{\delta^+}{R}\dots\overset{\delta^-}{X}$	$\overset{+}{R} + X$	No change	Dispersed	Retard
6	S_N^2 (c)	$\bar{Y} + \overset{+}{R}X$	$\overset{\delta^+}{Y}\dots R\dots\overset{\delta^-}{X}$	$YR + X$	Decrease	—	Retard

proportion of water in a water–alcohol solvent. Thus for the triethyl sulphonium bromide, we can write

$$(C_2H_5)_3S^+ \longrightarrow \overset{+\delta}{C_2H_5} \ldots \overset{+\delta}{(C_2H_5)_2S} \longrightarrow C_2H_5^+ + (C_2H_5)_2S, \quad (7.71)$$

$$C_2H_5^+ + Br^- \longrightarrow C_2H_5Br. \quad (7.72)$$

TABLE III

THE EFFECT OF INCREASED PERCENTAGE OF WATER ON THE SPECIFIC VELOCITY CONSTANT

EtOH% by vol.	90	80	70	60	50	40
[Halide] (M)	0.0755	0.0762	0.0824	0.0735	0.0810	0.0306
k_1' (hr^{-1})	0.00616	0.0329	0.145	0.453	1.32	4.66

Gleave et al. (95) found that the speed of the hydrolysis of the trimethyl sulphonium cation was decreased by increasing the amount of water in an aqueous alcoholic solvent, which result would be expected from prediction (c, 6). The mechanistic picture would be

$$HO^- + (CH_3)S\overset{+}{(CH_3)_2} \xrightarrow{\text{slow}} \overset{\delta-}{HO} \ldots (CH_3) \ldots \overset{\delta+}{S(CH_3)_2} \quad (7.73)$$
$$\longrightarrow CH_3OH + (CH_3)_2S.$$

There are many other examples of the rate of reaction being a function of the solvating power of the solvent (96, 97). We will mention a few of these, though we will not attempt to make a comprehensive list of such kinetic studies.

We might point out that not only rate but also the order and the mechanism of a reaction might be changed by changing solvent. Hughes and Ingold (80) and Hughes (98) have mentioned this point. In the hydrolysis of secondary and tertiary alkyl halides, Hughes et al. (99) recognized two mechanisms of this type of substitution. These are a bimolecular S_N2 mechanism involving attack by the hydroxide ions, and a unimolecular S_N1 mechanism kinetically dependent on the ionization of the alkyl halide. If R represents the alkyl radical and X represents the halide, these mechanisms can be represented by the equations

$$RX + OH^- \longrightarrow ROH + X^- \quad (S_N2),$$
$$RX \longrightarrow R^+ + X^-, \quad (7.74)$$

followed by

$$R^+ + OH^- \longrightarrow ROH \text{ (instantaneously)}. \quad (S_N1) \quad (7.75)$$

The changeover from the bimolecular to the unimolecular mechanism was found to depend on the medium and on the concentration as well as on the alkyl group involved. Hughes (98) found the changeover in dilute aqueous alcohol solutions to be between the ethyl and the isopropyl groups.

Benson (100) points out that if the attacking group Y can facilitate the departure of X, a case intermediate between S_N1 and S_N2 mechanisms can occur. If Y represents an ion such as a halide X^- or OH^- or RO^-, the displacement reaction is usually second order; when Y is a solvent molecule behaving as a nucleophilic reagent, but also simultaneously as an ionizing agent, intermediate cases arise in the kinetics. Reactions in which Y is a solvent water molecule is called hydrolysis. Benson feels that the complexity of the molecular systems in solvolysis reactions as contrasted with the over-simplification of the models used to explain them may be the cause of much of the controversy in the interpretations of these reactions.

The solvation of the leaving group is required in all nucleophilic displacements of anions from saturated carbon atoms according to Swain and Eddy (57). A solvation producing the quick formation of a coordinate covalent bond is most effective. This type of solvation is represented by the solvation of the bromine in methyl bromide by mercuric bromide, which in 0.05 M concentrations increases by sevenfold the termolecular displacement of bromine in methyl bromide by pyridine in benzene solution. Ionic bonding, for example, hydrogen bonding by phenols, is the next most effective type of solvation. Reaction (7.58) illustrates this phenomenon. Polarization solvation by polarizable solvent molecules may be important when only this type of solvation is possible. Swain and Eddy point out that the reaction in pure benzene of methyl bromide with pyridine is probably termolecular even though it is second order due to large excess of solvent. That this is the case is supported by the fact that neither this reaction, nor for that matter any nycleophilic displacement reaction, has been found to occur in the homogeneous gas phase. Also in hexane or cyclohexane, which resembles the gas phase strongly in that polarization forces are very weak, the reaction does not occur, except heterogeneously on the walls of the containing vessel or on the surface of precipitated crystals (101, 102).

The initial attack on the carbon, for example, by pyridine, is by formation of a ionic dipole–dipole bond which makes it easier for the old covalent bond to break. The ionic solvation bond with carbon is formed immediately, and by a slow process this ionic bond may form a covalent bond. The quantitative difference in the rate of this changeover may result in the intermediate "carbonium salt" in the case of methyl halides having a to-tal of nearly four covalent bonds on only a minute fraction of a unit positive charge on the carbon; while in the case of tertiary halides the carbonium ion has chiefly two ionic bonds and is relatively "free" and long-lived. In the teritary case the carbon-solvating reagent or solvent does not become to a great extent covalently bonded either in the rate-determining or in the subsequent product-determining step.

In the rate-determing step the attack on carbon may be of different types. Covalent bonding is involved in the case of primary halides. Solvation involving chiefly ionic or ion–dipolar bonding occurs with tertiary halides. The attack may even be on a carbon other than the one from which the leaving group is displaced, as, for example, in neopentyl chloride:

$$(7.76)$$

The whole process including the Wagner–Meerwein rearrangement can take place in one step.

Swain believes that the first-order racemizations in alcoholic or aqueous solutions may result from the rapid formation of disolvated carbonium ion, facilitated by the dissociation of the ion pair in these polar solvents and by the high concentration of hydroxylic reagent present. In aqueous solutions, due to the hydroxyl ion's large solvation shell, it is sterically hindered from approaching closely to the central carbon in tertiary halides, and under these circumstances the ion does not accelerate these reactions.

Satchell (103) divides the mechanisms of acylation into two types, without respect to the finer details. The types, without indication of the relative rates, which will depend on the system concerned, may be written

$$R \cdot COX \underset{\longleftarrow}{\overset{a}{\rightleftharpoons}} R \cdot CO^+ + X^- \text{ (or } R \cdot CO^+X^-) \xrightarrow[b]{s} \text{ Products}, \quad (7.77)$$

$$R \cdot COX + S \longrightarrow \text{ Products} . \quad (7.78)$$

In reaction (7.77) when reaction (a) alone is rate determining and especially when other molecules of A distinct from those actually acelated may assist ionization, then only on the basis of particular definitions of molecularity and bonding change can a rigid distinction be made between mechanisms

(7.77) and (7.78). Doubtful distinctions involving borderline cases really involve only definitions (*33*). An electron-attracting group X which possesses some stability as an ion X^- will enhance the plus charge on the acelium ion RCO^+ and promote cleavage. An electron-repelling group will have the opposite effect. Less well-defined variations of the leaving group achieved within a given type of reagent, as well as the electron-attracting or electron-repelling properties, are consistent with the usual qualitative assignment of polar properties to the group (*104*). The effect of changes in R on reactivity are not so clearly defined as are changes in X. If R is an electron donor it will reduce the charge on the carbonyl carbon atom in both the acelium ion and in the polarized reagent. If R attracts electrons it will increase this charge and therefore hinder ionization. Thus whether the substrate reacts primarily with the ionized or unionized acelylating agent determines the effect of changes in R. Of importance too, in this respect, is which step in the over-all process is rate controlling. Bender (*105*), Ingold (*104*), Gould (*106*), Hine (*33*), Streitwieser (*107*), and Satchell (*108*) have discussed the factors involved in acylation and other nucleophilic substitutions.

In many examples of mechanism (7.78) there may be direct addition of substrate to the carbonyl group rather than the synchronous displacement of X by S (*103, 105*). Satchell (*103*) writes for the substrate water the mechanism

$$R \cdot COX + H_2O \rightleftharpoons RC(OH)_2X \longrightarrow RCO_2H + HX \qquad (7.79)$$

but points out that the individual steps of this scheme may be made to fit varied situations. The concurrent occurrence of exchange of oxygen and acelation has been the main reason for postulating such intermediates, but such a postulation has not proved necessary in all examples—especially those involving acid catalysis (*109*).

A few examples of solvation involving the solvent water have been reported recently that appear to be of especial interest. Hammett and Dyrup (*110*) correlated rates of reactions and a factor H_0, which they termed an acidity function, up to 100% sulfuric acid. In measuring the acidity function, Hammett and Dyrup used the *step-method* of the relative basicity of indicators. The expression for H_0 was written

$$H_0 = - \log a_{H^+} \frac{f_B}{f_{BH^+}}, \qquad (7.80)$$

where a_{H^+} is the activity of the hydrogen ion and f_B and f_{BH^+} are the respective activity coefficients of the nonionized or neutral base indicator and its conjugate acid, respectively. In dilute aqueous solutions, where $f_B = f_{BH^+}$ = 1,

$$H_0 = - \log a_{H^+} = pH .\qquad (7.81)$$

Sometimes the function h_0 is used. It is defined as

$$h_0 = a_{H^+} \frac{f_B}{f_{BH^+}} .\qquad (7.82)$$

Therefore,

$$H_0 = - \log h_0 .\qquad (7.83)$$

Prichard and Long (111) found that, in the acid-catalyzed hydrolyses of substituted ethylene oxides, logarithms of the first-order rate constants for the acid-catalyzed hydrolysis of ethylene oxides were linear with $- H_0$.

Zucker and Hammett (112) studied the acid-catalyzed iodination of acetophenone in aqueous solutions ranging from 0.2 to 3.6 M perchloric acid. For this acidity range they found the reaction to be much more nearly proportional to concentration of hydrogen than to the h_0 acidity function. Considering this in conjunction with the known mechanism of the reaction caused them to propose that rates of acid-catalyzed reactions which included a water molecule in the transition state would depend on concentration of hydrogen ion rather than the h_0 acidity function.

Schwartzenbach and Wittwer (113) and Long and Bakule (114, 115) have studied the keto–enol transformation of 1,2-cychlohexanedione. The former investigators found the reaction to be sufficiently slow to be easily measurable by conventional means up to acidities of about 6 or 7 M mineral acid. The reaction was found to reach a measurable equilibrium in aqueous solutions. The equilibrium was complicated by the fact that the ketone in aqueous solutions was almost entirely in the form of its monohydrate. Combination of the kinetic and equilibrium data permits the determination of the first-order rate constants for both the ketonization and enolization reactions. The first-order specific velocity constant for the acid-catalyzed approach to equilibrium was given by the equation

$$k'_{H^+} = k'_{obs} - k^w_{obs} ,\qquad (7.84)$$

where k'_{obs} was observed specific velocity constant for approach to equilibrium and k^w_{obs} represents the water contribution to the rate. For the three catalyst acids, perchloric, sulfuric, and hydrochloric in the range 1 to 7 M,

Long and Bakule found that $\log k_{H^+}'$ was linear with $-H_0$, thus

$$\log k_{H^+}' = -0.64\,H_0 - 5.14\,, \tag{7.85}$$

also if k_f' and k_r' are the first-order specific velocity constants for the acid-catalyzed conversion of the enol to ketone and of the ketone to enol, respectively, then

$$k_H' = k_f' + k_r'\,, \tag{7.86}$$

and plots of $\log k_f'$ versus $-H_0$ and of $\log k_r'$ versus $-H_0$ are linear with slopes ranging from 0.5 to 0.7 and from 0.8 to 1.0, respectively, for the two types of plot depending on the catalyst acid. The second-order specific velocity constant k_e', for the reaction of enol with hydronium ion is related to k_f' by the equation

$$k_f' = k_e'C_{H_3O^+}\,, \tag{7.87}$$

and k_k' the second-order specific velocity constant for the reaction of the ketone with hydronium ion is related to k_r' by the equation

$$k_r' = k_k'C_{H_3O^+}\,. \tag{7.88}$$

Plots of $\log k_e'$ and $\log k_k'$ versus $1/T$ gave linear plots and straight lines, from which the values of the energies and entropies of activation were obtained (Table IV).

Since only the ketone is significantly hydrated a spectrum of transition states was written as follows

$$\tag{7.89}$$

The main reaction is from unhydrated enol to hydrated ketone. Thus the upper hydration equilibrium lies to the left and the lower one to the right.

TABLE IV

ENERGIES AND ENTROPIES OF ACTIVATION FOR BIMOLECULAR REACTION WITH HYDROGEN
ION, k' IN LITERS PER MOLE-SECOND

Specific velocity constant	Energy of activation, E^* (kcal/mole)	Entropy of activation, s^* (e. u.)
k_e' (ketonization)	17.6 ± 1	-26 ± 2
k_k' (enolization)	24.4 ± 1	-4 ± 2

The two extremes of carbonyl hydration in the transition state, one appropriate to reaction via the left-hand side of this equilibrium diagram and the other appropriate to the reaction via the right-hand side, were written

$$(7.90)$$

Unhydrated transition state   Hydrated transition state

Now the entropy of activation for a reaction of ordinary unhydrated ketones is about -10 e.u., which is the expected value for a second-order reaction when k' is in milliliters mole^{-1} sec.$^{-1}$. For a normal keto–enol reaction the over-all change in entropy is close to zero, and the S^* value for the reverse enol–ketone reaction should also be normal. Thus the S^*-value for the reaction of 1,2-cyclohexanedione is markedly abnormal. For the enol–ketone transformation, S^* is 15 to 20 e.u. more negative than normal, and for the reverse keto–enol transformation S^* is more positive by 5 or 6 e.u. than normal. Long and Bakule explain these phenomena on the basis that the transition state possesses a comparatively firmly bound water of hydration. Thus the transition state, according to these authors, corresponds closely to the right-hand hydrated one listed above. Because of this somewhat unexpected conclusion, it is suggested that further examination of entropies of activation for reactions of carbonyl compounds might be profitable.

Consistent with the expected slow proton transfer mechanism for the enol, the reaction of the enol is faster in H_2O than in D_2O.

It was concluded by Long and Bakule that the rates of the slow proton transfers to normal unhydrated enols will also approximately correlate with h_0, probably with slightly higher slopes of log k' versus $-H_0$ than in the present case. This is especially interesting since the enol reaction at lower acidities exhibited general acid catalysis. The reaction, thus, apparently shows general acid catalysis at low acidities but follows h_0 at higher acidities.

Thus solvation of the 1,2-cyclohexanedione ketone by a solvent water molecule makes the dependence of the rate of the establishment of the acid-catalyzed keto–enol equilibrium of this compound distinctly different from the normal keto–enol reaction.

Bunnett's (*116*) approach to acid catalysis did not add significantly to the understanding of the acid-catalyzed keto–enol equilibrium of 1,2-cyclohexanedione relative to a consideration of the mechanistic implications of the hydration equilibrium and the Arrhenius parameters. Bunnett proposed to introduce the activity of water as a somewhat necessary variable and to characterize a reaction by one or the other of two empirical factors w or w^*. The factor w is the slope of the plot of $\log(k'_{H^+}/h_0)$ versus log a_{H_2O}, and w^* is the slope of $\log(k'_{H^+}/C_{H^+})$ versus log a_{H_2O}. Bunnett's idea was that, for any given reaction, one or the other of these plots would be approximately linear. Neglecting the nonlinear plots, w approaches zero when the slope of log k_{H^+} versus $-H_0$ is about unity, is negative when the slope is greater than unity, and is positive when the slope is less than unity. In like manner, w^* is nearly zero when the plot of log k_{H^+} versus log C_{H^+} is about unity. Thus the parameters w and w^* permit a more complete classification of acid-catalyzed reactions than do the functions h_0 and C_{H^+} alone. According to Bunnett, in the reactions of ketones to form enols, the w plots are usually linear, with slopes ranging from 2 to 7. The w^* plots are occasionally linear with slopes of the order of -1.5. The same slope might be expected for ketonization of enols if approximate independence on keto–enol equilibrium with electrolyte concentration is assumed. Bunnett, whose correlations of w and w^* with mechanisms depend largely on results from ketones, prepares a chart containing conclusions on these correlations. From the chart it can be seen that a larger w than 3.3 and a larger w^* than -2 signifies that water is acting as a proton-transfer agent in the slow step.

In the case of the 1,2-cyclohexanedione, the ketone reaction gives w plots which are straight, but for which w is low and is in the expected range of water acting as a nucleophile. For the forward and reverse reactions, the slopes are so different as to lead one to believe that water was behaving as a nucleophile for one of the directions, but as a proton-transfer agent for the reverse. Long and Bakule point out that this is an unreasonable conclusion and indicate that since the keto–enol transformation of 1,2-cyclohexanedione involves a monohydrated ketone and perhaps also a monohydrated transition state, there is no particular reason why this reaction should behave identically like those of other ketones.

The indicator acidity functions H_0, treated originally as empirical concepts, have recently been rationalized by Bascombe and Bell (*117, 118*). These investigators showed that strong hydration of the proton in concentrated acid solutions are chiefly responsible for the high acidities of these solutions.

Yagil and Anbar (*119*) using the theory of Bascombe and Bell, and making calculations on the molar, mole fraction, and volume fraction scales, proved the hydration number of the hydroxide ion to be three. Their concept of the structure of the hydrated hydroxide ion ($H_7O_4^-$) can be depicted as

$$\left[\begin{array}{c} \quad\quad\ \overset{\displaystyle H}{\underset{|}{}} \quad\quad\ \overset{\displaystyle H}{\diagup} \\ O-H\cdots O\cdots H-O \\ \diagup \quad\quad \vdots \\ H \quad\quad\ \underset{|}{H} \\ \quad\quad\ O \\ \quad\quad\quad \diagdown H \end{array} \right]^- \qquad (7.91)$$

Anbar *et al.* (*120*) studied the effect of ionic hydration on the kinetics of base-catalyzed reactions in concentrated hydroxide solutions. Equations derived showed the relation between the observed rate constants of various base-catalyzed reactions and the indicator basicity of the solution expressed by the H_--function. Employing the trihydrated hydroxide ion the equations were expressed also in terms of C_{OH^-}, the stoichiometric concentration of the hydroxide ion, and of C_{H_2O} the concentration of "free" water. It was concluded that the rate dependence in concentrated alkaline solution reveals the change in the transition state with respect to the number of molecules of water involved in its formation. The rate expressions for eight types of base-catalyzed reactions were classified into three groups. Two of these followed H_- or $C_{OH^-}/C_{H_2O}^4$; five of these followed $H_- + \log C_{H_2O}$ or $C_{OH^-}/C_{H_2O}^3$, and one followed $H_- + 2 \log C_{H_2O}$ or C_{OH^-}/C_{H_2O}. In principle the three groups can be distinguished by these criteria; however, complications may make this distinction difficult in practice. For example, water may be involved in the formation of the transition state in a way not considered in these three groups. For example, consider the unimolecular decomposition of a conjugate base in which the transition state is stabilized by four water molecules of hydration not present in the hydration shell of the substrate SH. There would be no dependence on water concentration, as can be seen from the following rate expression:

$$-\frac{dC_{\text{SH}}}{dt} = k''C_{\text{S}}\text{-}C_{\text{H}_2\text{O}}^4 = k''K_cC_{\text{SH}}\,\frac{C_{\text{OH}^-}}{C_{\text{H}_2\text{O}}^4}\,C_{\text{H}_2\text{O}}^4$$
$$= k''K_cC_{\text{OH}^-}C_{\text{SH}}\,,\tag{7.92}$$

where K_c is the equilibrium constant for the reaction of the substrate with the hydrated hydroxide ion,

$$\text{SH} + \text{OH}(\text{H}_2\text{O})_3{}^- \;\rightleftharpoons\; \text{S}^- + 4\text{H}_2\text{O}\,.\tag{7.93}$$

In a proton-abstraction reaction or in nucleophilic substitution by a non-hydrated hydroxide, a rate law obeying $C_{\text{OH}^-}/C_{\text{H}_2\text{O}}^3$ was derived. The breakdown of serine phosphate in alkaline solution was shown to involve a mechanism of proton abstraction, and the hydrolysis of ethyl iodide, a S_N2 reaction, was used to illustrate the nucleophilic substitution by a nonhydrated hydroxide.

In their mechanism of base-catalyzed reactions, Anbar *et al.* use the definition of H_- as

$$H_- = -\log\frac{C_{\text{BH}}}{C_{\text{B}^-}}\,K_{\text{BH}}\,,\tag{7.94}$$

where K_{BH} is the acid dissociation constant of indicator from the equation

$$\text{BH} \;\rightleftharpoons\; \text{B}^- + \text{H}^+\,,\tag{7.95}$$

also $H_- = \log h_-$, and, therefore,

$$h_- = \frac{C_{\text{BH}}}{C_{\text{B}}}\,K_{\text{BH}} = \frac{f_{\text{B}^-}}{f_{\text{BH}}}\,a_{\text{H}^+} = \frac{f_{\text{B}^-}a_{\text{H}_2\text{O}}}{f_{\text{BH}}a_{\text{OH}^-}}\,K_w\,.\tag{7.96}$$

The dissociation of BH in alkaline solution can be written

$$\text{BH} + \text{OH}^- \;\rightleftharpoons\; \text{B}^- + \text{H}_2\text{O}\,,\tag{7.97}$$

hence a quantity b_- can be defined as

$$b_- = K_w/h_- = f_{\text{BH}}a_{\text{OH}^-}/f_{\text{B}^-}a_{\text{H}_2\text{O}}\,.\tag{7.98}$$

In dilute alkaline solutions, where the activity coefficients approach unity,

b_- approaches C_{OH^-}. The quantity b_- is directly related to the OH^- activity. Thus b_- is the alkaline counterpart of h_0 in acid solution. The relation

$$f_{BH}/f_{B^-} = b_- \frac{a_{H_2O}}{a_{OH^-}} \qquad (7.99)$$

can be written as a consequence of Eq. (7.98). This equation is basic to the formulation of the rate expressions for the base-catalyzed reactions.

Let us consider specifically case II presented by Anbar *et al.* For this case, the substrate SH is in rapid preequilibrium with its conjugate base S^-, thus

$$SH + OH^- \rightleftharpoons S^- + H_2O, \qquad (7.100)$$

but subsequently S^- forms products in a bimolecular rate-determining step with another reactant Y,

$$S^- + Y \xrightarrow{\text{slow}} \text{Products}. \qquad (7.101)$$

The rate expression is then

$$-\frac{dC_{SH}}{dt} = k' \frac{f_Y f_{S^-}}{f^*} C_Y C_{S^-}. \qquad (7.102)$$

Solving for C_{S^-} from Eq. (7.100) combined with a step in which S^- is converted to products in a unimolecular rate determining step thus

$$S^- \xrightarrow{\text{slow}} \text{Products}. \qquad (7.103)$$

one obtains

$$C_{S^-} = C_{SH} \frac{K_{SH}}{K_w} \frac{f_{SH}}{f_{S^-}} \frac{a_{OH^-}}{a_{H_2O}}, \qquad (7.104)$$

and this value of C_{S^-} substituted into Eq. (7.102) yields

$$-\frac{dC_{SH}}{dt} = k' \frac{K_{SH}}{K_w} \frac{f_Y f_{SH}}{f^*} \frac{a_{OH^-}}{a_{H_2O}} C_Y C_{SH}. \qquad (7.105)$$

If now the Y component in the transition state interacts similarly as in the free state with the environment, then

$$f_{S^-}^* = \frac{f^*}{f_Y} \qquad (7.106)$$

is the coefficient of the B^--like part of the transition state, if it can be assumed that

$$\frac{f_Y f_{SH}}{f^*} = \frac{f_{BH}}{f_{B^-}}. \tag{7.107}$$

Using this assumption and Eq. (7.99), the equation

$$k'_{obs} = -\frac{1}{C_Y C_{SH}}\frac{dC_{SH}}{dt} = \text{const } b_- \tag{7.108}$$

can be derived from Eq. (7.105). Applying the actual preequilibrium equation (7.93), one obtains, with no further assumptions, the expression

$$-\frac{dC_{SH}}{dt} = k'' C_S - C_Y = k'' K_c \frac{C_{OH^-}}{C_{H_2O}^4} C_Y C_{SH}, \tag{7.109}$$

which yields

$$k'_{obs} = \text{const}\, \frac{C_{OH^-}}{C_{H_2O}^4} = \text{const } b_-, \tag{7.110}$$

which corresponds exactly to Eq. (7.108). This equation is exactly that obtained in case I, where the same preequilibrium Equation (7.100) was postulated but where S^- is converted to reaction products in a unimolecular rate-determining step, as shown in Eq. (7.103).

The formation of hydrazine from chloramine (NH_2Cl) and ammonia in highly alkaline aqueous solutions is an example of case II. This reaction was found to be first order in b_-, and the slope was 0.90 (*121*). A plot of the data for this reaction is given in Fig. 2. The k'_{corr} in Fig. 2 was given by the relation

$$k'_{corr} = k'_{obs} - k_n', \tag{7.111}$$

where k'_{obs} was the observed pseudounimolecular specific velocity constant at high alkalinities obtained since ammonia was in large excess, and k'_n was the alkaline independent pseudounimolecular rate constant obtained when the pH was 13.7 or less. From Eq. (7.94), (7.96), (7.98), (7.110), and (7.111), $\log k'_{obs}$ or $\log k'_{corr}$ would be linear with $\log b_-$, $\log h_-$, and with H_-.

A series of reports on pulse radiolysis studies of organic and aquo-organic systems has recently appeared in the literature (*122–126*). Using a linear accelerator giving 2-μsec pulses of approximately 1.8-Mev electrons, and with the peak current in the pulse being variable up to 0.5 amp, Hart and

Boag (*122*) observed a transient absorption band with a peak at 7000 Å in deaereated water and in various aqueous solutions. The band was attributed to the hydrated electron. These authors studied the rate of decay of the absorption and the effect of various anions, catious, and dissolved gases. In concentrated solutions of ammonia and of methylamine, similar absorption-spectra-produced irradiations have been found. These absorption spectra resemble the known absorption spectra of solvated electrons in liquid ammonia or in liquid methylamine.

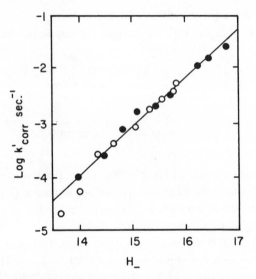

FIG. 2. Base-catalyzed chloramine-ammonia reaction. $[NH_2Cl] = 2 \times 10^{-3}$, $[NH_3] = 1.0$, $t = 27.3°$, $\bigcirc$, sodium hydroxide; $\bullet$, potassium hydroxide.

Dorfman and Taub (*126*) studied the kinetics of the elementary reactions of two transient species, the hydrated electron and the α-ethanol radical, in irradiated deaereated aqueous ethanol solution using the pulse radiolysis technique. The kinetics of the hydrated electron were measured after using a 0.4-μsec, 15-Mev electron pulse from a linear accelerator. They found that, in both acid and neutral solutions, the hydrated electron e_{aq} in the bulk of the solution in these high intensity experiments disappeared according to the equation

$$e_{aq} + H_{aq}^+ \longrightarrow H \, . \tag{7.112}$$

The absolute rate constant for the reaction represented by Eq. (7.112) was determined by observing the decay kinetics of the hydrated electron at 577–579 mμ. The solution was 0.5 M in ethanol so that the hydroxyl radical, which along with the hydrogen atom are formed as transient species in water, is removed by the equation

$$\text{OH} + \text{C}_2\text{H}_5\text{OH} \longrightarrow \text{CH}_3 \cdot \text{CHOH} + \text{H}_2\text{O} \tag{7.113}$$

and therefore cannot enter into competitive reaction for the hydrated electron. The hydrogen atoms are also removed by the reaction

$$\text{H} + \text{C}_2\text{H}_5\text{OH} \longrightarrow \text{CH}_3 \cdot \dot{\text{C}}\text{HOH} + \text{H}_2 . \tag{7.114}$$

The hydrogen ion concentration and the pulse current were regulated so that $[H_{aq}^{+}]/[e_{aq}^{-}] \gg 10$, and thus the reaction depicted by Eq. (7.112) was pseudo-first order. The absolute second-order rate constant obtained from the slope of the plot of the logarithm of the optical density versus time and the known hydrogen ion concentration, corrected for the small additional amount formed by the pulse itself, was found at 23° to be $(2.26 \pm 0.21) \times 10^{10}$ M^{-1} sec^{-1}. This compares to a rate constant of 1.3×10^{11} M^{-1} sec^{-1} found by Eigen and DeMaeyer (127) for the hydrated proton reacting with the hydrated hydroxide ion

$$\text{H}_{aq}^{+} + \text{OH}_{aq}^{-} \longrightarrow \text{H}_2\text{O} . \tag{7.115}$$

They have explained this high rate constant on the basis of a proton jump mechanism resulting in a phenomenalogical interaction distance of 8 Å. Dorfman and Taub were unable to offer an explanation of the difference in the rate constants for the reactions represented by Eqs. (7.112) and (7.115). The α-ethanol radicals formed in the reactions depicted in Eqs. (7.113) and (7.114) disappear in a bimolecular reaction discussed by Taub and Dorfman (125).

At much lower intensities for steady-state radiolysis depending on the intensity, the reaction represented by Eq. (7.112) will make a smaller contribution, and the first-order decay of the electron represented by the reaction

$$e_{aq}^{-} + \text{H}_2\text{O} \longrightarrow \text{H} + \text{OH}^{-} \tag{7.116}$$

will become more important. The first-order rate constant was obtained by taking the slope of the plot of the logarithm of the optical density versus time shown in Fig. 3. Decay curves were observed at pulse currents as low as 100 mamp. The value of k' was found to be given by $k' \leq 4.4 \times 10^4$

sec^{-1} at 23°. Only a upper limit for k' was given since, at the low initial concentration of the hydrated electron of less than 1 μM, trace impurities such as oxygen at concentrations less than 0.1 μM might make a significant contribution to the rate.

In basic solution, depending upon pH and upon intensity, the reaction represented by Eq. (7.112) will be very small or negligible in amount, and the reactions represented by Eq. (7.116) and by the equation

$$e_{aq}^- + e_{aq}^- \longrightarrow H_2 + 2OH^- . \tag{7.117}$$

will predominate. The reaction represented by Eq. (7.117) is bimolecular, and a plot of the reciprocal of the optical density versus time is a straight

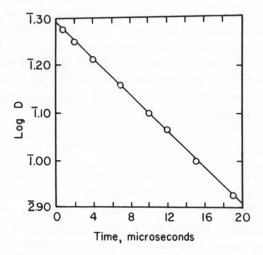

Fig. 3. Log D versus time for the first-order rate of disappearance of the hydrated electron at low concentration in a deaerated, strongly basic aqueous ethanol solution. Temperature, 23°C.

line for the disappearance of at least 65% of the initially formed hydrated electrons. Only the upper limit of the specific velocity constant can be obtained from the straight line portion of the curve, since there may be a small contribution to the rate of electron disappearance due to other processes such as the reaction of the electron with the small amount of hydrogen peroxide formed as a residual molecular yield and the small contribution under the conditions used of the unimolecular decay of the hydrated electron represented by Eq. (7.116). From the slope of the straight portion of the line for the second-order reaction one obtains at 23°

$$2k' \leq (1.32 \times 10^6)\, \varepsilon_{577}, \tag{7.118}$$

where ε_{577} is the molar extinction coefficient of the hydrated electron at the indicated wavelength. The value of ε_{577} may be found from the initial optical densities read from the rate curves in basic solution together with the appropriate yield of the electron. Dorfman and Taub (*126*) give the value of 2.7 $\pm$ 0.2 molecules/100 ev for $G_{e_{aq}}$, the hydrated electron portion of the total yield of reducing species produced in the bulk of the solution after correcting the initial observable optical densities for the electron disappearance during the pulse. This correction was small, since the half-life of the hydrated electron in basic solution is an order of magnitude greater than the pulse width. The value of ε_{577} was found to be (9.7 $\pm$ 1.5) $\times 10^3$ M^{-1} sec^{-1}. The upper limit for the specific velocity constant for the reaction represented by Eq. (7.117) is therefore, at 23°,

$$2k' < (1.3 + 0.2) \times 10^{10}\ M^{-1}\,\text{sec}^{-1}, \tag{7.119}$$

where k' is defined by the differential equation for the second-order rate expression

$$-\frac{d[e_{aq}^-]}{dt} = 2k'\,[e_{aq}^-]^2. \tag{7.120}$$

Baxendale *et al.* (*128*) determined absolute rate constants for the reactions of some metal ions with the hydrated electron by following the decay of the absorption of the hydrated electron at 7000 Å. The absorption was produced by 2 μsec pulse of 4-Mev electrons in the presence of the metal ions, and its decay was followed using a sensitive photomultiplier/oscilloscope technique. In the presence of the ions the decay rate is considerably increased. Conditions were obtained where the decay of the hydrated electron in pure water is negligible in comparison with the decay in the presence of 50–200 μM metal ions. The decay in the presence of the metal ions at these concentrations is a first-order process. The bimolecular rate constants k_2' were obtained for the electron–ion reaction and had the values of 0.17×10^{10}, 1.35×10^{10}, 2.3×10^{10}, 3.0×10^{10}, 5.8×10^{10}, $> 9 \times 10^{10}$, respectively, for the ions Zn^{2+}, Co^{2+}, Ni^{2+}, Cu^{2+}, Cd^{2+}, and $Co(NH_3)_6^{3+}$. The Mn^{2+} reaction was found to be an order of magnitude slower than the slowest measured.

Transient absorption was observed at $\sim$ 3000 Å in these solutions and was ascribed to the unstable valency states produced in these reactions, for example, Zn^+. This was confirmed using the photomultiplier technique. Thus it was found that these absorptions build up in the time period observed for the enhanced electron decay. Also, the order of the reactivity of the ions as listed above were confirmed by the buildup of the hydrated electron absorption during the pulse. It was suggested that the transient ultraviolet absorptions are the charge-transfer spectra of the reduced ions, the molar extinction coefficient of the Zn^+ species being at least 1.3×10^3.

It is evident from the material presented in this section that solvation and ionization phenomena are interwoven. The mechanism of ionization involves, and is impossible without, strong solvent interactions. Necessarily, however, ionization does not occur in all instances of solvation. Perforce, however, the two phenomenon are advantageously discussed together.

From what has been said in this section it can also be seen that solvation and ionization influence rates and mechanisms in various manners. In fact, solvation and ionization rank high among the important influences of solvents on rates and mechanisms of reactions in solution. Many of the concepts of the influence of the solvent exerted through solvation and ionization on rates and mechanisms have been verified experimentally and explained theoretically, but many of these effects are conjectural though reasonable. There remain differences of opinion on the explanations of some of the phenomena. It was therefore decided by the author to present the material in as unbiased a manner as possible and permit the reader to weigh and evaluate the data and their interpretations in a manner satisfactory to himself. Happy is the investigator who can reach sane, sound, and unprejudiced conclusions on the material pertinent to his own scientific line of endeavor.

Solvolysis and the Effect of Solvent on Rates

Certain types of solvolyses have already been discussed. For example, the first-order nucleophilic substitution or S_N1 type reactions in which the attacking group Y is a solvent molecule is classified as a solvolysis reaction. If water is the solvent, the reaction is termed a hydrolysis reaction.

Thus for the hydrolysis of alkyl or aryl halides in water solvent, Hughes et al. (129) and Ingold (130) have discussed a mechanism in which the ionization of the organic compound is the slow rate-determining step which is followed by a fast reaction with the nucleophilic reactant. If RX represents the alkyl or aryl halide, the hydrolysis may be written

$$RX \underset{k_2'}{\overset{k_1'}{\rightleftarrows}} R^+ + X^-, \qquad (7.121)$$

$$R^+ + H_2O \underset{k_4'}{\overset{k_3'}{\rightleftarrows}} ROH_2^+ \underset{k_6'}{\overset{k_5'}{\rightleftarrows}} ROH + H^+. \qquad (7.122)$$

The rate of the ionization with specific velocity constant k_1' is so slow, and the rate of the reaction of R^+ with water is so much faster than the rate of the reaction of R^+ with the halide X^-, that the rates of these hydrolyses reactions can be represented simply by the equation

$$-\frac{d(RX)}{dt} = k_1'(RX). \qquad (7.123)$$

The rates of such solvolysis reactions are strongly dependent on the organic group R and on the solvent. They do not take place in the gas phase, which is additional evidence that the mechanism involves ionic or polar intermediates.

Cram (131) studied the acetolysis of the p-toluenesulfonates of the stereo-isomers of 3-phenyl-2-butanol (3-phenyl-2-butyl tosylate) in glacial acetic acid. There was almost complete retention of optical activity in the case of the acetolysis of the erythro compound, however optically active threo tosylate gave the racemic threo acetate. The results are interpreted in terms of an intermediate cyclic carbonium ion in a Wagner–Meerwein rearrangement. The carbonium ion was symmetric in the case of the acetolysis of the erythro compound, but was asymmetrical in the case of the acetolysis of the threo compound. Cram proposed a simultaneous loss of protonated HOTs$^-$ group and the backside bridging of the phenyl group to form a carbonium ion. Thus for the erythro compound the attachment of an ester group to either of the two bridged carbon atoms in the carbonium ion will produce the identical optically active erythro product. Thus

$$(7.124)$$

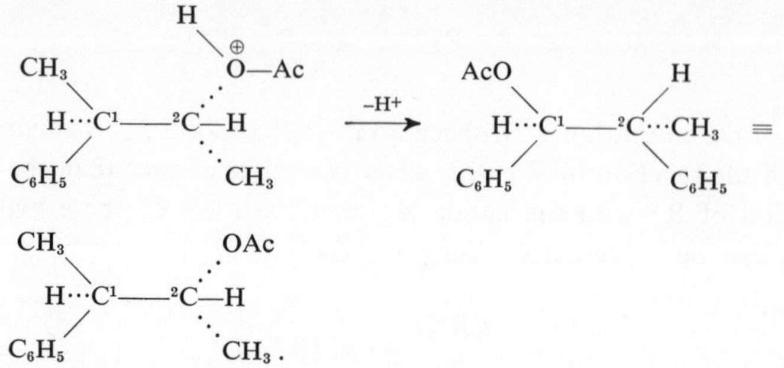

(7.124)

While with the thero tosylate the attachment of an ester group in a statistical distribution will result in a racemic threo product.

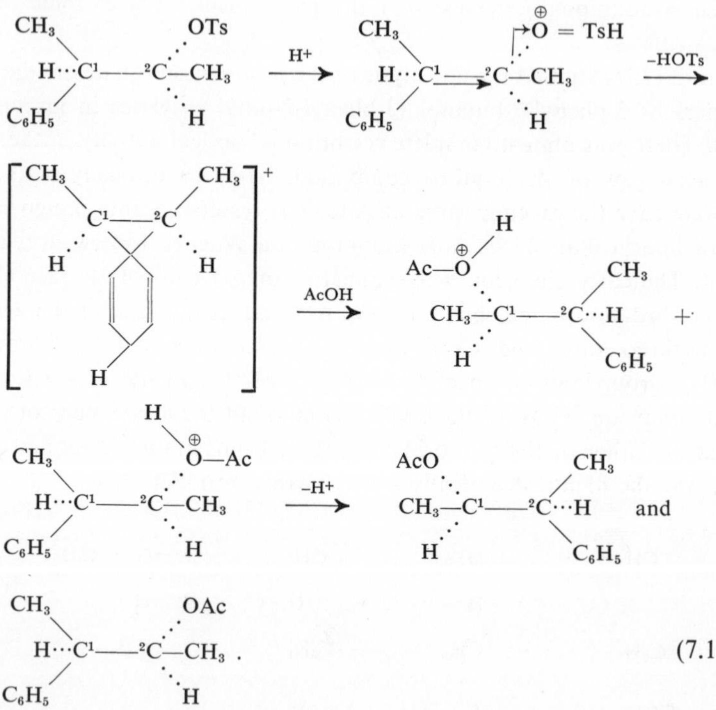

(7.125)

The hydrolyses of ethylene halohydrins, CH_2OHCH_2X, and of ethylene oxide, $\underset{\text{O}}{CH_2\text{—}CH_2}$ have been studied extensively (*132–137*). The reactions of ethylene chlorohydrin, ethylene bromohydrin, ethylene iodohydrin, and ethylene fluorohydrin with pure water are negligibly slow with respect to their rates of reaction with hydroxide ion; also ethylene chlorohydrin is less reactive than ethylene bromohydrin or ethylene iodohydrine but is more reactive than the fluorohydrin (*134*). Brønsted *et al.* (*133*) found that ethylene oxide in aqueous hydrochloric acid disappeared by four different paths. These are

$$\begin{array}{c} CH_2 \\ | \quad\quad\!\!>O + H_2O \\ CH_2 \end{array} \longrightarrow \begin{array}{c} CH_2OH \\ | \\ CH_2OH , \end{array} \qquad (7.126)$$

$$\begin{array}{c} CH_2 \\ | \quad\quad\!\!>O + H_2O \\ CH_2 \end{array} \xrightarrow{\ H_3O^+\ } \begin{array}{c} CH_2OH \\ | \\ CH_2OH , \end{array} \qquad (7.127)$$

$$\begin{array}{c} CH_2 \\ | \quad\quad\!\!>O + Cl^- + H_2O \\ CH_2 \end{array} \longrightarrow \begin{array}{c} CH_2Cl \\ | \quad\quad\quad + OH^- , \\ CH_2OH \end{array} \qquad (7.128)$$

$$\begin{array}{c} CH_2 \\ | \quad\quad\!\!>O + Cl^- + H_3O^+ \\ CH_2 \end{array} \longrightarrow \begin{array}{c} CH_2Cl \\ | \quad\quad\quad + H_2O . \\ CH_2OH \end{array} \qquad (7.129)$$

The first path is a direct hydrolysis and the second path is a hydronium ion catalyzed hydrolysis. The rate equation for the entire process causing the disappearance can be written

$$-\frac{1}{[C_2H_4O]} \frac{d[C_2H_4O]}{dt} = k_0' + k_0'[H_3O^+] + k_2'[Cl^-]$$

$$+ k_2'[Cl^-][H_3O^+] . \qquad (7.130)$$

If nucleophilic reagents other than water or chloride ion are present, each will react in a manner similar to the reactions depicted for water and chloride ion. If, as in the case of the acetolysis of 3-phenyl-2-butyl tosylate, there is first the addition of a hydrogen ion then, supposedly, an oxonium complex

is formed. The nucleophilic reagent will react from 500 to 10,000 times as fast with this complex as with the neutral oxide, since the weaker base

ROH rather than the strong base RO⁻ is being displaced from its bonding to carbon. The nucleophilic character of the attacking agent rather than its basic strength determines the rate. The more nucleophilic the attacking agent, the faster the rate.

Prichard and Long (111) studied the acid-catalyzed hydrolysis of substituted ethylene oxides at 0°C, using a dilatometric procedure. Aqueous perchloric acid ranging up to 3.5 molar was used. Substitution caused a wide variation in rates. The rate for isobutylene oxide was about 10^4 times greater than that for epibromohydrin at constant acid concentration. Since the logarithm of the first-order rate constant for the hydrolysis was found to be linear with $-H_0$, and on the basis of O^{18} studies in the basic hydrolysis, it was concluded that the reaction occurred by an A — 1 mechanism involving a carbonium ion intermediate in the following manner

$$R_1R_2\overset{\displaystyle O}{\overset{\displaystyle \triangle}{C}}\text{---}CH_2 + H^+ \rightleftharpoons R_1R_2\overset{\displaystyle OH^+}{\overset{\displaystyle \triangle}{C}}\text{---}CH_2 \qquad \text{(equilibrium)}, \qquad (7.131)$$

$$R_1R_2\overset{\displaystyle OH^+}{\overset{\displaystyle \triangle}{C}}\text{---}CH_2 \longrightarrow M^{*+} \longrightarrow R_1R_2\overset{\displaystyle O}{\overset{\displaystyle |}{C}}{}^+\text{---}CH_2 \quad \text{(slow)}, \qquad (7.132)$$

$$R_1R_2\overset{\displaystyle O}{\overset{\displaystyle |}{C}}{}^+\text{---}CH_2 + H_2O \xrightarrow{\underset{\text{complex}}{\overset{\text{activated}}{}}} R_1R_2\underset{\displaystyle \underset{\displaystyle OH}{|}}{\overset{\displaystyle OH}{\overset{\displaystyle |}{C}}}\text{---}CH_2 + H^+ \quad \text{(fast)}. \qquad (7.133)$$

If the mechanism had been A — 2 the rate-controlling step would have been an attack on the conjugate acid by a water molecule with the resulting inclusion of a water molecule in the activated complex. In A — 2 mechanisms experiments verify that rates closely parallel hydrogen ion concentration and not the acidity function. However, experience and theory indicate that, in an A — 1 mechanism, the rate will parallel the acidity function. In the A — 1 type mechanism the solvalytic action of the solvent is not rate controlling but is critical in the product-forming step.

For the present case the rate of reaction was written

$$\text{Rate} = k_1' \, C_{\text{oxide}}$$

$$= k_{bi}' \, C_{\text{oxide}} \, C_{\text{H}^+}$$

$$= k_{bi}^{0\prime} \, C_{\text{oxide}} \frac{C_{\text{H}^+} f_{\text{H}^+} f_{\text{oxide}}}{f_{\text{M}^{*+}}}, \qquad (7.134)$$

where $k_{bi}^{0'}$ is a true constant and the f's are the activity coefficients of the indicated species. Therefore,

$$\log k_1' = \log \frac{C_{H^+} f_{H^+} f_{\text{oxide}}}{f_M^*} + \text{constant},\qquad(7.135)$$

and the addition of Eqs. (7.80) and (7.135) yields

$$\log k_1' = -H_0 + \log \frac{f_{\text{oxide}}}{f_{M^{*+}}} - \log \frac{f_B}{f_{BH^+}} + \text{constant}.\qquad(7.136)$$

Since all activity coefficients become unity in dilute solutions, the experimental fact that $\log k_1'$ is linear in $-H_0$ with unit slope leads to the conclusion that

$$\frac{f_{\text{oxide}}}{f_{M^{*+}}} = \frac{f_B}{f_{BH^+}},\qquad(7.137)$$

which for an A — 1 mechanism is quite reasonable, since the activated complex and the conjugate acid of the reactant oxide are structurally very similar.

Long and Prichard (136) found from the acidic, basic, and neutral hydrolyses of propylene and isobutylene oxides in O^{18}-labeled water that the base-catalyzed reaction in the case of the propylene oxide occurs predominantly at the primary carbon atom, and exlcusively so for the isobutylene oxide. On the other hand, the acid-catalyzed reaction occurs exclusively at the branched chain for isobutylene oxide and predominantly so for propylene oxide.

The base-catalyzed hydrolysis would involve nucleophilic attack by the hydroxide ion, predominately for steric reasons, at the primary carbon atom of the epoxide ring. Thus for basic hydrolysis the mechanism would be

$$\text{(slow)},\qquad(7.138)$$

$$\text{(fast)}.\qquad(7.139)$$

The mechanism of the acid-catalyzed hydrolysis was discussed above.

The acid and basic hydrolyses of esters have been studied extensively. In Chapter II, where the theory of ion–dipolar molecule reactions was presented, these reactions were discussed at some length. Further discussion of these reactions will not be presented here.

Ammonolysis resembles hydrolysis. We may illustrate ammonolysis by the reaction (138)

$$
\begin{array}{c}
\underset{\overset{|}{H}}{\overset{OR}{C}} = O + NH_3 \rightleftarrows HC\overset{OR}{-}OH \rightleftarrows ROH + HC = O . \\
\underset{NH_2}{} \qquad\qquad NH_2
\end{array} \qquad (7.140)
$$

The ammonolysis may become very slow for higher esters. Thus, Hurd (139) reports that isobutrate reacts at room temperature only after about eight weeks.

Salts of weak acids and weak bases in the aqua series are hydrolyzed by water, and similarly in the ammonia series, ammonolysis takes place (140). In the case where the metal atom in a compound is not exercising its maximum coordination number, coordination with one or more molecules of water usually precedes hydrolysis. Fowles and Pollard (141) and Emeleus (142) believe that, in liquid ammonia, ammonolysis takes place in a manner involving similar steps. Thus Barnett and Wilson write the more familiar hydrolysis involving silicon tetrachloride as follows:

$$
SiCl_4 + 2H_2O = \overset{H_2O}{\underset{H_2O}{\diagdown}} SiCl_4 \longrightarrow 2HCl + Si(OH)_2Cl_2 , \qquad (7.141)
$$

$$
Si(OH)_2Cl_2 + 2H_2O = \overset{H_2O}{\underset{H_2O}{\diagdown}} Si(OH)_2Cl_2 \longrightarrow 2HCl + Si(OH)_4 . \qquad (7.142)
$$

Fowles and Pollard (141) found that in the Group IVA halides the degree of ammonolysis decreases regularly down the series $SiCl_4$, $TiCl_4$, $ZrCl_4$, and $ThCl_4$. Emeleus indicated that $SiCl_4$ underwent complete ammonolysis to $Si(NH_2)_4$, while according to Fowles and Pollard, $TiCl_4$ and $ZrCl_4$ form the amidochlorides $Ti(NH_2)_3Cl$ and $Zr(NH_2)Cl_3$ with liquid ammonia while $TiCl_4$ gives the hexaammoniate $TiCl_4 \cdot 6NH_3$. Hydrolysis (141) and alcoholysis (142) demonstrate a similar order of reactivity. The mechanism of the alcoholosis and the ammonolysis is perhaps similar. Thus for the alcoholysis (141, 143)

$$\text{ROH} + \text{MCl}_4 \longrightarrow$$

$$\begin{array}{c}\text{R} \\ \overset{\delta-}{\diagdown} \overset{\delta+}{\diagup} \\ \text{O--MCl}_3 \\ \diagup \uparrow \\ \text{H} \\ \overset{}{\scriptstyle \delta+}\end{array} \overset{\delta-}{\underset{}{\text{Cl}}} \longrightarrow \begin{array}{c} \text{R} \\ \diagdown \\ \text{O--M--Cl}_3 + \text{H}^+ + \text{Cl}^- \end{array} \qquad (7.143)$$

and for the ammonolysis

$$\text{NH}_3 + \text{M} \longrightarrow \begin{array}{c} \overset{\delta+}{\text{H}} \\ \diagdown \\ \text{H--N--MCl}_3 \\ \diagup \\ \text{H} \end{array} \overset{\delta-}{\underset{}{\text{Cl}}} \longrightarrow \text{NH}_2\text{--MCl}_3 + \text{H}^+ + \text{Cl}^-. \qquad (7.144)$$

It would be expected that the replacement of chlorine atoms by amino groups should occur most readily when the M–Cl bond is markedly ionic in character, but fractional ionic character for the Group IVA halides increases regularly from silicon to thorium tetrachloride. The specific surface charge intensity, however, calculated by taking into consideration the radius of the different atoms, is in the reverse order. This specific surface charge intensity determines the coordination energy available for the ionization of the M–Cl and N–H bonds, and hence the observed orders of reactivity and degree of displacement are as should be anticipated.

An alternate mechanism is offered by Fowles and Pollard in which the amide ion reacts with the halide molecule:

$$\text{NH}_2^- + \text{MCl}_4 \longrightarrow \left[\text{NH}_2 \ldots \underset{\text{Cl}_3}{\text{M}} \ldots \text{Cl} \right]^- \longrightarrow \text{NH}_2 \cdot \text{MCl}_2 + \text{Cl}^- \qquad (7.145)$$

followed by discharge of the chloride ion and the retention of the amide group. This mechanism is supported by the increased ammonolysis when amide ions in the form of alkali metal amides are added to the solution, though somewhat opposed to this theory is the fact that the variation of temperature, which should influence the autoionization, seems to have no effect. Both schemes predict the ionization of the M–Cl bond. The mechanism of Wardlaw (143) involves the initial coordination of an ammonium molecule followed by the ionization of the M–Cl and N–H bonds, while Fowles and Pollard's mechanism predicts the initial coordination of an amide ion. The amide ion then transmits its charge to the chlorine atom which splits off as an ion.

Ammonolysis and amination are synonymous in that both result in an amide group replacement of another atom or group, for example, a halogen atom in a molecule. Bergstrom et al. (144) in 1936 first aminated aromatic halides using alkali metal halides in liquid ammonia. They converted chloro-, bromo-, and iodo-benzene into aniline and several byproducts, including diphenyl amine, triphenylamine, and p-aminobiphenyl. They obtained no reaction with fluorobenzene. Using the procedure of the above authors, Gilman and Avakian (145) prepared meta-amino ethers form ortho-halogenated ethers. Benkeser and co-workers (146) demonstrated that the entering amide displaces a hydrogen atom ortho to the halogen atom. They also found that ether solutions of aryl halides were attacked by lithium dialkylamides. Bunnett and Zahler (147) reviewed the reaction in 1951. They proposed a mechanism for the rearrangement in which the migration of the hydride ion from carbon 2 to carbon 1 and its consequent displacement of the halide ion is forced by the attack by the amide ion on carbon 2 of a halobenzene. Thus

$$\text{(7.146)}$$

Roberts et al. (148) suggested a likely intermediate to be benzyne in the amination of unsubstituted halobenzenes, since almost equal amounts of aniline-1-C^{14} and aniline-2-C^{14} were obtained when chlorobenzene-1-C^{14} was aminated with potassium amide in liquid ammonia. Thus,

$$\text{(7.147)}$$

$$\text{(7.148)}$$

Roberts and co-workers (149) observed the same results with chlorobenzene-1-C^{14}, bromobenzene-1-C^{14}, and iodobenzene-1-C^{14}. They also found that in deuterated bromobenzene the percentage of deuterium increased with percentage of reaction. Thus the undeuterated material underwent amination faster than the deuterated material, the ratio of the specific velocity constants k_H'/k_D' for the amination being 5.5. Chlorobenzene-2-D in dilute solution showed a small increase in deuterium content after partial reaction, while in concentrated solutions it showed a decrease in deuterium content.

This was interpreted to mean that chlorobenzene-2-D exchanged deuterium with the solvent at a rate comparable with the amination. For chlorobenzene-2-D, k_H'/k_D' was found to be 2.70. In the case of fluorobenzene-2-D, deuterium was exchanged with the solvent very rapidly (see Dunn *et al.,* *150*). For the amination and exchange of chlorobenzene the mechanism given below was written:

$$(7.149)$$

The calculation of k_H'/k_D' was made using the following procedure. The rates of disappearance of chlorobenzene-2-D (D) and ordinary chlorobenzene (H) is given by

$$- \frac{d[D]}{dt} = [k_D' + k_H'] \, [D] \, [NH_2^-]^n, \qquad (7.150)$$

$$- \frac{d[H]}{dt} = 2k'_H [H] \, [NH_2^-]^n . \qquad (7.151)$$

If n is taken as unity, Eq. (7.150) divided by Eq. (7.151), the quotient rearranged, and the resulting expression integrated, k_H'/k_D' becomes

$$k_H'/k_D' = \tfrac{1}{2} \, \{\ln \, ([D]_t/[D]_0)/\ln \, ([H]_t/[H]_0)\} - 1. \qquad (7.152)$$

In making this derivation it was assumed that all the reactions were constant volume processes, the reactions were first order with respect to halobenzene, only hydrogen or deuterium *ortho* to the halonges are involved in the over-all amination reactions, and the reaction of benzene to give products is not rate determining or reversible.

A kinetic expression for mechanism (7.149) to account for the change in the deuterium content of chlorobenzene during amination was derived by Roberts and co-workers (*149*). The rate of disappearance of chlorobenzene-2-D is

$$- \frac{d[D]}{dt} = (k_H' + k_D') [D] [NH_2^-] - k_{-1} [\text{anion IV}] ; \quad (7.153)$$

using the steady-state assumption gives

$$[\text{anion IV}] = \frac{k_H' [D] [NH_2^-]}{k_{-1}' + k_2'} . \quad (7.154)$$

Substituting Eq. (7.154) into Eq. (7.153) gives

$$- \frac{d[D]}{dt} = (k_H' + k_D') [NH_2^-] - \frac{k_{-1}' k_H'}{k_{-1}' + k_2'} [D] [NH_2^-]. \quad (7.155)$$

In like manner Eq. (7.156) is obtained for the rate of disappearance of ordinary chlorobenzene:

$$- \frac{d[H]}{dt} = 2k_H' [H] [NH_2^-] - \frac{k_{-1}'}{k_{-1}' + k_2'} (2k_H' [H] [NH_2^-]$$
$$+ k' [D] [NH_2^-] . \quad (7.156)$$

Letting $F = k_{-1}'/(k_{-1}' + k_2') = $ fraction of the intermediate anion returning to initial reactant and the isotope effect $i = k_H'/k_D'$ then

$$\frac{d[D]}{d[H]} = \frac{[i(1 - F) + 1] [D]}{2i(1 - F)[H] - F[D]} . \quad (7.157)$$

If Eq. (7.157) is divided by [D] and if it is assumed that $[i(1 - F) + 1]$, $2i(1 - F)$, and $-F$ are constants, we obtain, after integration,

$$\ln [D] \Big]_{D_0}^{D_t} + \frac{1 + (1 - F)i}{1 - (1 - F)i} \ln \left[(-1 + (1 - F)i) \frac{[H]}{[D]} - F \right]_{[H]_0/[D]_0}^{[H]_t/[D]_t} = 0.$$
$$(7.158)$$

Dunn *et al.* (*150*) write the symbol d for the fraction of the chlorobenzene deuterated at time t and the symbol P for the fraction of starting material which has reacted in time t. Using these symbols the Eq. (7.158) can be written

$$\ln \frac{d}{d_0}(1-P) = \frac{i(1-F)+1}{i(1-F)-1} \ln \frac{(1/d)-1-[F/\{i(1-F)-1\}]}{(1/d_0)-1-[F/\{i(1-F)-1\}]}.$$

(7.159)

The above equations apply to the change in concentration of deuterium in unreacted chlorobenzene. Dunn *et al.*, in order to calculate the aniline composition, predicted by mechanism (7.149), let A_D and A_H represent deuterated and undeuterated aniline, respectively. Then

$$\frac{d[A_D]}{dt} = k_2 [\text{anion IV}],$$

(7.160)

and dividing Eq. (7.160) by (7.153) yields

$$-\frac{d[A_D]}{d[D]} = \frac{k_2' [\text{anion IV}]}{(k_H' + k_D')[D][NH_2] - k_{-1}'[\text{anion IV}]}.$$

(7.161)

Substituting the steady-state expression for [anion IV] and introducing i and F where appropriate gives

$$-\frac{d[A_D]}{d[D]} = \frac{i(1-F)}{i(1-F)+1}.$$

(7.162)

Integration under the condition that $[A_D] = 0$ when $[D] = [D_0]$ yields

$$[A_D] = \frac{i(1-F)}{i(1-F)+1}([D]_0 - [D]).$$

(7.163)

Assuming $P = ([A_D] + [A_H])/([D]_0 + [H]_0)$ and $1-P = ([D] + [H])/([D]_0 + [H]_0)$, Eq. (7.163) becomes

$$\frac{[A_D]}{[A_D] + [A_H]} = \frac{1}{P}\frac{i(1-F)}{i(1-F)+1}\frac{[D]_0}{[D]_0 + [H]_0}$$
$$- (1-P)\frac{[D]}{[D]+[H]}.$$

(7.164)

To simplify this equation let a equal the fraction of the aniline which is deuterated and d the fraction of the chlorobenzene which is deuterated, then

$$a = \frac{i(1 - F)}{i(1 - F) + 1} \; \frac{d_0 - d(1 - P)}{P}. \qquad (7.165)$$

Dunn et al. subjected mixtures of ordinary chlorobenzene acid and chlorobenzene-2-D to partial amination by sodamide in liquid ammonia. They analyzed both the unreacted starting material and the product aniline for deuterium. They found the deuterium in the aniline to be distributed almost equally between the *ortho* and *meta* positions. Their results give strong support to mechanism (7.149), in which the slow step is the formation of an intermediate such as benzyne symmetrical with respect to carbon atoms 1 and 2. Table V presents the data of Dunn et al., where a was calculated using Eq. (7.165).

In the ammonolysis of ethylene chlorohydrin (I), Krishnamurthy and Rao (151) found that the yield varied directly as the NH_3/I ratio, with the optimum at approximately at 8. The optimum temperature appeared to be 90°C. For any NH_3/I ratio the reaction was independent of the concentration of ammonia. Aqueous ammonia was found to be a far more effective reagent in the ammonolysis than liquid ammonia alone.

Szychlinski (152) studied the photochemical ammonolysis of p-chlorophenetol, p-bromophenetol, o-chloroanisole, o-bromoanisole, and p-bromoanisole in 25% methanol at 30–40° using a mercury arc lamp and passing a stream of ammonia at a pressure through the reaction flask. The reaction was zero order and yielded mainly primary aromatic amines. The reaction was the same in a nitrogen as in an ammonia stream, which indicated that ammonolysis was a secondary process, and that the primary process was photolysis.

By treating solutions of $Na_3P_3O_9$ with concentrated aqueous ammonia, Quimby and Flautt (153) prepared $Na_4P_3O_9NH_2$. The formation of $Na_4P_3O_9NH_2$ was followed by nuclear magnetic resonance measurements at 26°C when the $Na_3P_3O_9$ was 0.27 M and the NH_3 was 5.8 M. The first-order rate constant was $10^{-2}/min$. Acidified solutions of $Na_4P_3O_9NH_2$ yielded NH_4^+ and the cyclic P_3O_9 was regenerated.

Among other reactions in liquid ammonia is one in which nitrogen from ammonia is exchanged with the nitrogen in carboxylic acid amides for example (154). In the reference cited the aminolysis of p-nitrobenzamide, p-$O_2NC_6H_4CONH_2$ in liquid ammonia was followed kinetically, and the first-order rate constant was found at 20°C to be $k' = 1.27 \times 10^{-8}$ sec^{-1}. The reaction is catalyzed by NH_4Cl without which higher temperatures are

TABLE V

AMINATION OF MIXTURES OF CHLOROBENZENE AND CHLOROBENZENE-2-D

d_0[a]	d[b]	P[c]	F_{calc}	a_{calc}[d]	a_{obs}
0.407 ± 0.003	0.350 ± 0.008	0.452 ± 0.006	0.844 ± 0.011	—	Not enough product for analysis
0.816 ± 0.019	0.414 ± 0.014	0.718 ± 0.006	0.870 ± 0.005	0.416 ± 0.004	0.441 ± 0.000
0.594 ± 0.014	0.290 ± 0.005	0.819 ± 0.012	0.868 ± 0.001	0.283 ± 0.004	0.326 ± 0.005
0.396 ± 0.009	0.227 ± 0.009	0.783 ± 0.007	0.874 ± 0.005	0.185 ± 0.005	0.204 ± 0.003

[a] d_0 = mole fraction of chlorobenzene-2-D at zero time.

[b] d = mole fraction of chlorobenzene-2-D at time t.

[c] P = mole fraction of starting material converted to chloride ion at time t.

[d] a = mole fraction of aniline-D at time t. The maximum value was calculated from Eq. (7.165) using the minimum value cf d and the maximum values of d_0, P and F. The maximum value of F was calculated from Eq. (7.159) using the minimum value of d and the maximum value of d_0 and P; the reverse combination gave the minimum value.

required. In the instance cited above, the p-nitrobenzamide was 0.13 M, the ammonium chloride was 3.33 M, and 12.23% of the nitrogen in the p-nitrobenzamide was nitrogen fifteen. The reaction is a form of ammonolysis in which a NH_2, and a hydrogen from ammonia is added to the amide of the acid, and was pictured by Heyns, Brockmann, and Roggenbuck as follows:

$$(7.166)$$

Thus in ammonolysis and amination, the composition of the media is an important factor. For example the presence of water or a salt as NH_4Cl influence the effectiveness and the speed of the reaction. As in other reactions, the solvent or the media in general is indeed influential in the ammonolysis and amination reactions.

Solvolysis could be discussed at much greater length, but only one or two other examples will be mentioned here to illustrate how solvolysis may change the nature of the reactant species and thus determine the mechanism and rate of a reaction.

Harkness and Halpern (155) studied the kinetics of the reaction

$$U(IV) + Tl(III) \longrightarrow U(VI) + Tl(I) \qquad (7.167)$$

in aqueous perchloric acid solution and found them to comply with the equation

$$-\frac{d[U(IV)]}{dt} = [U^{4+}] [Tl^{3+}] (k_1'[H^+]^{-1} + k_2'[H^+]^{-2}) . \qquad (7.168)$$

The two rate constants were identified with the following reaction paths involving activated complexes of the compositions $(U \cdot OH \cdot Tl)^{6+}$ and $(UOTl)^{5+}$:

$$U^{4+} + Tl^{3+} + H_2O \; \underset{}{\overset{k_1'}{\rightleftharpoons}} \; (U \cdot OH \cdot Tl)^{6+} + H^+, \tag{7.169}$$

$$U^{4+} + Tl^3 + H_2O \; \underset{}{\overset{k_2'}{\rightleftharpoons}} \; (U \cdot O \cdot Tl)^{5+} + 2H^+. \tag{7.170}$$

Jones and Amis (156) studying the same reaction in 25% by weight methanol found that their rate data could be reproduced using the equation

$$-\frac{d[U(IV)]}{dt} = k_1'[U^{4+}] [Tl^{3+}] [H^+]^{-1} + k_2'[U^{4+}] [Tl^{3+}] [H^+]^{-2}, \tag{7.171}$$

which can be put in the form

$$-\frac{d[U(IV)]}{dt} \left[\frac{([H^+]+K_U) ([H^+]+K_{Tl})^2}{[U(IV)] [Tl(III)]^2 [H^+]} \right] = k_1' \left[\frac{([H^+] + K_{Tl})}{[Tl(III)]} \right] + k_2'. \tag{7.172}$$

A plot of the left-hand side of Eq. (7.172) versus $([H^+] + K_{Tl})/[Tl(III)]$ yielded a straight line from which, at a given temperature, k_1' was obtained as the slope and k_2' as the intercepts. K_U and K_{Tl} were hydrolysis constants of U^{4+} and Tl^{3+} whose values were available in the literature. The activation process in 25 wt % methanol was written

$$U^{4+} + Tl^{3+} + H_2O \; \underset{}{\overset{k_1'}{\rightleftharpoons}} \; (U \cdot OH \cdot Tl)^{6+} + H^+, \tag{7.173}$$

$$U^{4+} + 2Tl^{3+} + 2H_2O \; \underset{}{\overset{k_2'}{\rightleftharpoons}} \; (Tl \cdot HO \cdot U \cdot OH \cdot Tl)^{8+} + 2H^+. \tag{7.174}$$

The kinetic data are reproduced in Table VI. Corresponding data for water media, as given by Harkness and Halpern (155) are presented for comparison in Table VII. It is clear from the rate expressions and tables of data that changing the solvent from water to 25% MeOH–75% H_2O by weight changed the rates, the mechanisms, and the thermodynamic quantities for the chemical reaction. The difference in solvolytic processes for the two cases, Eqs. (7.169) and (7.170) in the one case and Eqs. (7.173) and (7.174) in the other case, are basic in explaining the differences in rates, mechanisms, and thermodynamic quantities. Increasing the weight percent of methanol in the media caused still greater changes in the kinetic factors of this reaction. It is reasonable to suppose that, as the dielectric constant of the medium is decreased by adding increasing amounts of methanol, the increased forces

TABLE VI

KINETIC DATA FOR THE OXIDATION OF U(IV) BY Tl(III), IONIC STRENGTH IS 2.9, IN 25% METHYL ALCOHOL–75% WATER MEDIA

Path [a]	15°C	20°C	25°C	Z	ΔH^* (cal/mole)	ΔS^* (e. u.)	ΔF^* (cal/mole)
1	0.0037	0.0040	0.0041	0.79	3112	−59.0	20,694
2	0.75	0.85	0.98	1871	4474	−43.6	17,467

[a] Path 1 refers to k_1' and path 2 to k_2'. Units of k_1' and k_2' are sec⁻¹.

TABLE VII

KINETIC DATA FOR THE OXIDATION OF U(IV) BY Tl(III), IONIC STRENGTH IS 2.9, IN WATER MEDIA

Path [a]	k'			ΔH^* (cal/mole)	ΔS^* (e. u.)	ΔF^* (cal/mole)
	16°C	20°C	25°C			
1	0.57	1.02	2.11	24,600	16	19,700
2	0.67	1.17	2.13	21,700	7	19,700

[a] Path 1 refers to k_1' and path 2 to k_2'. Units of k_1' and k_2' are sec⁻¹.

among oppositely charged ions, for example, OH^- and U^{4+} and OH^- and Tl^{3+}, will favor the formation of more complex intermediates. This is what is actually implied in the two activation processes in the two media, water and water–methanol.

Nitration of Aromatic Compounds and the Solvent

The nitration of aromatic compounds will be mentioned in relation to solvent effect upon rates and mechanisms because the reaction has been so extensively studied, more than any other reaction of aromatic compounds, because the solvent plays such an important role in the rate and mechanism of the reaction, and because its mechanism has been rather completely elucidated.

Euler (*157*) in 1903 suggested that the nitronium ion, NO_2^+, was the reactive species in nitration. That a positive ion or a strong dipole was an active intermediate which reacts with the benzene ring in nitration was indicated by the fact that substitution in the ring of electron-withdrawing groups such as the nitro group reduced the readiness with which nitration occurred. Hantzsch (*158*), based on an incorrect value of about 3 for the van't Hoff factor, *i*, concluded that $H_2NO_3^+$ and $H_3NO_3^{2+}$ were the ions present in nitric acid dissolved in concentrated sulfuric acid. He based his conclusions on ultraviolet absorption, conductance, and freezing-point depression data. The positive-ion mechanism was supported by Ri and Eyring (*159*) on orientation by substituents. These investigators on the basis of electrostatic interactions caluclated fairly accurately the *ortho–para* to *meta* ratio for the nitration of several substituted benzenes, assuming a positive ion reactant and using bond distance and dipole moment data. Chedin (*160*) found two Raman spectrum lines at 1050 cm^{-1} and 1400 cm^{-1} in mixtures of nitric acid and sulfuric acid which were not present in the Raman spectra of the pure substances. The 1400-cm^{-1} line was also present in perchloric and selenic acid solutions of perchloric acid (*161*). This line must be common to a substance present in all strong acid solutions of nitric acid. The two lines, therefore, do not come from a common molecular source. That the source of the frequency 1400 cm^{-1} has no other line in its spectrum was proved by a detailed study of the spectra. Thus the source is unambiguously the nitronium ion, since no other structure formed from the elements in nitric acid would yield the observed spectral characters. There are several other lines in the Raman spectrum of the 1050-cm^{-1} frequency as produced by the use of sulfuric acid. They were found to identify the hydrogen sulfate ion HSO_4^-. It was found (162) from freezing points that nitric acid reacts with solvent sulfuric acid quantitatively according to the equation

$$HNO_3 + 2H_2SO_4 \longrightarrow NO_2^+ + H_3O^+ + 2HSO_4^- , \qquad (7.175)$$

except that since water is only semi-strong base there is a recombination of a fraction of the H_3O^+ and HSO_4^- ions as depicted below:

$$H_3O^+ + HSO_4^- \rightleftharpoons H_2O + H_2SO_4 , \qquad (7.176)$$

so that the van't Hoff factor i is not 4 as would be expected if Eq. (7.175) were completely quantitative, but was observed to be 3.77, expressed as a mean over the concentrations used.

If further proof of the existence of the nitronium ion is needed, there are data on the existence of its stable salts (161, 163–165). Even N_2O_5 has been identified as nitronium nitrate NO_2NO_3 from the Raman spectra of the solid material.

This rather lengthy discussion of the existence of the nitronium ion in strong acid solutions of nitric acid has been given in support of Euler's early suggestion that nitronium ion was the reactive species in nitration and in confirmation of recent evidence on this role of the nitronium ion with respect to nitration. Nitration can be achieved by using either strong acid solutions of nitric acid or nitronium salts as a source of the nitration reagent, that is, of nitronium ion. It has been found that rates of nitration by nitric acid of aromatic compounds in sulfuric acid–water mixtures increase with increasing sulfuric acid up to about 90% sulfuric acid (166), after which the rates decrease with increasing percent of sulfuric acid.

Westheimer and Kharasch (167) pointed out that the ionization of nitric acid to form the nitronium ion in a sulfuric acid medium is analogous to the ionization of a triarlycarbinol indicator. They also showed that the rate of nitration of nitrobenzene increased by a factor of about 3000 when the concentration of sulfuric acid is increased from 80 to 90%, and that this increase in rate paralled the ionization of trinitrotriphenylcarbinol in the same media. The rate of nitration did not parallel, however, the ionization of anthraquinone which does not ionize in sulfuric acid in a manner similar to that of nitric acid when the latter forms the nitronium ion.

Nitric acid, trinitrophenylcarbinol, and anthraquinone ionize, respectively, as shown below, in the region of 90% sulfuric acid (162):

$$HNO_3 + 2H_2SO_4 \longrightarrow NO_2^+ + H_3O^+ + 2HSO_4^- , \qquad (7.177)$$

$$(O_2NC_6H_4)_3COH + 2H_2SO_4 \longrightarrow (O_2NC_6H_4)_3C^+ + H_3O^+ + 2HSO_4^-, \quad (7.178)$$

$$+ HSO_4^-. \quad (7.179)$$

It is evident that the ionization of nitric acid and of trinitrotriphenylcarbinol are entirely similar with the nitronium ion in the case of nitric acid being replaced by the triphenylcarbonium ion in the case of trinitrotriphenylcarboninol. However, there is no similarity in the ionization of nitric acid and anthraquinone. Now the rate of nitration with increasing sulfuric acid concentration paralleled the ionization of trinitrotriphenylcarbinol, but not that of the ionization of anthraquinone. Hence, it is logical to assume that the rate of nitration paralleled the increase in nitronium ion concentration. At the concentration of sulfuric acid where the ionization to nitronium ion is complete, the rate of nitration reaches a maximum. With further increase of sulfuric acid concentration there is no further increase in rate. The decrease in rate after the maximum has been explained by Bennett and co-workers (168) who assumed that the removal of the proton was part of rate-controlling step. Thus

$$NO_2^+ + ArH + SO_4H^- \longrightarrow ArNO_2 + H_2SO_4. \quad (7.180)$$

This would indicate that the composition of the medium in which the rate of nitration is a maximum represents a balance between the acid function of the solvent and its basic function as a proton acceptor. There is a difference of opinion about the removal of the proton being a part of the rate step. See, for example, the work of Melander (169) and of Gillespie et al. (170). Accepting the rate-controlling step to involve the removal of a proton as represented in Eq. (7.180), then as the composition of the acid becomes richer than 90% in sulfuric acid the concentration of HSO_4^- should decrease rapidly since the bisulfite ion is produced chiefly by the loss of a proton by sulfuric acid to water to produce hydronium ion and bisulfite ion. As a consequence of this theory addition of bisulphite ion should accelerate the rate of nitration in solutions rich in sulfuric acid, but should decelerate the rate of nitration in solutions more dilute in sulfuric acid than the optimum amount. These effects have been noted (167, 168). Westheimer and Kharasch showed that the determining factor for nitration in sulfuric acid is the acidity of the medium as measured by the H_0 function. Thus, the influence of the bisulfite ion on the rate of nitration is understandable on the basis of the hypothesis that the maximum rate occurs at a definite value of H_0, the acidity, rather than a definite composition of acid. Thus, bisulfite ion which lowers the acidity will cause a decrease in the rate when the acidity is less than the optimum but an increase in the rate when the acidity is already greater than the optimum. The rate is not appreciably affected by substances, such as dinitrobenzene and phosphorous pentoxide, which do not materially affect the acidity of the medium.

Frost and Pearson (14) mention several objections to the supposition that the removal of the proton is involved in the rate-determining step. For one thing, such an hypothesis would predict that the maximum in the rate would be a function only of the solvent and independent of the aromatic compound used. This is not quite the case experimentally. Another consequence of such a supposition would be the constancy, independent of solvent composition, of the ratio of rate constants for two compounds. This is far from true. Also Melander (169) showed that in nitration tritium was displaced from the ring as readily as was hydrogen. This is more direct proof that removal of the proton is not part of the rate-determining step, since if such were the case, tritium ion's being removed less readily than the proton would cause the rate of nitration to be slower in the tritium-substituted compound. Since there is no exchange of the hydrogen with the acids and the water the existence of a dissociation equilibrium cannot be the reason. Melander concludes that the addition of nitronium ion is the rate-determining step, and that this is followed by a rapid splitting off of the proton.

The acceptable explanation of the decrease in rate in higher percentages of sulfuric acid is apparently the formation of unreactive positively charged or hydrogen-bonded species (14). The positively charged species $C_6H_5NO_2H^+$ and the hydrogen-bonded species $C_6H_5NO_2H_2SO_4$ have been described by Gillespie (171) in solutions of nitrobenzene in sulfuric acid. Cherbuliez (172) and Masson (173) discuss the properties of the compound $C_6H_5NO_2 \cdot H_2SO_4$. Other positively charged specied, such as $CH_3C_6H_4NO_2H^+$ and $CH_3NO_2H^+$, are described by Gillespie. Frost and Pearson suggest that there is a relation between the basicity of the organic molecule and the rate of fall in rate of nitration after the maximum in rate is reached. The more basic a compound is the steeper the maximum. They cite the work of Baker and Hey (174), who explained the increase of *meta* nitration for benzaldehyde, acetophenone, and ethyl benzoate with increase of sulfuric acid in the nitrating mixture as arising from the conversion of these substances to positively charged ions, thus increasing the tendency toward *meta* orientation.

Frost and Pearson also argue that sulfuric acid is more polar than water, that there would be a spreading out of the charge in the activated complex for a rate-determining step between a neutral molecule and a positive ion in going to the more polar solvent, and that, therefore, there would be a decrease in rate with increasing sulfuric acid. If the rate-controlling step involved a destruction of charge, there would be a more marked decrease in rate in going to a more polar solvent.

A mechanism proposed by Gillespie *et al.* (*170*), for nitration of aromatic compounds when nitronium ion is formed from nitric acid can be written

$$2HNO_3 \underset{k_2'}{\overset{k_1'}{\rightleftharpoons}} H_2NO_3{}^+ + NO_3{}^- \qquad \text{(fast)}, \qquad (7.181)$$

$$O_2N \cdot \overset{+}{O}H_2 \underset{k_4'}{\overset{k_3'}{\rightleftharpoons}} O_2N^+ + OH_2 \qquad \text{(slow)}, \qquad (7.182)$$

$$NO_2{}^+ + ArH \overset{k_5'}{\longrightarrow} \overset{+}{Ar} \overset{\displaystyle H}{\underset{\displaystyle NO_2}{\big\backslash}} \qquad \text{(slow)}, \qquad (7.183)$$

$$\overset{+}{Ar} \overset{\displaystyle H}{\underset{\displaystyle NO_2}{\big\backslash}} \overset{k_6'}{\longrightarrow} ArNO_2 + H^+ \qquad \text{(fast)}. \qquad (7.184)$$

This mechanism seems in agreement with the facts to date. Unlike the mechanism represented in Eq. (7.180) this mechanism does not show a high solvent sensitivity. The rate represented by the rate constant k_2' Eq. (7.181), would be solvent sensitive as would be the rate of the reaction represented by Eq. (7.180), but in the mechanism proposed by Gillespie *et al.* the reverse step in Eq. (7.181) is not rate-controlling. The reaction represented in Eq. (7.183) shows a slight solvent effect due to spreading of the charge in the transition state, and this is observed to be the case. In Eq. (7.181) one of the molecules of nitric acid may be replaced by a molecule of any strong acid, the anion of which would replace the nitrate anion in the reaction represented by Eq. (7.181). Another observation which supports the mechanism given above is the hyperbolic repression of the rate of nitration by nitrate ions (rate $\propto C^{-1}$), as would be required by reverse reaction represented in Eq. (7.181), without disturbing the reaction order, even where the order is zero.

The rate expressions represented by Eqs. (7.181) to (7.184) cannot be solved exactly. However, Frost and Pearson (*14*) have given the solutions for the limiting cases in which k_5' is much less than k_3' or k_4' and in which k_5' is much greater than k_3' or k_4'.

The reverse step in Eq. (7.182) takes place only in the presence of relatively large amounts of water, and then the zeroth-order reaction goes over to a first-order reaction in which the kinetic effect of water on first-order nitration is comparable to nitrate ion (*175*). The acceleration by sulfuric acid and the specific retardation by nitrate ions from ionizing nitrates of aromatic nitration in nitric acid, or in nitromethane or acetic acid, with nitric acid in constant excess, obey linear limiting laws and occur alike with zeroth- and first-order reactions, always without disturbance to the reaction order. Thus one nitrate ion is eliminated and one proton used in the reversible conversion of nitric acid to nitronium ion. These steps precede the step which produces nitronium ion, because a zeroth-order nitration only maintains its kinetic order where the latter step is not reversed.

Aromatic nitration is kinetically second-order in sufuric acid, first-order in nitric acid, and is zeroth-order for sufficiently reactive aromatic compounds in either nitromethane or acetic acid with nitric acid in constant excess. The zeroth-order kinetics in organic solvents, and its transition to first-order kinetics with decreasing aromatic reactivity, proves conclusively the formation and effectiveness of the nitronium ion, NO_2^+. All of the observations support the mechanism presented in Eqs. (7.181) to (7.184).

It would be well to conclude a book on the effects of solvents on reaction rates and mechanisms with a reaction, such as aromatic nitration, which shows such a varied dependence of rates and mechanisms on the solvent. Hydrogen bonding, complex formation, polarization, and ionization are among the influences which solvents exert on this type of reaction. The solvent can influence mechanism to the extent of causing reversibility in at least one step of the mechanism and of changing the order of the reaction. This book will not conclude with aromatic nitration, however, since some other topics deserve attention and will be presented in a final chapter.

REFERENCES

1. E. S. Amis, "Kinetics of Chemical Change in Solution," Chapter 4. Macmillan, New York, 1949.

2. K. J. Laidler and H. Eyring, *Ann. N. Y. Acad. Sci.* **39**, 303 (1940).

3. L. J. Minnick and M. Kilpatrick, *J. Phys. Chem.* **43**, 259 (1939).

4. H. S. Harned and N. D. Embree, *J. Am. Chem. Soc.* **57**, 1669 (1935).

5. H. S. Harned and G. L. Kazanjia, *J. Am. Chem. Soc.* **58**, 1912 (1936).

6. E. S. Amis and V. K. LaMer, *J. Am. Chem. Soc.* **61**, 905 (1939).

7. W. A. Cowdrey, E. D. Hughes, C. K. Ingold, S. Masterman, and A. D. Scott, *J. Chem. Soc.* **1937**, 1267.

8. E. S. Lewis and C. E. Boozer, *J. Am. Chem. Soc.* **74**, 308 (1952).

9. V. K. LaMer, *J. Franklin Instit.* **225**, 709 (1938).

10. E. S. Amis, *Anal. Chem.* **27**, 1672 (1955).

11. F. D. Chattaway, *J. Chem. Soc.* **101**, 170 (1912).

12. J. Walker, *Trans. Chem. Soc.* **67**, 746 (1895); **71**, 489 (1897).

13. L. M. Lowry, *Trans. Faraday Soc.* **30**, 375 (1934).

14. A. A. Frost and R. G. Pearson, "Kinetics and Mechanism," pp. 307–315. Wiley, New York, 1961.

15. E. S. Lewis and C. E. Boozer, *J .Am. Chem. Soc.* **74**, 308 (1952).

16. C. E. Boozer and E. S. Lewis, *J. Am. Chem. Soc.* **75**, 3182 (1953).

17. D. J. Cram, *J. Am. Chem. Soc.* **75**, 332 (1953).

18. S. Winstein and A. H. Fainberg, *J. Am. Chem. Soc.* **79**, 5937 (1957).

19. E. S. Amis, *J. Phys. Chem.* **60**, 428 (1956).

20. P. A. Landskroener and K. J. Laidler, "Effects of Solvation on Hydrolysis," Ph. D. dissertation, Catholic University of America. Catholic University of America Press, Washington, D. C., 1954.

21. N. Goldenberg and E. S. Amis, *Z. physik. Chem.* [N. S.] **31**, 145 (1962).

22. J. Franck and E. Rabinowitch, *Trans. Faraday Soc.* **30**, 120 (1934).

23. E. Rabinowitch and W. C. Wood, *Trans. Faraday Soc.* **32**, 1381 (1936).

24. M. V. Smoluchowski, *Z. physik. Chem.* **92**, 129 (1917).

25. R. A. Ogg, Jr., P. A. Leighton, and F. W. Bergstrom, *J. Am. Chem. Soc.* **56**, 1754 (1933).

26. E. Warburg and W. Rump, *Z. Physik* **47**, 305 (1928); **58**, 291 (1929); *Berl. Akad. Ber.*, **1918**, (1928).

27. R. M. Noyes, *J. Chem. Phys.* **22**, 1349 (1954).

28. R. M. Noyes, *J. Am. Chem. Soc.* **77**, 2042 (1955).

29. R. M. Noyes, *J. Am. Chem. Soc.* **78**, 5486 (1956).

30. R. M. Noyes, *J. Am. Chem. Soc.* **79**, 551 (1957).

31. S. W. Benson, "The Foundations of Chemical Kinetics." McGraw-Hill, New York, 1960.

32. D. F. DeTar and C. Weis, *J. Am. Chem. Soc.* **79**, 3045 (1957).

33. J. Hine, "Physical Organic Chemistry." McGraw-Hill, New York, 1962.

34. A. Rembaum and M. Szwarc, *J. Am. Chem. Soc.* **76**, 5975 (1954).

35. M. Szwarc, *J. Polymer. Sci.* **16**, 367 (1955).

36. A. C. Harkness and J. Halpern, *J. Am. Chem. Soc.* **81**, 3526 (1959).

37. J. Halpern and J. G. Smith, *Can. J. Chem.* **34**, 1419 (1956).

38. W. C. E. Higginson and J. W. Marshall, *J. Chem. Soc.* **1957**, 447.

39. J. Halpern, *Can. J. Chem.* **37**, 148 (1959); A. M. Armstrong, J. Halpern, and W. C. E. Higginson, *J. Phys. Chem* **60**, 1661 (1956); A. M. Armstrong and J. Halpern, *Can. J. Chem.* **35**, 1020 (1957); H. N. Halvorson and J. Halpern, *J. Am. Chem. Soc.* **78**, 5562 (1956).

40. F. A. Jones and E. S. Amis, *J. Inorg. Nucl. Chem.* **26**, 1045 (1964).

41. G. C. Pimentel and A. L. McClellan, "The Hydrogen Bond." Freeman, San Francisco, California, 1960.

42. L. Hunter, *J. Chem. Soc.* **1945**, 806.

43. A. Sherman, *J. Phys. Chem.* **41**, 117 (1937).

44. G. V. Tsitsishvili, *J. Phys. Chem. (U.S.S.R.)* **15**, 1082 (1941).

45. A. R. Ubbelohde, *J. chim. phys.* **46**, 429 (1949).

46. K. Wirtz, *Z. Naturforsch.* **2a**, 264 (1947).

47. W. Gordy and S. C. Stanford, *J. Chem. Phys.* **9**, 204 (1941).

48. W. Gordy and S. C. Stanford, *J. Chem. Phys.* **8**, 170 (1940).

49. Y. Sato and S. Nagakura, *Sci. Light* (Tokyo) **4**, 120 (1955).

50. H. Tsubomura, *Bull. Chem. Soc. Japan*, **27**, 445 (1954).

51. H. Hart, F. A. Cassis, and J. J. Bordeaux, *J. Am. Chem. Soc.* **76**, 1639 (1954).

52. W. J. Svirbely and J. F. Roth, *J. Am. Chem. Soc.* **75**, 3106 (1953).

53. M. Szwarc and J. Smid, *J. Chem. Phys.* **27**, 421 (1957).

54. S. Palit, *J. Org. Chem.* **12**, 752 (1957).

55. P. S. Rees, D. P. Tong, and G. T. Young, *J. Chem. Soc.* **1954**, 662.

56. C. G. Swain, *J. Am. Chem. Soc.* **70**, 1119 (1948).

57. C. G. Swain and R. W. Eddy, *J. Am. Chem. Soc.* **70**, 2989 (1948).

58. G. Aksnes, *Acta Chem. Scand.* **14**, 1526 (1960).

59. H. Euler and A. Olander, *Z. physik. Chem.* **131**, 107 (1928).

60. S. Ono, K. Hiromi, and Y. Sano, *Bull. Chem. Soc. Japan* **36**, 431 (1963).

61. K. Hiromi, *Bull. Chem. Soc. Japan* **33**, 1251 (1960).

62. K. Hiromi, *Bull. Chem. Soc. Japan* **33**, 1264 (1960).

63. W. H. R. Shaw and D. G. Walker, *J. Am. Chem. Soc.* **79**, 4329 (1957).

64. J. L. Rabinowitz and E. C. Wagner, *J. Am. Chem. Soc.* **73**, 3030 (1951).

65. N. L. Miller and E. C. Wagner, *J. Am. Chem. Soc.* **76**, 1847 (1954).

66. M. E. Runner, M. L. Kilpatrick, and E. C. Wagner, *J. Am. Chem. Soc.* **69**, 1406 (1947).

67. M. E. Runner and E. C. Wagner, *J. Am. Chem. Soc.* **74**, 2529 (1952).

68. G. Vavon and P. Montheard, *Bull. Soc. chim. France* **7**, 551 (1940); **7**, 560 (1940).

69. M. J. Astle and M. J. McConnell, *J. Am. Chem. Soc.* **65**, 35 (1943).

70. M. J. Astle and M. P. Cropper, *J. Am. Chem. Soc.* **65**, 2395 (1943).

71. M. J. Astle and S. T. Stephenson, *J. Am. Chem. Soc.* **65**, 2399 (1943).

72. M. J. Astle, *Trans. Electrochem. Soc.* **92**, 473 (1947).

73. L. Pekkarinen, *Ann. Acad. Sci. Fennicae*: Ser A. II, No. 62, 77 (1954).

74. D. G. Knorre and N. M. Emanuel, *Voprosy Khim. Kinetiki, Kataliza, i Reaktsoinnoi Sposobnosti, Akad. Nauk SSSR, Otd.* 1955, 106.

75. D. G. Knorre and N. M. Emanuel, *Zh. Fiz. Khim.* **26**, 425 (1952).

76. D. G. Knorre and N. M. Emanuel, *Dokl. Akad. Nauk SSSR* **91**, 1163 (1953).

77. D. G. Knorre and N. M. Emanuel, *Usp. Khim.* **24**, 275 (1955).

78. L. Grossman, D. Stollar, and K. Herrington, *J. chim. phys.* **58**, 1078 (1961).

79. L. G. Augenstine, C. A. Gihron, K. L. Grist, and R. Mason, *Proc. Natl. Acad. Sci. U.S.* **47**, 1733 (1961).

80. E. D. Hughes and C. K. Ingold, *J. Chem. Soc.* 1935, 255.

81. E. D. Hughes, *J. Chem. Soc.* 1935, 255.

82. C. A. L. de Bruyn and A. Steger, *Rec. trav. chim.* **18**, 41, 311 (1899).

83. E. Bergmann, M. Polanyi, and A. Szabo, *Z. physik. Chem.* **B20**, 161 (1933).

84. N. Menschutkin, *Z. physik Chem.* **5**, 589 (1890); **6**, 41 (1890).

85. G. Carrera, *Gazetta*, **24i**, 180 (1894).

86. A. Hemptinne and A. Bekaert, *Z. physik. Chem.* **28**, 225 (1899).

87. H. von Halban, *Z. physik Chem.* **84**, 128 (1913).

88. H. E. Cox, *J. Chem. Soc.* **119**, 142 (1921).

89. J. A. Hawkins, *J. Chem. Soc.* **121**, 1170 (1922).

90. G. E. Muchin, R. B. Ginsberg, and Kh. M. Moissejera, *Ukr. Khim. Zh.* **2**, 136 (1926).

91. H. McCombie, H. A. Scarborough, and F. F. P. Smith, *J. Chem. Soc.* 1927, 802.

92. H. Essex and O. Gelormini, *J. Am. Chem. Soc.* **48**, 882 (1926).

93. H. von Halban, *Z. physik. Chem.* **67**, 29 (1909).

94. E. D. Hughes and C. K. Ingold, *J. Chem. Soc.* 1933, 1571.

95. J. L. Gleave, E. D. Hughes, and C. K. Ingold, *J. Chem. Soc.* 1935, 236.

96. S. C. J. Oliver and A. P. H. Weber, *Rec. trav. chim.* **53**, 869 (1934).

97. S. C. J. Oliver, *Rec trav. chim.* **53**, 891, 1934.

98. E. D. Hughes, *J. Am. Chem. Soc.* **57**, 708 (1935).

99. E. D. Hughes, C. K. Ingold, and C. S. Patel, *J. Chem. Soc.* 1933, 526.

100. S. W. Benson, "The Foundations of Chemical Kinetics," Chapter XVI. McGraw-Hill, New York, 1960.

101. N. J. T. Pickles and C. N. Hinshelwood, *J. Chem. Soc.* 1936, 1353.

102. E. A. Moelwyn-Hughes, and C. N. Hinshelwood, *J. Chem. Soc.* 1932, 231.

103. D. P. N. Satchell, *Quart. Rev.* (*London*) **17**, No. 2, 160 (1963).

104. C. K. Ingold, "Structure and Mechanism in Organic Chemistry," Bell and Sons, London, 1953.

105. M. L. Bender, *Chem. Rev.* **60**, 53 (1960).

106. C. W. Gould, "Mechanism and Structure in Organic Chemistry." Holt, New York, 1959.

107. A. Streitweiser, Jr., *Chem. Rev.* **56**, 571 (1956).

108. D. P. N. Satchell, *J. Chem. Soc.* **1951**, 5404.

109. J. P. M. Satchell, *J. Chem. Soc.* **1963**, 555.

110. L. P. Hammett and A. J. Dyrup, *J. Am. Chem. Soc.* **54**, 2721 (1932).

111. J. G. Prichard and F. A. Long, *J. Am. Chem. Soc.* **78**, 2667 (1956).

112. L. Zucker and L. P. Hammett, *J. Am. Chem. Soc.* **61**, 2791 (1939).

113. G. Schwartzenbach and C. Wittwer, *Helv. Chim. Acta* **30**, 663 (1947).

114. F. A. Long and R. Bakule, *J. Am. Chem. Soc.* **85**, 2313 (1963).

115. R. Bakule and F. A. Long, *J. Am. Chem. Soc.* **85**, 2309 (1963).

116. J. F. Bunnett, *J. Am. Chem. Soc.* **83**, 4956, 4968, 4973, 4978 (1961).

117. K. N. Bascombe and R. P. Bell, *Discussions Faraday Soc.* **24**, 158 (1957).

118. R. P. Bell, "The Proton in Chemistry," p. 81 ff. Cornell Univ. Press, Ithaca, New York 1959.

119. G. Yagil and M. Anbar, *J. Am. Chem. Soc.* **85**, 2376 (1963).

120. M. Anbar, M. Bobtelsky, D. Samuel, B. Silver, and G. Yagil, *J. Am. Chem. Soc.* **85**, 2380 (1963).

121. G. Yagil and M. Anbar, *J. Am. Chem. Soc.* **84**, 1797 (1962).

122. E. J. Hart and J. W. Boag, *J. Am. Chem. Soc.* **84**, 4090 (1962).

123. J. W. Boag and E. J. Hart, *Nature* **197**, 45 (1963).

124. L. M. Dorfman, I. A. Taub, and R. E. Bühler, *J. Chem. Phys.* **36**, 3051 (1962).

125. I. A. Taub and L. M. Dorfman, *J. Am. Chem. Soc.* **84**, 4053 (1962).

126. L. M. Dorfman and I. A. Taub, *J. Am. Chem. Soc.* **85**, 2370 (1963).

127. M. Eigen and L. DeMaeyer, *Z. Elehtrochem.* **59**, 986 (1955).

128. J. H. Baxendale, E. M. Fielden, and J. P. Keene, *Proc. Chem. Soc.* **1963**, 242–243.

129. E. D. Hughes, C. K. Ingold, and C. S. Patel, *J. Chem. Soc.* **1933**, 526.

130. C. K. Ingold, "Structure and Mechanism in Organic Chemistry." Cornell Univ. Press, Ithaca, New York, 1953.

131. D. J. Cram, *J. Am. Chem. Soc.* **71**, 3863 (1949); **74**, 2149 (1952).

132. S. Winstein and R. B. Henderson, chapter on "Ethylene and Triethylene Oxides" in "Heterocyclic Compounds" (R. C. Elderfield, ed.), Vol. I. Wiley, New York, 1950.

133. J. N. Brønsted, M. L. Kilpatrick, and M. Kilpatrick, *J. Am. Chem. Soc.* **51**, 428 (1929).

134. C. L. McCabe and J. C. Warner, *J. Am. Chem. Soc.* **70**, 4031 (1948).

135. S. Weinstein and H. J. Lucas, *J. Am. Chem. Soc.* **61**, 1576 (1939).

136. F. A. Long and J. G. Prichard, *J. Am. Chem. Soc.* **78**, 2663 (1958).

137. R. E. Parker and N. S. Isaacs, *Chem. Revs.*, **59**, 737 (1959).

138. F. C. Whitmore, "Organic Chemistry," p. 339. Van Nostrand, Princeton, New Jersey, 1937.

139. C. D. Hurd, "Pyrolysis of Carbon Compounds," A. C. S. Monograph. Reinhold, New York, 1929.

140. E. de B. Barnett and C. L. Wilson, "Inorganic Chemistry," Longmans, Green, New York, 1958.

141. G. W. A. Fowles and F. H. Pollard, *J. Chem. Soc.* 1953, 4128.

142. H. J. Emeleus, *Chem. & Ind. (London)* 1937, 813.

143. D. C. Bradley, F. M. Abd-el Halim, and W. Wardlaw, *J. Chem. Soc.* 1950, 3450.

144. F. W. Bergstrom, R. E. Wright, C. Chandler, and W. A. Gilkey, *J. Org. Chem.* 1, 170 (1936); R. E. Wright and F. W. Bergstrom, *J. Org. Chem.* 1, 179 (1936).

145. S. Gilman and S. Avakian, *J. Am. Chem. Soc.* 67, 349 (1945).

146. R. E. Benkeser and W. E. Buting, *J. Am. Chem. Soc.* 74, 3011 (1952); R. A. Benkeser and G. Schroll, *ibid.* 75, 3196 (1953); R. A. Benkeser and R. G. Severson, *ibid.* 71, 3838 (1949).

147. J. F. Bunnett and R. E. Zahler, *Chem. Rev.* 49, 373 (1951).

148. J. D. Roberts, H. E. Simmons, Jr., L. A. Carlsmith, and C. W. Vaughan, *J. Am. Chem. Soc.* 75, 3290 (1953).

149. J. D. Roberts, D. A. Semenow, H. E. Simmons, Jr., and L. A. Carlsmith, *J. Am. Chem. Soc.* 78, 601 (1956).

150. G. E. Dunn, P. J. Krueger, and W. Rodewald, *Can. J. Chem.* 39, 180 (1961).

151. V. A. Krishnamurthy and M. R. A. Rao, *J. Indian Inst. Sci.* 40, 145 (1958).

152. J. Szychlinski, *Roczniki Chem.* 33, 443 (1959).

153. O. L. Quimby and T. J. Flautt, *Z. anorg. u. allgem. Chem.* 296, 294 (1958).

154. K. Heyns, R. Brockman, and A. Roggenbuck, *Ann. Chem.* 614, 97 (1958).

155. A. C. Harkness and J. Halpern, *J. Am. Chem. Soc.* 81, 3526 (1959).

156. F. A. Jones and E. S. Amis, *J. Inorg. & Nucl. Chem.* 26, 1045 (1964).

157. H. Euler, *Ann. Chem.* 330, 280 (1903).

158. A. Hantzsch, *Z. physik. Chem.* 61, 257 (1907); 65, 41 (1908); *Ber.* 58 B, 941 (1925).

159. T. Ri and H. Eyring, *J. Chem. Phys.* 8, 433 (1940).

160. J. Chedin, *Compt. rend.* 200, 1397 (1935).

161. C. K. Ingold, D. J. Millen, and H. G. Poole, *J. Chem. Soc.* 1950, 2576.

162. R. J. Gillespie, J. Graham, E. D. Hughes, C. K. Ingold, and E. A. R. Peeling, *J. Chem. Soc.* 1950, 2504.

163. D. R. Goddard, E. D. Hughes, and C. K. Ingold, *J. Chem. Soc.* 1950, 2559.

164. G. Olah, S. Kuhn, and A. Mlinkò, *J. Chem. Soc.* 1956, 4257.

165. D. J. Millen, *J. Chem. Soc.* 1950, 2606.

166. R. J. Gillespie and D. J. Millen, *Quart. Rev. (London)* 2, 227 (1948).

167. F. H. Westheimer and M. S. Kharasch, *J. Am. Chem. Soc.* 68, 1871 (1946).

168. G. M. Bennett, J. C. D. Brand, D. M. James, T. G. Saunders, and G. Williams, *J. Chem. Soc.* 1947, 474.

169. L. Melander, *Nature,* 163, 599 (1949).

170. R. J. Gillespie, E. D. Hughes, C. K. Ingold, D. J. Millen, and R. I. Reed, *Nature*, **163**, 599 (1949).

171. R. J. Gillespie, *J. Chem. Soc.* **1950**, 2542.

172. E. Cherbuliez, *Helv. Chim. Acta* **6**, 281 (1923).

173. J. Masson, *J. Chem. Soc.* **1931**, 3200.

174. J. W. Baker and L. Hey, *J. Chem. Soc.* **1932**, 1227, 2917.

175. E. D. Hughes, C. K. Ingold, and R. I. Reed, *J. Chem. Soc.* **1950**, 2400.

CHAPTER VIII

FURTHER CONSIDERATION OF SOLVENT EFFECTS

Some Specific Ion Effects

Instead of rates of reactions between ionic reactants obeying either Eq. (1.100) or Eq. (1.101) in all cases, many such reactants are found to obey an equation of the form

$$\log k' = \log k_0' + \frac{2Az_A z_B \sqrt{\mu}}{1 + \sqrt{\mu}} + B. \tag{8.1}$$

Hoppé and Prue (*1*) found that the rate data on the alkaline hydrolyses of potassium ethyl malonate and oxalate in the presence of alkali cations followed this equation, where the A for water at 25°C is 0.509.

Indelli and Amis (*2*) studied the salt effects in the reaction between bromate and iodide ions. Equation (8.1) with $2A = 2.036$ and $B = 1.1$ accounted for the rate data when the nitrates of potassium, magnesium, and lanthanum were used. The chlorides gave higher results, presumably due to a reaction between bromate and chloride ions. When the reduction in the hydrogen ion is taken into account in the case of sulfates, these results are very similar to those for the nitrates at equal cation concentration. The reduction of hydrogen ion resulted from the formation of HSO_4^- with a dissociation constant of 1.03×10^{-2} given by Davies *et al.* (*3*). The reaction was first order with respect to BrO_3^- and I^- and second order with respect to H^+, and hence, was very sensitive to change in hydrogen ion concentration. It can be assumed that reaction takes place through some intermediate steps such as

$$BrO_3^- + 2H^+ \rightleftharpoons BrO_2^+ + H_2O \qquad \text{rapid equilibrium}$$
$$BrO_2^+ + I^- \longrightarrow BrO^+ + IO^- \qquad \text{rate determining}$$

followed by the rapid reactions of the BrO^+ and IO^- ions with H^+ and I^-.

The increased H^+ coming from the hydrolysis of uranyl ion, UO_2^{2+}, according to the equation

$$2UO_2^{2+} + 2H_2O \longrightarrow (UO_2OH)_2^{2+} + 2H^+ \tag{8.2}$$

and using the value of the hydrolysis constant of 1.90×10^{-6} mole/liter (4), could not account for the acceleration of the rate of this reaction by the UO_2^{2+}. From the above value of the hydrolysis constant a H^+ concentration of 0.0027 M is calculated for a solution containing 0.0025 M HNO_3 and 0.02 M $UO_2(NO_3)_2$. A value of 8.5 (liter/equivalent)^3 is calculated for the corresponding rate constant for the reaction of 8.33×10^{-4} M $KBrO_3$ and 2.50×10^{-3} M KI neglecting the salt effect. The corresponding experimental value is 70 (liter/equivalent)^3. A specific catalysis by the UO_2^{2+} ion must therefore be present. A complex (5) of the UO_2^{2+} with the iodide, UO_2I^+, which reacted with the BrO_3^- ion in a much more rapid way, but to yield the same products as the original reactants, probably accounts for the observed results. The reaction between UO_2I^+ and BrO_3^- would be more rapid than between I^- and BrO_3^- from simple electrostatic considerations.

In this reaction there was no great difference in the specific effects of the cations even for different valence of cations, when the anion was the nitrate ion. However, the rate constants in the presence of chlorides is always greater than in the presence of the corresponding nitrates. This was attributed to a slower but similar reaction between the chloride and bromate ions as compared to the reaction between the iodide and bromate ions. Skrabal and Webertsch (6) reported that chloride ion accelerates the analogous reaction between bromide and bromate ions.

The Olson-Simonson Criticism of the Brønsted-Bjerrum Equation (7)

Olson and Simonson (8) have discussed in some detail the plot, given in Fig. 3 of Chapter 1, of log k'/k_0' versus $\sqrt{\mu}$ for various ionic charge type reactants. They have provided supplementary data of their own on the two reactions involving the bromopentamine cobaltic ion. This plot was generally accepted as convincing proof of the Brønsted-Bjerrum treatment. They conclude that the effect of the addition of inert salts for reactions between ions of the same charge sign "is caused almost exclusively by the concentration and character of salt ions of charge sign opposite to that of the reactants," and that "the rate is not dependent upon the ionic strength of the solution." "The [salt] effects are quantitatively interpretable in terms of an ion association constant and specific rate constants for the associated and nonassociated reactants. The further introduction of activity coefficients is not necessary."

In the light of the precise knowledge long available on association constants derived from extrakinetic sources (9), it is obvious that, in reactions between ions, allowance must be made for the role played by ion pairs. It has become equally clear that, when quantitative allowance is made for their role, the concept of activity coefficients must indisputably be retained. The role of ion pairs in ionic reactions may now be accepted as a supplement to, rather than as a replacement of, the role of activity coefficients.

Let us, to illustrate the situation, compare the effect of two bivalent ions on the rate of two ionic reactions,

$$
\begin{matrix} CHBr \cdot COO^- \\ | \\ CHBr \cdot COO^- \end{matrix} + OH^- \longrightarrow \begin{matrix} CH(OH) \cdot COO^- \\ | \\ CHBr \cdot COO^- , \end{matrix} + Br^- \qquad (8.3)
$$

$$[CO(NH_3)_5Br]^{2+} + OH^- \longrightarrow [CO(NH_3)_5OH]^{2+} + Br^- , \qquad (8.4)$$

using the simple electrostatic principle that, *ceteris paribus*, the rate of reaction between particles A and B has a normal value when either A or B is uncharged, and that, when the charges on A and B are of the same or opposite sign, the rate is, respectively, less or greater than their normal value. When the divalent cations Ba^{2+} are added to a solution in which reaction (8.3) takes place, the rate of reaction is increased. If Ba^{2+} ions form molecules or ion pairs, such as

$$
\begin{matrix} CHBr \cdot COO \\ \diagdown \\ | \qquad \qquad Ba , \\ \diagup \\ CHBr \cdot COO \end{matrix}
$$

they will, being uncharged, react more rapidly with OH^- than the organic anions do. Following the work of Holmberg (10), this effect was once referred to as "cationic catalysis." When the divalent anions SO_4^{2-} are added to solutions in which reaction (8.4) takes place, the rate of reaction is lowered. The molecule or ion pair here formed is

$$[CO(NH_3)_5Br]SO_4$$

which, being again uncharged, reacts less rapidly with the hydroxyl ion than does the complex cation.

These qualitative explanations, being satisfactory on electrostatic principles, may now be accepted as correct following their quantitative extension, by means of association constants obtained from conductimetric and solubility data unrelated to reaction kinetics. We shall consider only one reaction, which we shall treat in an oversimplified form, omitting a relevant step for which quantitative allowance has also been made. The reaction is that between sodium bromacetate and inorganic thiosulphates in aqueous solution, which, assuming both salts to be completely ionized, would be written stoichiometrically as

$$CH_2Br \cdot COO^- + S_2O_3^{2-} \longrightarrow CH_2S_2O_3 \cdot COO^{2-} + Br^- \qquad (8.5)$$

and kinetically as

$$\frac{d[S_2O_3^{2-}]}{dt} = k_i[CH_2Br \cdot COO^-] [S_2O_3^{2-}] \left(\frac{f_A f_B}{f_X}\right), \qquad (8.6)$$

where the terms in the square brackets are concentrations, and the f terms are activity coefficients, the subscripts of which denote the net ionic charge of the bromacetate ion (f_A), the thiosulphate ion (f_B), and the complex formed by their union (f_X). In the presence of a divalent cation, M^{2+}, we must allow for the equilibrium

$$MS_2O_3 \rightleftarrows M^{2+} + S_2O_3^{2-}, \qquad (8.7)$$

governed by the constant

$$K = \frac{[M^{2+}] [S_2O_3^{2-}]}{MS_2O_3} f_B^2. \qquad (8.8)$$

The net rate of reaction now becomes

$$-\frac{d[S_2O_3^{2-}]}{dt} = k_i[CH_2Br \cdot COO^-] [S_2O_3^{2-}] \left(\frac{f_A f_B}{f_X}\right) + k_m[CH_2Br \cdot COO^-]$$
$$[MS_2O_3] \frac{f_0 f_A}{f_A}. \qquad (8.9)$$

It is usually assumed that the activity coefficient f_0 of an uncharged solute is unity. Hence,

$$-\frac{d[S_2O_3^{2-}]}{dt} = k_i[CH_2Br \cdot COO^-] [S_2O_3^{2-}] \left(\frac{f_A f_B}{f_X}\right)$$
$$\left\{1 + \frac{k_m}{k_i} \frac{[M^{2+}]}{K} \frac{f_B f_X}{f_A}\right\}. \qquad (8.10)$$

By measuring the rate of reaction at various ionic strengths and concentrations of the divalent cation, Davies and his collaborators have determined the two rate constants, and the rate constant for another step which, as stated, has been omitted from the present scheme. In water at 25°C, the bimolecular constants, in liters per mole-second, are as follows:

Reactants	k
$S_2O_3^{2-} + CH_2Br \cdot COO^-$	0.247
$MS_2O_3 + CH_2Br \cdot COO^-$	1.23
$S_2O_3^{2-} + CH_2Br \cdot COOM^+$	7.63

Davies has also shown that, in agreement with Eq. (1.101) the gradients of $d \log_{10} k / d\sqrt{\mu}$ have the values of $+2$, 0, and -2. Other reactions have also been shown to conform with the Brønsted-Bjerrum theory, after allowance has been made for the role of ion pairs.

The joint roles of the activity coefficient and ion pairing may be formulated in a slightly different way as follows. Let C_A be the total concentration of bromacetate ions and C_B the total concentration of MS_2O_3, dissociated and associated. Then, in terms of the degree of dissociation α of MS_2O_3. Eq. (8.8) may be written as

$$K = \frac{C_B \alpha^2}{1 - \alpha} f_B^2,$$ (8.11)

and

$$[CH_2Br \cdot COO^-] = C_A$$
$$[MS_2O_3] = C_B(1 - \alpha)$$
$$[S_2O_3^{2-}] = M^{2+} = C_B \alpha.$$

The rate equation (8.10) can now be expressed as follows:

$$-\frac{dC_A}{dt} = C_A C_B \left[k_m + \left(k_i \frac{f_A f_B}{f_X} - k_m \right) \alpha \right],$$ (8.12)

which is an amended form of the equation of Robertson and Acree (11). When α is zero, we have the rate equation between the bromacetate ion and the ion pair, and when α is unity, we recover Eq. (1.101). Ion pairing will be discussed more fully later.

Relative Solvation and Salt Effects

Indelli and Amis (12) studied the temperature coefficients of the iodide ion–persulfate ion reaction using 5° increments over the 25°–50°C temperature range. The object was to try to detect nonelectrostatic effects from measured energies of activation of the reaction when various salts over ranges of concentrations were used to vary the ionic strength. From Eq. (7.8) for the energy of activation, ΔE_{In}, arising from the ionic strength effects, namely,

$$\Delta E_{In} = 2.303 \, RTZ_A Z_B \sqrt{\mu} \qquad (8.13)$$

it would be expected, for reactants of like charge sign, that ΔE_{In} and therefore the over-all activation energy would increase with increasing ionic strength. In Fig. 1 it is apparent that ΔE_{app} increases when a small amount

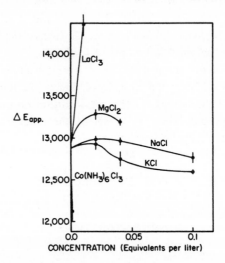

Fig. 1. The experimental activation energies (and their errors indicated by vertical lines) versus concentration.

of salt is added to the reaction system, but that at higher concentrations of salt a definite decrease is observed in the cases of magnesium chloride and sodium chloride, and more so for potassium chloride. The vertical lines indicate the errors in these energies. A decrease in ΔE_{app} by potassium salt is observed from other data in the literature (13, 14). Indelli and Amis explained the decrease in energy of activation by the polarization which the cation exerts on the electrons of the anion when an ion pair is formed or when a cation is near enough to the activated complex. For both polarization and electrostatic effects to be present, it is only necessary that an ion of opposite sign be within a certain distance of a reacting ion or an activated complex. The magnitude of this critical distance is somewhat arbitrary, as has been shown by Brown and Prue (15) and by Guggenheim (16) for thermodynamic properties of solutions of by-bivalent electrolytes.

In the case of the iodide ion–persulfate ion reaction there must be the transfer of an electron from the iodide to the persulfate ion. An ion pair such as

$$+ \delta \; \overset{\displaystyle \overset{O^{+\delta}}{\|}}{\underset{\underset{O^-}{\displaystyle |{-2\delta}}}{O=S}}-O-O-\overset{\displaystyle \overset{O^{+\delta}}{\|}}{\underset{\underset{{}^-O}{\displaystyle |{-2\delta}}}{S=O}} + \delta$$

$$M^+$$

formed between the metal and persulfate ions should be a better electron acceptor than the free persulfate ion. Due to the particular structure of the $S_2O_8^{2-}$ ion, cations of small radius can, for equal charge on the cations, produce a greater negative charge toward the center of the persulfate ion than cations of larger radius. Unambiguous determination of ionic radius in solution is difficult (17). This is where the solvent enters the picture. In the same solvent, at the same temperature, and in the same concentration range, different ions are solvated to a different extent. Solvation decreases with increasing concentration and with increasing temperature. In many solvation determinations only relative solvation of the ions of an electrolyte, and not the absolute values of the solvations of the individual ions, are found. Wear and Amis (18) found from transference number and solvation number considerations, using the Hittorf procedure involving an inert reference substance, that sodium ion has a greater solvation than potassium ion in aqueous solutions at like concentration and temperature. This means that in aqueous solutions sodium ion would be more bulky than potassium ion, although the unsolvated sodium ion has a radius of about 0.9 Å and the unsolvated potassium ion a radius of 1.3 Å. Therefore, in, our model, the hydrated sodium ion would cause less polarization of the persulfate ion than would potassium ion. Indelli and Amis assumed that ion conductivity might be used as a reasonable indication of the ionic radius in a given solvent. These authors compared the salt effects on the rate of reaction between persulfate and iodide ions with the ionic conductances of cations at infinite dilution using several salts. The results are presented in Table I for different salts at a concentration of 0.04 equivalents/liter in the kinetic runs at 25°C. For cations of like valence there is a marked parallelism between the decrease in the rate constant and the decrease in the limiting conductance of the cations at infinite dilution. The only exception being the inversion between Li^+ and $N(CH_3)_4^+$.

TABLE I

COMPARISON OF THE SALT EFFECT ON THE RATE OF REACTION BETWEEN PERSULFATE AND
IODIDE AND THE LIMITING EQUIVALENT CONDUCTANCE AT INFINITE DILUTION OF THE CATION

Salt	10^3k	λ^0
KCl	2.70	73.5
NaCl	2.12	50.1
LiCl	1.91	38.6
$N(CH_3)_4Br$	1.89	44.9
$N(C_2H_5)_4Br$	1.48	32.6
$MgCl_2$	2.89	53.0
$Co(NH_3)_6Cl_3$	38.7	101.9
$LaCl_3$	~ 6.5 [a]	69.7

[a] Extrapolated.

Other Approaches to Electrostatic Effects among Ions in Solution

a. *Conductance and Electrostatics*

Figure 1 in Chapter VII was used to illustrate the point that ions tend to be selectively solvated by the more polar component of a mixed solvent even when this component is present in statistically unfavorable proportions. If it is assumed that the actual radius r_i of a solvated ion is the radius r_i^0 of the nonsolvated ion plus a term proportional to the electrostatic force F between the ion and the dipolar molecules of solvent, then r_i can be represented by the equation (*19*)

$$r_i = r_i^0 + KF, \tag{8.14}$$

where K is a constant. Now if r_i^0 is small compared to the distance at which electrostatic forces can be exerted, that is, if $r_i^0 < KF$, then

$$r_i \cong KF. \tag{8.15}$$

Neglecting higher order effects, the electrostatic force between an ion and a molecule is given by the equation

$$F = \frac{2z_i\varepsilon\mu \cos\theta}{Dr_s^3} \tag{8.16}$$

where z_i is the valence of the ion, ε the charge on the electron, μ the dipole moment of the molecule, r_s the distance between the ion and the dipole at which the force between them is F, and θ the angle which a line drawn from the ion to one of the centers of charge of the dipole makes with the line determined by the centers of charge of the dipole. For head-on alignment of the ion and the dipole

$$F = \frac{2z_i \varepsilon \mu}{Dr_s^3} \tag{8.17}$$

and

$$r_i = KF = \frac{2Kz_i \varepsilon \mu}{Dr_s^3} = \frac{2K\varepsilon \mu}{Dr_s^3} \tag{8.18}$$

for univalent ions.

The equation [Eq. (7.20)] for Walden's rule for uni-univalent electrolytes where the radii r_0^+ and r_0^- of the positive and negative ions are assumed to be equal ($r_0^+ = r_0^- = r_0$) can be written

$$\Lambda_0 = \frac{\varepsilon \mathscr{F}}{900\pi\eta r_0} \tag{8.19}$$

where Λ_0 is the limiting equivalent conductance of the electrolyte, $\mathscr{F}$ the faraday, η the viscosity of the medium, and the other terms are as defined previously.

Substituting the values of r_i from Eq. (8.18) for r_0 in Eq. (8.19), and solving for r_s yields

$$r_s = \left(\frac{1800\pi\eta K\mu\Lambda_0}{D\mathscr{F}} \right)^{1/3}. \tag{8.20}$$

The value of r_s has for like-charge sign of ions a much greater probability of remaining constant than does the r from Walden's equation, since in Eq. (8.20) when the dielectric constant D which occurs in the denominator increases, the dipole moment μ which occurs in the numerator likewise decreases and *vice versa*, and since r_s involves the distance of electrostatic interaction of ions with dipoles which should be independent of the actual size of like-charge ions.

For potassium chloride in water; in 20.2, 40.2, 60.7, and 80.7% methanol in water; and in 100% methanol, $r_s = 12.61 \pm 0.35$ at 25°, 35°, and 45°C. When the data for tetramethylammonium chloride and tetraethylammonium picrate in 100% methanol and ethanol at 25°C were included, $r_s = 12.72 \pm 0.36$. Sodium chloride in water and in methanol gave somewhat abnormal values for r_s, but in ethanol this salt gave a normal value of r_s (12.38).

A mathematical expression corresponding to Walden's rule would, from Eq. (8.20), be

$$(\Lambda_0 \eta \mu / D)^{1/3} = \text{constant.} \qquad (8.21)$$

Further correlation of the theory of conductance and electrostatics is of course evidenced by the fundamental conductance theory of Onsager (20), whose time of relaxation effect is in essence an electrostatic one. The electrophoretic effect arises from a sort of twofold electrostatic effect. There is first the formation of the ion atmosphere around a central ion due to electrostatic forces; then there is the movement of the solvent enclosed by the atmosphere along with, or accompanying, the atmopshere which is moving in a direction opposite to the central ion, so that the central ion finds itself swimming upstream against its own solvent. The solvent tends to accompany the atmosphere at least partially because of the strong ion–dipolar molecule attraction between the ions of the atmosphere and the dipolar molecules of solvent.

The numerous applications of the Onsager and related theories to electrolyte and ionic conductances is thus an application of electrostatic theory to ions in solution.

The temperature coefficient for the conductance process developed by Pedersen and Amis (21) is based on the Onsager equation and as such involves electrostatic considerations as well as solvent influences. The solvent influence arises of course from the dependence of the dielectric constant and of the viscosity of the solvent on temperature.

Conductance can be used to determine the fraction of solute which is unassociated (22) and hence can be used directly in the interpretation of equilibrium and kinetic data involving ions. See section on Olsen-Simonson criticism of Brønsted-Bjerrum equation discussed previously.

b. *Ionic Association Equilibrium*

Moelwyn-Hughes (23) points out that the picture of an ion surrounded by a diffuse and continuous cloud of electricity of opposite sign to the central ion has mathematical advantages but physicochemical deficiencies. For a close distance of approach of two ions the "cloud" surrounding either ion is simply the charge on the other ion, and the region occupied by the ion pair is on the average neutral.

Bjerrum (24) recognized that the theory of complete ionization had to be modified to account for ion pairs. His theory did not account for the variation of the association constant K_A with the dielectric constant of the solvent, and this, together with the artificiality of one of his assumptions, led to criticism of his theory and to its being eventually superceded by a theory (25, 26) more closely in agreement with data and mathematically less complex.

Extrapolated to infinite dilution, the equilibrium constant K_A for the association of univalent ions,

$$A^+ + B^- \; \rightleftharpoons \; A^+B^-, \tag{8.22}$$

can be written in terms of concentration

$$K_A = \frac{C_{A^+B^+}}{C_{A^+}C_{B^+}}. \tag{8.23}$$

Independently Denison and Ramsey and Fuoss and Kraus have derived the equation

$$K_A = K_A{}^0 \exp \left\{ \frac{\varepsilon^2}{DakT} \right\}, \tag{8.24}$$

where ε is the electronic charge, a is the distance of separation of the charges in the ion pair, D is the dielectric constant of the medium, k is the Boltzmann gas constant, T the absolute temperature, and $K_A{}^0$ is the association constant for a pair of uncharged particles. It is readily seen that the term $(\varepsilon^2/DakT)$ is the ratio of the electrostatic energy to the thermal energy of the ion pair.

Equation (8.24) predicts that log K_A plotted versus $1/D$ should yield a straight line with a slope $\varepsilon^2/2.303akT$, from which values of a having molecular dimensions should be found. Moelwyn-Hughes (23) has proven the validity of the predictions of the equation both as to the predicted plot and the values of a for substances ranging from silver nitrate in water, methanol, and ethanol to tetraisoamylammonium nitrate in dioxane-water mixtures of different compositions, with dielectric constants ranging form 2 to 80.

Denison and Ramsey (25) found that points fell off the straight line plot of log K_A versus $1/D$ for a given solute in different pure solvents when solvents, such as ethylene chloride, that exist in two forms of different polarity are used. He suggested that this divergence from theory arose from the enhanced effective dielectric constants as compared to the macroscopic dielectric constants of such solvents. Inami et al. (27) calculate the effective dielectric constant D_{eff} of etnylene chloride solvent in the presence of picrate solute by assuming the best fit of the plot of the log K_A versus $100/D$ of the $(n—C_4H_9)_4NPi$ in various solvents (Fig. 2) was given by the points

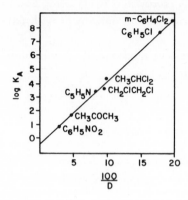

FIG. 2. Association constants of tertiary butylammonium picrate at 25°C in the various solvents indicated in the figure.

for the six solvents other than ethylene chloride. The slope and the intercept for the line were determined and inserted in the logarithmic form of Eq. (8.24) as follows:

$$\log K_A = \text{slope} \left(\frac{100}{D_{eff}}\right) + \text{Intercept} \tag{8.25}$$

or

$$3.64 = 0.464 \left(\frac{100}{D_{eff}}\right) - 0.470 \tag{8.26}$$

and D_{eff} was found to be 11.2 compared to the macroscopic dielectric constant of 10.23.

Equation (8.24) applies only to salts in which the ions are electrically symmetrical and in which the primary stabilizing force of the ion pair is considered to be the charge-charge interaction. If either of the ions in the salt is electrically asymmetrical an additional stabilizing force arises from charge-dipole interaction. Picrates of electrically symmetrical cations, frequently used in conductance work, require this additional stabilizing force in ion-pair formation. Accascina et al. (28, 29) have replaced the logarithmic form of Eq. (8.24) by the equation

$$\log K_A = \log K_A^o + \frac{\varepsilon^2}{2.303aDkT} + \frac{\mu\varepsilon}{2.303DkTd^2} \tag{8.27}$$

where μ is the dipole moment of the anion and d is the distance from the center of the cation to the center of the dipole of the anion when cation and anion exist as an associated ion pair.

An approximate expression for $K_A{}^0$ which allows an independent calculation of a can be made as follows. Consider a solution containing, per cm³, n_A spheres of radius r_A and n_B spheres of radius r_B. The probability that a sphere of radius r_A shall contact a sphere of radius r_B is the ratio of the sum of the volume of the shells $\frac{4}{3}\pi[(r_A + r_B)^3 - r_B{}^3]N_B$ to the total volume V of the solution. Here N_B is the total number of B particles in the solution. But $N_B/V = n_B$ and $r_A + r_B = a$. Per unit volume the number of contacting pairs is n_A times as great as given above, and is therefore $\frac{4}{3}\pi[a^3 - r_B{}^3]n_A n_B$. Calculating the number of contacting pairs in the other order would yield $\frac{4}{3}\pi[a^3 - r_A{}^3]n_A n_B$. The mean number of contacting pairs is one-half the sum of the above two expressions, and is

$$n_{AB} = \tfrac{2}{3}\pi a^3 \left[1 + \frac{3r_A r_B}{a^2} \right] n_A n_B. \tag{8.28}$$

Therefore the association $K_A{}^0$ is given by

$$K_A{}^0 = \frac{n_{AB}}{n_A n_B} = \frac{2}{3}\pi a^3 \left[1 + \frac{3r_A r_B}{a^2} \right] \quad \text{cm}^3/\text{molecule} \tag{8.29}$$

$$= \frac{2}{3}\frac{1000}{N_0}\pi a^3 \left[1 + \frac{3r_A r_B}{a^2} \right] \quad \text{liters/mole} \tag{8.30}$$

and for $r_A = r_B$

$$K_A{}^0 = \frac{7}{6}\frac{N_0}{1000}\pi a^3 \tag{8.31}$$

Moelwyn-Hughes calculated the ion-pair separations using both the slopes and separately the intercepts of Eq. (8.24). The results of these calculations are given in Table II. The agreement between the two calculations is as reasonable as could be expected. The use of Eq. (8.29) would perhaps yield greater internal consistency, but would require the knowledge of at least one ionic radius, and there is now some doubt about the authenticity of radii obtained from Stokes law (19, 28).

Induction energies and still further terms, in addition to the ion-dipole term, will no doubt have to be added to Eq. (8.24) in order to obtain a

refined theory capable of accounting for all the facts. Moelwyn-Hughes
(23) introduced the intrinsic repulsion in the simple statistical manner given
below.

TABLE II

CALCULATION OF THE DISTANCE a FROM THE SLOPE AND FROM THE INTERCEPT OF A PLOT
OF $\log K_A$ VERSUS $1/D$ AT 25°C

Electrolyte	$\log K_A{}^\circ$	Slope	a (Å)		
			From slope	From intercept	Mean
AgNo$_3$	$\bar{2}.802$	88.5	2.75	3.06	2.91±0.16
(C$_4$H$_9$)$_4$NBr	$\bar{1}.041$	56.4	4.31	3.68	4.00±0.31
(C$_4$H$_9$)$_4$NI	$\bar{1}.892$	43.9	5.55	7.06	6.31±0.76
(C$_4$H$_9$)$_4$NClO$_4$	$\bar{1}.755$	50.2	4.85	6.37	5.61±0.76
(C$_4$H$_9$)$_4$NPic	$\bar{1}.543$	46.0	5.29	5.41	5.35±0.06
(iso C$_5$H$_{11}$)$_4$NNO$_3$	$\bar{1}.549$	41.7	5.83	5.43	5.63±0.20

It is assumed that ions repel each other with a force varying as the inverse $(s + 1)$th power of the distance a separating them. Therefore, the
energy of interaction φ of a pair of univalent ions of unlike charge signs in
a medium of dielectric constant D is

$$\varphi = \frac{A}{a^s} - \frac{\varepsilon^2}{Da} \tag{8.32}$$

which has a minimum value φ_e given by

$$\varphi_e = -\frac{\varepsilon^2}{Da_e}\left(1 - \frac{1}{s}\right) \tag{8.33}$$

when $a = a_e$. A harmonic vibration frequency ν_e accompanies displacements
from equilibrium separation. Thus

$$\nu_e = \frac{\varepsilon}{2\pi a_e}\left[\frac{s-1}{m^* Da_e}\right]^{1/2}, \tag{8.34}$$

where m^* is the reduced mass of the ion pair. The translational components
of the total partition functions of the ions is all that need be considered,
if it is assumed that the internal motions of the ions are the same in the

free and associated states. The partition functions f_+ and f_- of the plus and minus ions, respectively, are given by the equations

$$f_+ = \frac{(2\pi m_+ kT)^{3/2} Ve}{h^3 N_+} \exp\left(-u_+/kT\right) \tag{8.35}$$

$$f_- = \frac{(2\pi m_- kT)^{3/2} Ve}{h^3 N_-} \exp\left(u_-/kT\right) \tag{8.36}$$

where m represents mass, N the numbers in total volume V, and u the average potential energies of the ions of the charges indicated by the subscripts. The masses may be those of either the free or solvated ions.

The motions of the ion pair can be resolved into translation throughout the volume V, rotation about its center of gravity, and vibration along the line of centers. Then, when the vibration frequency v_e is such that $hv_e < kT$ so that the classical partition function can be used for this motion, the total partition function for the ion pair can be written

$$f_\pm = \frac{[2\pi(m_+ + m_-)kT]^{3/2} Ve}{h^3 N_\pm} \frac{8\pi I kT}{h^2} \frac{kT}{hv_e} \exp\left(-u_\pm/kT\right), \tag{8.37}$$

where I is the moment of inertia of the ion pair and equals $m^* a_e^2$, and the rest of the terms have already been defined or have their usual significance.

For equilibrium the sum of the chemical potentials of the ions must equal the chemical potential of the ion pair, that is:

$$\mu_+ + \mu_- = \mu_\pm, \tag{8.38}$$

where the chemical potential of the species i is given by the equation

$$\mu_i = -kT\left[\ln f_i + \frac{d\ln f_i}{d\ln N_i}\right]_{T,V}. \tag{8.39}$$

Thus it follows that the association constant K_A is

$$K_A = \frac{n_\pm}{n_+ n_-} = 4\pi a_e^3 \left[\frac{2\pi D a_e kT}{(s-1)\varepsilon^2}\right]^{1/2} \exp\left[-\frac{u_\pm - u_+ - u_-}{kT}\right]. \tag{8.40}$$

But from Eq. (8.33), the algebraic sum of the potential energies in the exponential term in Eq. (8.40) is given by $(\varepsilon^2/D a_e)(s-1)/s$, and converting the units of K_A from cm³/molecule to liters/mole, Eq. (8.40) becomes

$$K_A = \frac{4N_0 \pi a_e^3}{1000} \left[\frac{2\pi D a_e kT}{(s-1)\varepsilon^2}\right]^{1/2} \exp\left[\frac{\varepsilon^2}{D a_e kT}\left(\frac{s-1}{s}\right)\right]. \tag{8.41}$$

In applying the above equation to data Moelwyn-Hughes wrote it in the logarithmic form as follows:

$$\log \frac{K_A}{D^{1/2}} = \overline{4}.905 + \log\left[\frac{\mathring{a}_e^{7/2}}{(s-1)^{1/2}}\right] + 243\left[\frac{s-1}{s\mathring{a}_e}\right]\frac{1}{D}$$

$$= \overline{4}.905 + \log \beta + 243\gamma\,\frac{1}{D}, \qquad (8.42)$$

where β and γ are specific constants which may be obtained from the intercept and slope, respectively, of the isothermal plot of $\log (K_A/D^{1/2})$ versus $1/D$. This treatment also permits the calculation of s and a_e. The value of s is calculated from the equation

$$s = \beta^2\gamma^7\left(1 - \frac{1}{s}\right)^6. \qquad (8.43)$$

Uncertainties in β and γ cause fluctuations in values of s. The value of s ranges from 5 for silver nitrate to 50 for tetraisoamylammonium nitrate.

These values are reasonable in themselves and when compared with the repulsive integers found for crystals and gases.

In this treatment the concept of ionic radius has not been emphasized. The quantity a_e is the average separation of charges in the ion pair when it is in the state of lowest potential energy. The root-mean-square displacement about the average separation is given by the equation

$$\frac{\bar{x}}{a_e} = \left[\frac{Da_e kT}{(s-1)\varepsilon^2}\right]^{1/2}. \qquad (8.44)$$

Plots of $\log K_A$, rather than $\log (K_A/D^{1/2})$, versus $1/D$ yield not a constant gradient, but one that increases with increasing $1/D$. This is true since

$$\frac{d\log K_A}{d(1/D)} = \frac{1}{2.303}\left[\frac{\varepsilon^2}{a_e kT}\left(\frac{s-1}{s}\right) - \frac{D}{2}\right]. \qquad (8.45)$$

Bodenseh and Ramsey (30) have noted deviations from linearity in this direction.

The Fuoss-Shedlavsky theory (31) has frequently been used to determine the ion-pair association constant from conductance data. This theory was found by Roach and Amis (32) to be very discriminating with respect to the charge type of electrolyte to which the data apertains. These authors were able to determine the primary ionic species of uranium from uranium

(IV) chloride in pure ethanol and in various weight per cents of ethanol in water. The ion association constants and their temperature coefficients, for the different hydrolyzed uranium ion species, were obtained.

The solvent effect on ion-pair equilibrium is of interest kinetically since application of the electrostatic theory of the salt effect to kinetic data requires correction for the extent of ion pairing. Knowledge of the equilibrium constant for ion-pair formation in different solvents make possible the correction of kinetic data for ion association.

c. *Electromotive Force and Electrostatics*

Harned and co-workers (*33, 34, 35*) have written the electromotive force E of a cell as the sum of the standard potential E^o and an activity quotient term, involving the ratio of the products of the activities of products to product of the activities of reactants each activity raised to an exponent represented by the coefficient of the particular substance in the stochiometric equation for the cell reaction. The activities were then substituted for in terms of activity coefficients and actual concentrations of the substances involved in the cell reaction. The activity coefficients were in turn written in terms of the Debye-Hückel theory.

A recent application (*36*) of this approach was the study of the cell

$$Cd\text{–}Cd_x Hg\ (11\%)\ |\ CdCl_2\ (m),\ (XMeOH),\ (YH_2O)\ |\ AgCl\text{–}Ag,$$

where X represents the weight per cent water and Y the weight per cent methanol in the solvent of the cell. The equation for the cell potential was written

$$E = E^o - k \log 4m^3 \gamma_\pm^3, \tag{8.46}$$

where k is $2.303RT/2F$, F is the Faraday, m is the molality of the cadmium chloride, and $\gamma_\pm$ is the mean activity coefficient of the salt.

From the Debye-Hückel theory

$$\log \gamma_\pm = - S_f \sqrt{m} + Bm. \tag{8.47}$$

This value of $\log \gamma_\pm$ substituted into Eq. (8.46) gave

$$E + k \log 4m^3 - 3kS_f \sqrt{m} = E^o - 3kBm = E_H. \tag{8.48}$$

In the above equations S_f is the limiting slope of the Debye-Hückel theory and E_H is a function that should be linear with m in dilute solutions when the Debye-Hückel theory holds, and provided the electrolyte is completely dissociated. For cadmium chloride, dissociation is incomplete and a plot of E_H versus $E + k \log 4m^3 - 3kS_f\sqrt{m}$ is not linear as required by Eq. (8.48).

If the ionization of $CdCl_2$ takes place in the two steps

$$CdCl_2 \longrightarrow CdCl^+ + Cl^- \tag{8.49}$$

$$CdCl^+ \rightleftharpoons Cd^{2+} + Cl^-, \tag{8.50}$$

if the first ionization step is complete; if one assumes $\gamma' = \gamma'_{Cl^-}$, where γ' is the activity coefficient; if the concentrations of $CdCl^+$, Cd^{2+}, and Cl^- are, respectively, $m(1 - \alpha)$, $m\alpha$, and $m(1 + \alpha)$; and if $\gamma'_{Cd^{2+}} = \gamma'' = (\gamma')^4$, then the equilibrium constant for the second ionization step can be written

$$K = \frac{(\alpha m\gamma'')\,(1 + \alpha)\gamma' m}{m(1 - \alpha)\gamma'} = \frac{m\gamma''\alpha(1 + \alpha)}{1 - \alpha}, \tag{8.51}$$

where α is the degree of dissociation of the $CdCl^-$ ion.

Equation (8.51) can be solved for α and yields

$$\alpha = \frac{1}{2}\left[-\left(1 + \frac{K}{m\gamma''}\right) + \sqrt{\left(1 + \frac{K}{m\gamma''}\right)^2 + \frac{4K}{m\gamma''}} \right]. \tag{8.52}$$

From the Debye-Hückel theory

$$\log \gamma'' = -\frac{4S_f\sqrt{\mu'}}{1 + A\sqrt{\mu'}} \tag{8.53}$$

where the constant A is given by

$$A = \frac{\sqrt{2}\,(35.57)\,\mathring{a} \times 10^8}{(DT)^{3/2}} \tag{8.54}$$

where $\mathring{a}$ is the distance between the centers of the ions when in the position of closest approach. In these equations the primed quantities are real, the unprimed are stoichiometric. Thus

$$\mu' = \tfrac{1}{2}\sum m_i' z_i^2, \tag{8.55}$$

where m_i' represents the actual ionic concentration. The actual ionic strength was calculated from the equation

$$\mu' = (2\alpha + 1)m. \tag{8.56}$$

In making calculations, tentative values of K, between 0.07 and 0.001, and a value of α, say 0.80, were assumed. The value of μ' was calculated using the assumed value of α and employing Eq. (8.56). The calculated value of μ' was inserted in Eq. (8.53) and a value of γ'' obtained. This value of γ'' and an assumed value of K made possible the calculation of a better value of α employing Eq. (8.52). This α inserted in Eq. (8.56) yielded a better value of μ' which using Eq. (8.53) resulted in a better value of γ'' to be used in Eq. (8.52) to give a better value of α. This reiteration was continued, for each concentration in each solvent, at each temperature until the same value of α, to within a few ten thousandths unit, was obtained twice.

Correcting the concentrations in Eq. (8.46) for the degree of ionization α there results

$$E = E^{\circ} - k \log \alpha (1 + \alpha)^2 m^3 \gamma_{\pm}'^3, \qquad (8.57)$$

where $\gamma_{\pm}'$ the real activity coefficient has replaced the stoichiometric activity coefficient $\gamma_{\pm}$. Subtracting E° from each side of Eq. (8.48) and substituting m in terms of μ gives

$$E_{\mathrm{H}} - E^{\circ} = E - E^{\circ} + k \log 4m^3 - 3k\, S_f' \sqrt{2\mu}. \qquad (8.58)$$

Substituting the value for $E - E^{\circ}$ from Eq. (8.57) into Eq. (8.58) there results

$$E_{\mathrm{H}} - E^{\circ} = k \log 4 - k \log \alpha (1 + \alpha)^2 - 3k \log \gamma_{\pm}' - 3kS_f' \sqrt{2\mu}. \quad (8.59)^{\dagger}$$

Computed values of $E_{\mathrm{H}} - E^{\circ}$ using the activity coefficients of cadmium chloride were plotted against $\sqrt{m}$ for various values of K. Values of E_{H} calculated from Eq. (8.48) were also plotted versus $\sqrt{m}$. The value of K which caused the curve of Eq. (8.59) to be most nearly symmetrical with the curve Eq. (8.48) was taken as the dissociation constant. The ordinates of the curve from Eq. (8.59) subtraced from the ordinates of the curve from Eq. (8.48), each difference being taken at the same value of $\sqrt{m}$, gave values of E°, which values should be independent of the molality. The calculations were made using an IBM 7074 computer. The program was stated so that different values of K and a selected value of $\mathring{a}$ was tried until a correct combination gave an average E° value with an average deviation of 1.5 mV. This amounts to the difference between the curves from Eq. (8.48) and Eq. (8.59) being a straight line with a slope of zero within 1.5 mV for a given solvent at a given temperature. From these values of

† Where $S_f' = \sqrt{3}\, S_f$ when μ is calculated using concentrations in molalities.

E^o at different temperatures, the thermodynamic functions for the cell reaction were calculated. The values of K, $\mathring{a}$, and the thermodynamic functions are listed in Table III. The cell reaction to which the thermodynamic functions apply is

$$Cd(s) + 2Ag^+(s) \longrightarrow Cd^{2+}(m) + 2Ag(s). \tag{8.60}$$

The value of $\mathring{a}$ could have been varied somewhat and the values of E^o in the different solvents would have been linear with the reciprocals of the dielectric constants of the solvent. However holding the value constant resulted in the slight curverature of E^o versus $1/D$ plots shown in Fig. 3.

Amis (37) derived an equation relating E^o to the dielectric constant of the medium. He assumed that the influence of the dielectric constant of

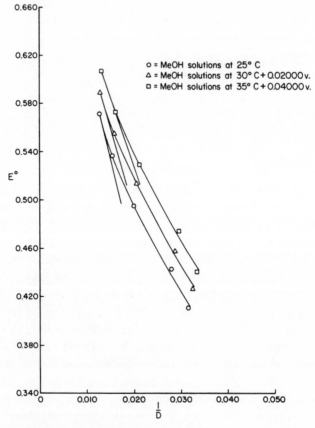

FIG. 3. Standard potentials of the Cd–Cd$_x$Hg (11%) | CdCl$_2$(m) | AgCl–Ag cell plotted against $1/D$ using methanol solutions.

TABLE III

TABULATED VALUES OF DERIVED QUANTITIES IN DIFFERENT SOLVENTS AND AT DIFFERENT TEMPERATURES

Solvent	T (°K)	D	K	$\mathring{a}$ (Å)	$E°$ international volts	Deviation in $E°$	$-\Delta F°$ calories per mole ±200 cal	$-\Delta H°$ calories per mole ±200 cal	$-\Delta S°$ e.u. ±0.6 e.u.
H₂O	298.16	78.5	0.07	5	0.57162	0.00130	26,400	33,800	25
H₂O	303.16	76.8	0.07	5	0.56890		26,200	32,400	20
H₂O / 30% MeOH	308.16 / 298.16	75.0 / 64.6	0.07 / 0.05	5 / 5	0.56670 / 0.53611	0.00137	26,100 / 24,700	27,300	8.7
30% MeOH	303.16	62.7	0.05	5	0.53502	0.00097	24,700	31,400	22
30% MeOH / 60% MeOH	308.16 / 298.16	61.3 / 50.1	0.05 / 0.02	5 / 5	0.53262 / 0.49510	0.00074 / 0.00152	24,600 / 22,800	27,000	22
60% MeOH	303.16	48.6	0.02	5	0.49362	0.00157	22,800	36,600	46
60% MeOH / 90% MeOH	308.16 / 298.16	47.1 / 35.8	0.02 / 0.01	5 / 5	0.48864 / 0.44253	0.00114 / 0.00184	22,500 / 20,400	26,900	22
90% MeOH	303.16	34.6	0.01	5	0.44019 [a]	0.00192	20,300	37,700	57
90% MeOH / 100% MeOH	308.16 / 298.16	33.7 / 31.6	0.01 / 0.001	5 / 5	0.43396 / 0.41033	0.00162 / 0.00183	20,000 / 18,900	30,900	40
100% MeOH	303.16	30.7	0.001	5	0.40598	0.00176	18,700	34,100	51
100% MeOH	308.16	29.9	0.001	5	0.40050	0.00183	18,500		

[a] Smoothed values.

the medium on the standard potential of a cell, arose from the difference in work at different dielectric constants necessary to bring the ionic species within a critical distance r of each other.

The work to bring a particle of charge z_A from infinity to within a distance r of an ion of charge z_B in a medium of dielectric constant D is given by the equation

$$w = -\frac{z_A z_B \varepsilon^2}{KDr} \tag{8.61}$$

and the work per mole of ions is

$$w = -\frac{N z_A z_B \varepsilon^2}{KDr} \tag{8.62}$$

where $K = 10^7$ ergs/joule.

In the cell reaction represented in Eq. (8.60), the total maximum work or free energy change necessary for the process at unit activities of reactants and products is $n \mathscr{F} E^o$, where n is the number of electrons involved in the electrode process, e.g., $n = 2$ in the reaction given above, $\mathscr{F}$ is the faraday (96,500 coulombs) and E^o is the standard potential in volts, i.e., the potential when reactants and products are at unit activity. This potential can be split into two parts: one part, $E_1{}^o$, arising from the reaction freed from charge effects, and the other part, $E_2{}^o$, related to the electrostatic contribution to the work. Hence per mole the free energy change for the cell process is

$$n \mathscr{F} E^o = n \mathscr{F} E_1{}^o + n \mathscr{F} E_2{}^o. \tag{8.63}$$

But

$$n \mathscr{F} E_2{}^o = -\frac{N z_A z_B \varepsilon^2}{KDr}. \tag{8.64}$$

Therefore

$$n \mathscr{F} E^o = n \mathscr{F} E_1{}^o - \frac{N z_A z_B \varepsilon^2}{KDr} \tag{8.65}$$

and [†]

$$E^o = E_1{}^o - \frac{N z_A z_B \varepsilon^2}{n \mathscr{F} Dkr}. \tag{8.66}$$

[†] It has been suggested to the author that the dielectric constant effect on the standard potential of the cell is related to the difference in free energy of solvation of the ions in the two media. Use the crude Born solvation free energy to calculate the effect of dielectric constant on E^o, that is, let nFE^o contain a term

$$-\frac{z_B{}^2}{2a_B}\left(1 - \frac{1}{D}\right) + \frac{z_A{}^2}{2a_A}\left(1 - \frac{1}{D}\right), \tag{i}$$

The derivation is general for all cells. From Eq. (8.65) if E^o is plotted versus $1/D$ a straight line should result with a slope equal to $(- Nz_A z_B \varepsilon^2/ n \mathscr{F} kr)$ and an intercept equal to E_1^o.

Corsaro and Stephens (38) studied the cell

$$\text{Zn–Hg} \mid \text{ZnCl}_2(m), \text{ (X), (Y)} \mid \text{AgCl}(s); \quad \text{Ag} \tag{8.67}$$

---

where a_A and a_B are the corrected radii of Latimer, Pitzer, and Slansky [*J. Chem. Phys.* 7, 108 (1939)]. Thus one obtains Eq. (8.66) with the last term modified as indicated

$$E^o = E_1^o + \frac{N\varepsilon^2}{nFDK} \left(\frac{z_B^2}{2a_B} - \frac{z_A^2}{2a_A} \right). \tag{ii}$$

Harned and Owen ("The Physical Chemistry of Electrolytic Solutions," 2nd ed., pp. 54 and 338. Reinhold, New York, 1950) use the electrical work of transferring the charges of n' molecules of electrolyte from water of dielectric constant D_1 to a solvent of dielectric constant D_2 in order to obtain an equation for the effect of dielectric constant on the standard potential of a cell. The transfer process results in the Born equation for the activity coefficient of an electrolyte. Replacing the natural logarithm of the activity coefficient by the corresponding expression for the electromotive force leads to an equation relating E^o to the reciprocal of the dielectric constant. The resulting equation for a 1–1 electrolyte is

$$\frac{(E^o - E^{o\prime})n\mathscr{F}}{RT} = 1.210 \times 10^2 \left(\frac{D_1 - D_2}{D_1 D_2} \right) \Sigma \frac{1}{b_j} \tag{iii}$$

and contains the thermal energy RT. The symbol b_j represents the radius of an ion considered as a sphere, E^o is the standard potential in water of dielectric constant D_1 and $E^{o\prime}$ is the standard potential in any mixture relative to unit activity coefficient of the electrolyte at infinite dilution in that solvent. The Harned and Owen approach involved the electrical work of transferring charges from one dielectric constant medium to another. This is the approach used in deriving Eq. (8.66).

The author is convinced that there is more than one effect operative in the dependence of E^o on $1/D$. Difference in solvation energy in the different media is no doubt important, but would be of reduced magnitude in instances, which now appear to be numerous, where selective solvation occurs to a marked extent. There is also an effect arising from the difference in energy required to remove an ion from its ionic environment of unit activity with respect to ionic atmosphere in a given mixed solvent series at a given temperature, when the dielectric constants of the solvents change at constant temperature. In other words, E^o has yet to be extrapolated to a reference state of dielectric constant, just as in kinetics double transitions of the rate constants have to be made to both standard reference states of ionic strength and dielectric constant. (See Amis, "Kinetics of Chemical Change in Solution," Chap. 4. Macmillan, New York, 1949.) Harned and Owen used the standard reference state of dielectric constant as that of water at a given temperature. Equation (8.66) involves infinity as the standard reference state of dielectric constant. At this dielectric constant of the medium, the ions of the electrolyte would be free from all Coulombic forces.

using water, water-methanol, water-ethanol, and water-dioxane solvents. They observed that the standard potential of the cell in all three water–organic component–solvent systems was linear with the reciprocal of the dielectric constant with a precision better than 0.1 mV. All plots extrapolated through the standard cell potential with water solvent.

Harned and Thomas (*39*) and Harned *et al.* (*40*) investigated the cell

$$\text{Pt, H}_2 \mid \text{HCl}(m), \text{ (X), (Y)} \mid ; \quad \text{AgCl; Ag} \tag{8.68}$$

where (X) is the weight per cent dioxane or methanol in (Y) weight per cent water. The plots of the standard potentials versus the reciprocals of the dielectric constants approached linearity in the higher dielectric regions of the solvents but became more pronouncedly curved in the lower dielectric regions. From Fig. 3 it will be recalled that the plots of the standard potentials of the cadmium chloride cell also showed some curvature.

In Table IV are presented the values of r calculated from the E^0 versus $1/D$ plots for the various cells involving various solvents and various temperatures. Equation (8.66) was used in the calculations. For those curves which showed curvature, the slopes were taken in the higher dielectric constant regions of the solvents. For example, in Fig. 3 the slopes of the straight lines drawn tangent to the curves were taken as the limiting slopes

TABLE IV

VALUES OF r CALCULATED FROM THE SLOPES OF E^0 VERSUS $1/D$ PLOTS FOR CELLS COMPOSED OF DIFFERENT ELECTROLYTES AND DIFFERENT ELECTRODE SYSTEMS

Salt	Solvent	Temp. (°C)	r (Å)
CdCl$_2$	H$_2$O–MeOH	25	0.98
		30	1.03
		35	1.18
CdCl$_2$	H$_2$O–EtOH [a]	25	1.22
		30	1.16
		35	1.31
ZnCl$_2$	H$_2$O–Dioxane	25	2.08
	H$_2$O–MeOH	25	0.94
	H$_2$O–EtOH	25	1.06
HCl	H$_2$O–Dioxane	25	2.77
	H$_2$O–MeOH	25	0.64

[a] This cell using this solvent was also studied by J. D. Hefley and E. Amis, *J. Phys. Chem.* **69**, 2082 (1965).

of the curves. Of course in the case of the cell studied by Corsaro and Stephens using various solvents no such extrapolation was necessary. The values of r presented in Table IV are gratifying if somewhat small, and support the authenticity of the theoretical approach. The values of r are reasonable and represent different electrolytes, at different temperatures in different solvents.

The values of E_1^o, could be obtained from calculations using Eq. (8.66) or from extrapolation of E^o versus $/1D$ plots to $1/D = 0$ or $D = \infty$. For example the E_1^o at 25°C for the cell depicted in the lower curve of Fig. 3 is found to be 0.7225 V. The relatively high voltage of the cell at infinite dielectric constant where there would be no electrostatic forces between ions indicates that the cell potentials of cells, freed from electrostatic effects among ions, would be considerably greater than the potentials observed in actual cases where electrostatic forces are extant. Much work is involved in separating the ions due to their charges, and this work does not appear as available work in the actual case.

The upward curvature of the E^o versus $1/D$ curves at lower dielectric constants of the solvent is reminiscent of curves obtained from kinetic data when logarithms of specific velocity constants are plotted as reciprocals of dielectric constants for certain reactions. The upward curvature in the potential curves at lower dielectric constants is probably due to the selective solvation of the ions by the higher dielectric constant, more polar component of the solvent, in this case water. This is probably true, especially in solutes furnishing the hydrogen ion due to strong hydrogen bonding and the formation of H_3O^+ and $H_9O_4^+$ ions in water (41). This selective solvation (or solvent binding) continues to be influential until only a small amount of water remains in the solvent (19). The behavior of the ions in the mixed solvents tends toward their behavior in the pure component with which they are most closely associated in the mixed solvent. This effect has been observed in rate phenomena (42). The water, being the more polar component of the solvent and being more intimately associated with the ions, reduces their attraction for each other and allows their more ready dissociation. Hence less work than is expected at the lower dielectric constants is required to free the ions of opposite charge from each other's influence, and, therefore, more work than anticipated becomes available for external use. This increase in work is manifested as a higher potential than would be expected from the electrostatic forces between the ions of opposite charge at the lower dielectric constants.

Harned and Owen (43) using the work content-activity coefficient approach derived an equation for the effect of dielectric constant on the standard potentials of cells which involved the thermal energy of the ions. They attributed the deviation from theory at low dielectric constants to be due to the replacement of hydronium ion H_3O^+ by (solvent) H^+ according to the reaction

$$H_3O^+ + \text{(solvent)} \; \rightleftharpoons \; H_2O + \text{(solvent)} \, H^+. \qquad (8.69)$$

They believe the phenomenon to be too complicated to be completely explained by so simple an electrostatic theory. In the author's opinion the data in Table IV exemplify the surprising ability of simple electrostatic theory to explain many chemical phenomena observed for ions in solution.

There are many other approaches for the determination of the association constants of ions, for example, polarographic, spectrometric, and spectrophotometric. However, such presentations would require excessive space and are not necessary for the purpose of illustrating the fact of ion association; how such is influenced by solvent, temperature, and concentrations; and the necessity of taking cognizance of such in interpreting reaction rates between ions in solution.

Further Notes on Electron Exchange Reactions

a. *Introduction*

Several recent reviews of electron exchange reactions have been published. Among these are two by Sutin (*44*) and Marcus (*45*), respectively. These reviews deal mainly with purely electron exchange reactions, in which a transfer of one or more electrons takes place between two different valence forms of the same element without the formation or rupture of chemical bonds. Sutin goes into some detail concerning experimental techniques, and the natures of the electron exchange processes in the transition and nontransition elements. He further presents general considerations with respect to electron exchange reactions, and deals particularly with the theories of Marcus (*46*), Halpern (*47*) and Hush (*48*).

While mentioning several treatments briefly, Marcus deals chiefly with the treatments of Hush; of Levich, Dogonadze, and Chizmadzhev; of Dogonadze; and of Marcus. Marcus points out the common assumptions and some of the differences in the treatments.

Both Sutin and Marcus present comparisons of the treatments with experiment. References (*44*) and (*45*), especially the latter, will be the source of the material presented in the following pages, up to page 292.

b. *General Reflection*

Most theories of exchange reactions are based upon the Franck-Condon principle, according to which internuclear distances and nuclear velocities do not change during the transition of an electron. But the equilibrium configuration of the reactants and products in an electron exchange reaction differ, since the equilibrium configurations of the coordination shells of an element depend upon its state of oxidation. Hence if electron transfer is to occur without a resulting distortion of the coordination shells of the products and a consequent higher potential energy of products than reactants, there must be a prior reorganization of the coordination shells of the reactants.

An N-dimensional surface in an $(N + 1)$-dimensional atomic configuration space can be used to represent the potential energy of reactants, where N is the number of independent variables necessary to define the separation of reactants and the configuration of their coordination shells. Similarly the potential energy of the products may be represented by a N-dimensional

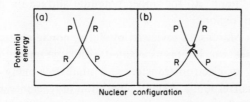

FIG. 4. Profile of N-dimensional potential energy surface of reactants R and products P plotted against the nuclear configuration of all the atoms in the system. (a) represents zero interaction and (b) finite interaction between the redox orbitals of the reacting species. The smaller the electronic interaction the smaller the separation of the two surfaces shown in (b), while the greater the interaction the greater the separation of the two surfaces. The separation is always twice the interaction energy, $2E_{RP}$.

surface in a $(N + 1)$-dimensional space. In Fig. 4(a) is shown the intersection of the profile of the reactant surface R and the product surface P on an $(N - 1)$-dimensional surface where there is zero interaction of the electronic orbitals of the two reacting species. If there is interaction of the electronic species there is a splitting of the surfaces as shown in Fig. 4(b). The extent of separation of the two surfaces depends on interaction between the redox orbitals of the reactants and equals twice the interaction energy of the system.

Sutin, in accordance with the theory of Marcus, using ΔG^* as the free energy for the change from the equilibrium distribution of the coordinates of the atomic reactants to those corresponding to the nonequilibrium distribution of the intersection region, wrote the equation for the specific velocity constant k' of the electron exchange reaction as

$$k' = \frac{pkT}{h} \exp\left(-\Delta G^*/RT\right), \qquad (8.70)$$

where p is the probability of reactants being converted to products in the intersection region.

Some of these reactions will be adiabatic and some nonadiabatic depending on how the system behaves on reaching the intersection region. When the system passes the intersection with low velocity it does not "jump" from the lower R surface to the upper R surface. There is usually time for electron transfer to take place, p is unity and all the systems remain on the lower potential energy surface on passing through the intersection region. Such reactions are called adiabatic reactions. In nonadiabatic reactions, the system passes through the intersection with high velocity so that there is little time for electron transfer. The p factor is less than unity and some of the systems "jump" to the upper potential surface on passing through the intersection region.

According to Marcus, as reviewed by Sutin, nonadiabatic reactions have two limiting cases: one limit is associated with large separation of reactants, low probability of electron transfer, and small free energy of activation; the other limit corresponds to small separation of reactants, high probability of electron transfer, and a large free energy of activation.

Reactions will be adiabatic when E_{RP} is larger than about 0.2 kcal mole^{-1} and p, therefore, is close to unity. These conditions will correspond to a wide separation of upper and lower energy surfaces in Fig. 4(b), so that all systems will tend to remain on the lower surface on passing through the intersection region. The calculation of E_{RP} is very difficult, but when E_{RP} is less than about 1 kcal mole^{-1}, it may be neglected in calculating ΔG^*.

c. Common Assumptions in the Principal Treatments and Some Differences

Marcus (45) lists some common assumptions and some differences characteristic of the principal treatments. The common assumptions are: (1) Analogous to the absolute rate theory and the distribution of the polarization oscillators of Levich et al., there is assumed to be thermal equilibrium between the classical or quantum mechanical reaction configurations or microscopic states and the remaining states (2) The interaction of the electronic orbitals of the two reactants is sufficiently weak so that the potential energy surface of the reactants is practically identical to that for zero electronic interaction except when this surface intersects the potential energy surface for which the electronic charge distribution is characteristic of the products. (3) The rate of the electron exchange reaction is set equal to the rate of "first passage" across the intersection of reactant and product surfaces, during which passage electron transfer is assumed to take place.

With respect to differences among the treatments, it might be pointed out that Marcus made the evaluation of the Arrhenius frequency factor to bear some of the difficulties in the statistical mechanical treatment whereby a surface integral has been expressed in terms of a volume integral. Hush in his treatment included the contribution of the inner coordination shell calculated from ion-dipole interactions and ligand field influences. Levich and Dogonadze (49) treated in an elegant manner microscopic polarization dynamics, but simplified their presentation by leaving out certain factors including dielectric dispersion and any variations in the inner coordination shell.

In the Marcus and in the Hush theories the total free energy ΔG^* for the electron exchange process is made up of three parts. One part ΔG_R^* arises from bringing the reactants from infinite distance apart to their distance R apart in the activated complex. A second part, ΔG_s^*, results from the necessity of reorganizing the solvent molecules surrounding the molecules, and a third part, ΔG_i^*, is a consequence of the reorganization of the inner coordination shells of the reactants. Thus

$$\Delta G^* = \Delta G_R^* + \Delta G_s^* + \Delta G_i^*. \qquad (8.71)$$

Only brief treatments of the theories of Marcus, of Levich, Dogonadze, and Chizmadzhev, of Dogonadze, and of Hush will be presented here since these have received rather complete treatment by either or both Sutin (44) and Marcus (45).

d. The Theory of R. A. Marcus

R. A. Marcus treated electron exchange reactions statistical mechanically introducing an "equivalent equilibrium distribution" to facilitate the calculation. In one of the last steps a substitution of the dielectric continuum equivalent was made for the contribution of the medium to the free energy of activation. Recently Marcus has been concerned with generalizing the statistical mechanical treatment so as to broaden the theoretical basis of the correlations. In addition to the three common assumptions and the classical treatment of the nuclei Marcus has made three additional assumptions in later work. These are (1) partial dielectric unsaturation outside the inner coordination shells, (2) harmonic forces in the inner coordination shell, and (3) practical independence of fluctuations of coordinates inside and outside the inner shell of the activated complex.

It was shown that, in obtaining the required functional form for the predicted correlations, every force constant in the reactant K_r and the corresponding one the in the product K_p could be replaced by a reduced force constant $2K_rK_p/(K_r + K_p)$ without causing significant error in the rate constant.

Introducing the quantities resulting from these assumptions into the equation for the rate constant k' gave the equation

$$k' = \varkappa \, pZ \exp\left(-\, \Delta G^*/RT\right). \tag{8.72}$$

In the above equation $\varkappa$ is the velocity-weighted average of the transition probability γ; p, which should have the value of about unity, is the ratio of the root-mean-square fluctuations in separation distance in the activated complex to the root-mean-square fluctuations of a coordinate for motion away from the intersection surface in the previously mentioned volume integral; Z is the number of collisions between two neutral species per unit volume of solution per unit time at the mean separation distance in the activated complex, and ΔG^* is defined by the equation

$$\Delta G^* = \frac{w_r + w_p}{2} + \frac{\lambda}{4} + \frac{\Delta G^\circ}{2} + \frac{(\Delta G^\circ + w_p - w_r)^2}{4\lambda}. \tag{8.73}$$

In Eq. (8.73) w_r is the polar and nonpolar work necessary to bring the reactants to within the mean separation distance R of each other; w_p is the negative of the polar and nonpolar work to separate the products from R to infinity; ΔG° is the "standard" free energy of reaction in the prevailing medium; and λ is the sum of the contribution, λ_i, from the changes in coordinates in each inner coordination shell, and of the contribution, λ_0, from changes in coordinates in each outer coordination shell. According to Marcus $(\Delta G^\circ + w_p - w_r)$ is the "standard" free energy in the medium in question at the separation distance R of the reactants.

For electrochemical reactions using metal electrodes Marcus states that Eq. (8.72) applies where Z is the number of collisions per unit area per unit time made by the uncharged reactants, and where ΔG^* is given by the equation

$$\Delta G^* = \frac{w_r + w_p}{2} + \frac{\lambda_{\text{el}}}{4} + \frac{n\varepsilon(E - E^\circ)}{2} + \frac{[n\varepsilon(E - E^\circ) + w_p - w_r]^2}{4\lambda_{\text{el}}}, \tag{8.74}$$

where E is the observed potential of the electrode, E^o is its formal potential, w_r is the polar and nonpolar work required to bring the reactant to within the mean separation distance R of the electrode, w_p is the negative of the polar and nonpolar work to remove the products from R to infinity, and λ_{el} again is composed of inner, $\lambda_{el\,i}$, and outer, $\lambda_{el\,o}$, contribution. These two contributions are only one-half those found for homogeneous electron exchange reactions when the mean separation distance between the reactant and electrode is only one-half that between two reactant particles.

Marcus writes the quantity λ_i discussed in connection with Eq. (8.73) and with Eq. (8.74) as a scalar product in terms of the intramolecular coordinates q_i of the inner coordination shell of one reactant in the electrode case and of two reactants in the case of the homogeneous reaction.

Marcus has applied his theory to experimental results by the calculation of rate constants and factors in the equation for the rate constant. See Table III of Chapter IV. He also has used his theory to predict correlations among experimental data. Several data have confirmed these predictions, but a few data (three cobalt reactions) do not agree with the predictions. For example, the prediction that, for small activation overpotential, the electrochemical transfer coefficient at metal electrodes is 0.5 has been found to hold in a substantial number of cases with few exceptions when measurements were made at high salt concentrations to minimize the work term (50).

e. *The Theory of Hush*

Hush, in addition to the three basic assumptions common to all treatments, makes the following additional assumptions. (1) The configurations of the coordinate shells are in equilibrium with the charges on the reactants throughout the electron exchange reaction. (2) The model of the inner coordinate shell is linear in the charge of the ion. (3) The mechanism is adiabatic. (4) The medium is an unsaturated dielectric continuum. (5) The range of distance of approach at which electron exchange can take place can be neglected. (6) It is not necessary to formulate a detailed mechanism of crossing the intersection surface. (7) An electron density parameter λ can be used as a measure of the probability of locating the electron on the oxidizing agent.

The form of the potential energy function assumed by Hush in order to calculate ΔG_i was given by

$$U = \frac{-n\,|\,z\,|\,\varepsilon\mu}{d^2} + \frac{\xi}{d^m}, \qquad (8.75)$$

where U is the potential energy of the inner coordination shell of the ion, n is the number of water molecules coordinated to the ion, $z\varepsilon$ is the charge on the ion, μ is the dipole moment of the water molecule, d is the ion-oxygen distance, and ξ and m are constants. This equation neglects the ligand field effects.

Sutin points out that if ξ and m are independent of charge over the range $z\varepsilon$ to $(z-1)\varepsilon$, then

$$d_{\varepsilon(z-1)} = d_{z\varepsilon} \left(\frac{z}{z-1} \right)^{1/m-2} \tag{8.76}$$

where $d_{z\varepsilon}$ is the equilibrium distance when the charge on the ion is $z\varepsilon$. The inner coordination shell reorganization free energy ΔG_i^* is given by

$$\Delta G_i^* = \frac{n\varepsilon\mu}{4(m-2)} \left[\frac{1}{z d_{z\varepsilon}^2} + \frac{1}{(z-1)(d_{z\varepsilon-1})^2} \right]. \tag{8.77}$$

Hush represented the energy of solvation of an ion as occurring in two steps. (1) The response of the medium through electronic polarization to the immersed ion and (2) the thermal equilibration of the medium with the ionic charge. The dielectric continuum theory free energy of solvation, ΔG_{si}^*, of an isolated ion of radius a, namely, $-z^2(1 - 1/D_s)/2a$, where D_s is the static dielectric constant, was written as the sum

$$\Delta G_{si}^* = - \frac{(z\varepsilon)^2}{2a} \left(1 - \frac{1}{D_0} \right) - \frac{(z\varepsilon)^2}{2a} \left(\frac{1}{D_0} - \frac{1}{D_s} \right), \tag{8.78}$$

where D_0 is the optical dielectric constant.

Analogously for the case of two ions of charges z_A and z_B, radii a_A and a_B, and probabilities of location of the electron on ion A of λ and on ion B of $(1-\lambda)$, the solvation contribution to the free energy ΔG_s, in going from the free ions to a complex with a separation distance R of the ions. was

$$\Delta G_s^* = - \left(\frac{1}{2a_A} + \frac{1}{2a_B} - \frac{1}{R} \right) \lambda(1-\lambda)\varepsilon^2 \left(\frac{1}{D_0} - \frac{1}{D_s} \right) + \frac{\lambda\varepsilon^2}{RD_s}(z_A - z_B - 1), \tag{8.79}$$

and the over-all free energy ΔG^* for the electron transfer was therefore

$$\Delta G^* = \Delta G_i^* + \Delta G_{si}^* + \lambda\Delta G^\circ, \tag{8.80}$$

where ΔG° is the same as defined above. This ΔG^* was calculated from the equation minimizing ΔG^* with respect to λ. By inserting ΔG^* in the activated complex rate equation the rate constants could be calculated. Marcus points out that in computing values of ΔG^* for insertion in the rate equation, the loss in translational entropy in forming the activated complex was not correctly accounted for, but that the consequent error in the rate constant was only a factor of ten.

Hush (48) applied his theory to the calculation of the rate constants of a number of different electron exchange reactions, with fairly satisfactory results. Marcus mentions that when data on force constants and bond lengths in the inner coordination shell become available, independent tests of some assumed properties of this shell can be made.

f. *Theory of Levich, Dogonadze, and Chizmadzhev*

In this brief summary we will mention the treatment of only chemical transfer reactions, although Levich, Dogonadze, and Chizmadzhev treated both these and also electrochemical transfer reactions both chemically and quantum mechanically. Neither changes in inner coordination shell angles nor in equilibrium bond lengths were present in the reactions considered, and the medium external to the inner coordination shell of a reactant was treated as a dielectric continuum. Nonadiabatic methods using perturbation theory were applied to rate calculations by assuming weak overlap of electronic orbitals of the reactants. The motion arising from polarization was treated using the Hamiltonian. In the one-electron approximation a Schrödinger equation was used which was so written as to account for the sensitivity of the electronic wave function to nuclear configuration in the intersection region of the reactant and product surfaces. Fourier components were used to represent the nuclear part of the Hamiltonian. Quantum treatment of the modes of polarization was used to formulate the rate expression.

There was a neglect of dielectric image effects and of a damping term in the macroscopic equation of polarization in the solvents; only an approximate treatment of the relative motion of reactants was made. The three basic assumptions common to all treatments were included.

For $\mathscr{H} \ll 1$, Levich, Dogonadze, and Chizmadzhev gave the expression for the probability p_{AB} of electron transfer at a separation distance R between reactants to be

$$p_{AB} = \sqrt{\frac{\pi}{kT\lambda_0}} \; \frac{\varepsilon^2_{AB}}{\hbar} \exp\left[-\left(\Delta G^{\circ}_R + \lambda_0\right)^2 / 4\lambda_0 kT\right], \qquad (8.81)$$

where ε_{AB} is half the splitting term at the intersection of the reactant and product surfaces; $\hbar$ is $h/2\pi$; ΔG°_R is the standard free energy of reaction at the separation distance R, i.e., it is the difference in the solvation free energy plus electronic free energy of the products and the sum of like energies with respect to the reactants; and λ_0 is given by the equation

$$\lambda_0 = \frac{(\Delta\varepsilon)^2}{2} \left(\frac{1}{2a_A} + \frac{1}{2a_B} - \frac{1}{R} \right) \left(\frac{1}{D_0} - \frac{1}{D_s} \right). \tag{8.82}$$

The expression for the specific velocity constant, k'_{AB}, was derived to be

$$k'_{AB} = \int_0^\infty p_{AB} \exp\left(- \frac{z_A z_B \varepsilon^2}{D_s R k T} \right) 4\pi R^2 \, dR. \tag{8.83}$$

Levich and Dogonadze (49), using the frequency ν_0 associated with polarization, obtained, at the low temperature end of the quantum expression, the equation

$$p_{AB} = \frac{2\pi}{\hbar} \frac{\varepsilon_{AB}^2 \exp\left(- \lambda_0/h\nu_0 \right)}{(\Delta G^\circ{}_R/h\nu_0) \, ! \, h\nu_0} \left(\frac{\lambda_0}{h\nu_0} \right)^{\Delta G^\circ{}_R/h\nu_0} \exp\left(- \frac{\Delta G^\circ{}_R}{kT} \right) \tag{8.84}$$

for a homogeneous electron transfer reaction. See reference (46) for the origin of the notation used here.

Dogonadze (51) derived an expression for the probability p_{AB} of transfer of an electron when $\mathscr{H} \cong 1$, that is, for an adiabatic electron exchange reaction at the separation distance R of the two participating particles. The equation for $\mathscr{H} = 1$ is

$$p_{AB} = \nu_0 \exp\left(- \frac{\lambda_0}{4kT} \right), \tag{8.85}$$

where

$$\lambda_0 = \frac{(\Delta\varepsilon)^2}{2} \left(\frac{1}{a} - \frac{1}{R} \right) \left(\frac{1}{D_0} - \frac{1}{D_s} \right) \tag{8.86}$$

since in this case $a_A = a_B = a$. See Eq. (8.82).

Marcus generalized this equation for $\mathscr{H} = 1$ to adiabatic electron transfer reactions. The resulting equation is

$$p_{AB} = \nu_0 \exp\left[- \frac{(\Delta G^\circ{}_R + \lambda_0)^2}{4\lambda_0 kT} \right]. \tag{8.87}$$

The expression for the specific velocity constant is identical with Eq. (8.83).

Levich and Co-workers applied their theory to the calculation of the rate of the ferrous-ferric electron exchange reaction.

g. *Tunneling*

The potential energy barrier to inner-shell reorganization will, in general, be high and narrow, since the differences in the radii of two reactants are of the order of a few tenths of an angstrom, while the energies of reorganization of inner shells are several kilocalories per mole. Thus the two reactants should exhibit a high probability of acquiring the same configurations in their inner coordination shells by tunneling through the potential barrier. This identity of configuration in the inner coordination shells is the configuration of the transition state. The barrier is nearly identical with a steep isosceles triangle at potential energies of the inner coordination shells above about 5 kcal/mole. Sutin terms such tunneling "nuclear tunneling." He gives equations for the nuclear transmission coefficient and for the nuclear tunneling factor (*52*).

Marcus believes that electron tunneling calculations should be used to evaluate the splitting ε_{AB} at the intersection of the reactant and product surfaces, and do not imply a nonadiabatic reaction. In the approximation of electron tunneling calculations,

$$\frac{\varepsilon_{AB}}{\pi\hbar} = p\nu_e \tag{8.88}$$

where ν_e is the number of times per second the electron in its orbit strikes the tunneling barrier and p is not simply a transmission coefficient, but involves the probability that an electron striking the tunneling barrier will form the final electronic configuration. Thus $p\nu_e$ is the number of times per second the final electronic configuration is formed.

h. *The Theory of Laidler and of Laidler and Sacher (43, 54, 55)*

Laidler in his theory assumed that electron exchange reactions are controlled by the rate of diffusion of the reacting species. He made the rate of such reaction to be the product of the rate of diffusion of reactants toward each other, an electron tunneling factor, and a Coulomb repulsion term.

Laidler and Sacher retained the electron tunneling factor, but replaced the other two terms by an Eyring rate expression containing a free energy of activation composed of three terms. The three terms were: (1) the electrostatic work required to bring the reactants together in infinitely dilute solution, (2) an inner coordination shell reorganization free energy, and (3) a change in solvation free energy due to change in ion size.

The new approach eliminated diffusion but still regarded p as a transmission coefficient. This treatment also neglected the contribution to the free energy of reorganization of the medium outside of the inner coordination shell. Marcus feels that the major weaknesses of this approach are the treatment of the p factor and the neglect of the contribution to the free energy described above. This free energy contribution arises from fluctuations in vibration-orientation polarization of the medium at any point external to the inner coordination shell. This polarization has a magnitude which depends on the charge of the ion and of the complex being considered. Laidler and Sacher did include the contributions to the free energy of reorganization arising from changes of bond lengths and angles in the inner coordination shell and the perhaps small contribution to the solvation free energy caused by change in ion size.

Marcus thinks that Laidler and Sacher should not include as a separate class those electron exchange reactions for which the mechanism is one of electron tunneling. Laidler and Sacher in addition to the above class of electron exchange reactions also designated the adiabatic type in which ε_{AB} is large and another type in which ε_{AB} is small.

i. *Other Complexities in Electron Exchange Reactions*

One such complexity was discussed in Chapter VII, pages 196 and 197. It was shown that the uranium (IV)–thallium (III) reaction occurred through a two-equivalent mechanism in solvent water, but by one electron step in solvent water-methanol. Some possible explanations of the differences in the mechanisms of the reaction in the two types of solvents were difference in strength of the cage walls in the two types of solvents, the difference in solvation of the ionic species in the different media, and the difference in the solvolytic reactant species in the different solvents.

Another complexity (56) observed by the author and his co-workers is the effect of a specific ion upon rate of an electron exchange reaction. Thus in the case of the neptunium (V)–uranium (IV) reaction chloride ion was found to accelerate the rate, though other anions used did not perceptibly influence the rate of the reduction. The mechanism of the neptunium (V)–uranium (IV) was postulated to be the same in both chloride and perchlorate media, since the rate data could be explained using the same mechanism. The over-all reaction between uranium (IV) and neptunium (V) was

$$U(IV) + 2Np(V) = U(VI) + 2Np(IV). \tag{8.89}$$

The rate of the reaction was determined in two ways. In one method the growth of the neptunium (IV) peak at 723 mμ, the decrease in the uranium

(IV) peak at 649 mμ, and the decrease in the neptunium (V) peak at 983 mμ were followed on a Beckman DU spectrophotometer or on a Cary 14 recording spectrophotometer, the cell compartments of which were maintained at the desired temperature. In the other method the neptunium (IV) was extracted with theonyltrifluoroacetone, and its concentration determined by radiometric assay. When excess uranium (IV) was present at lower acid concentration, the neptunium (IV) concentration reached a maximum indicating a further reduction of neptunium (IV) to neptunium (III) by uranium (IV). This further reduction of neptunium (IV) by uranium (IV) constituted a part of a further study (57). No photochemical or wall effects were observed.

The orders of the reaction with respect to various reactants were determined using the methods of initial rates and fractional rates (58). The two methods gave consistent results. The reaction was found to be first order with respect to uranium (IV), zero order with respect to neptunium (V), first order with respect to neptunium (IV), negative second order with respect to hydrogen ion and first order with respect to chloride ion in that part of the reaction which was chloride dependent. In the rate equation to be given later two specific velocity constants k_1' and k_2', are involved. Only one of these, k_2', was found to depend on chloride ion concentration. The dependence was exhibited only above a critical concentration of chloride ion, and was first order with respect to chloride ion concentration above this critical value of about 0.068 M, as can be seen from Fig. 5. Table V shows that k_1 was independent of chloride ion concentration.

The following mechanism based on experimental data and known reactions was written for constant hydrogen ion concentration:

$$Np(V) + U(IV) \xrightarrow{k_1'} Np(IV) + U(V), \qquad (8.90)$$

$$Np(IV) + U(IV) \xrightarrow{k_2'} Np(III) + U(V), \qquad (8.91)$$

$$Np(V) + Np(III) \xrightarrow{k_3'} 2Np(IV), \qquad (8.92)$$

$$Np(V) + U(V) \xrightarrow{k_4'} Np(IV) + U(VI), \qquad (8.93)$$

$$Np(IV) + U(V) \xrightarrow{k_5'} Np(III) + U(VI). \qquad (8.94)$$

From the above equations the following rate expressions can be written:

$$\frac{d[Np(IV)]}{dt} = k_1'[Np(V)]\,[U(IV)] - k_2'[Np(IV)]\,[U(IV)]$$
$$+ 2k_3'[Np(V)]\,[Np(III)] + k_4'[Np(V)]\,[U(V)] \qquad (8.95)$$
$$- k_5'[Np(IV)]\,[U(V)],$$

$$\frac{d[\text{Np(III)}]}{dt} = k_2'[\text{Np(IV)}]\,[\text{U(IV)}] - k_3'[\text{Np(V)}]\,[\text{Np(III)}] \tag{8.96}$$
$$+ k_5'[\text{Np(IV)}]\,[\text{U(V)}],$$

$$\frac{d[\text{U(V)}]}{dt} = k_1'[\text{Np(V)}]\,[\text{U(IV)}] + k_2'[\text{Np(IV)}]\,[\text{U(IV)}] \tag{8.97}$$
$$- k_4'[\text{Np(V)}]\,[\text{U(V)}] - k_5'[\text{Np(IV)}]\,[\text{U(V)}].$$

TABLE V

CONCENTRATION OF CHLORIDE ION AND k_1' FOR 25°C [a]

$[\text{Cl}^-]\,M$	$k_1\ M^{-1}\,\text{min}^{-1}$
0.0	0.019
0.0055	0.019
0.0119	0.019
0.0238	0.018
0.0358	0.020
0.0471	0.016
0.0598	0.021
0.0708	0.019
0.0822	0.020
0.1000	0.020
0.1204	0.020
0.1450	0.017

[a] $\mu = 0.15$, $[\text{H}^+] = 0.07\ M$, $[\text{Np(V)}] = 1.45 \times 10^{-3}\ M$, $[\text{U(IV)}] = 4.4 \times 10^{-3}\ M$.

By applying steady state approximations to Eqs. (8.96) and (8.97), the concentrations [Np(III)] and [U(V)] can be determined and substituted into Eq. (8.95) to yield

$$\frac{1}{2}\frac{d[\text{Np(IV)}]}{dt} = k_1'[\text{Np(V)}]\,[\text{U(IV)}] + k_2'[\text{Np(IV)}]\,[\text{U(IV)}]. \tag{8.98}$$

In the case of the perchlorate media the concentration of uranium (IV) was in large excess, and remained constant and equal to its initial concentration. Setting $[\text{Np(V)}] = [\text{Np(V)}]_0 - [\text{Np(IV)}]$, Eq. (8.98) can be integrated to yield

$$2(k_2' - k_1')\,[\text{U(IV)}]_0 t = -\ln 2k_1'[\text{Np(V)}]_0 + \ln\{2k_1'[\text{Np(V)}]_0$$
$$+ 2(k_2' - k_1')[\text{Np(IV)}]\}. \tag{8.99}$$

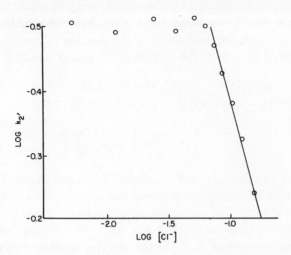

FIG. 5. A plot of $\log k_2'$ versus $\log [\text{Cl}^-]$ to determine effect of chloride concentration on k_2' at 25°C.

Assuming that $(k_2' - k_1') [\text{Np(IV)}] > k_1'[\text{Np(V)}]_0$ after a suitable length of time if $k_2' > k_1'$, Eq. (8.99) can be rearranged to give

$$\log \frac{[\text{Np(V)}]_0}{[\text{Np(IV)}]} = - \frac{2(k_2' - k_1')}{2.303} [\text{U(IV)}]_0 t + \log\left(\frac{k_2' - k_1'}{k_1'}\right) \quad (8.100)$$

which, with the hydrogen ion dependence being included in the k' values, has the same form as the empirical equation

$$\log \frac{[\text{Np(V)}]_0}{[\text{Np(IV)}]} = - \frac{k't[\text{U(IV)}]_0 [\text{H}^+]_0^{-2}}{2.303} + \log k'' \quad (8.101)$$

which reproduced the data and was actually instrumental in suggesting the mechanism proposed above.

Using the slopes and the intercepts from plots of $\log \{[\text{Np(V)}]_0/[\text{Np(IV)}]\}$ versus time and Eq. (8.100) at $[\text{H}^+] = 0.1\ M$, the apparent velocity constants were found to be $k_1' = 0.05\ M^{-1}\ \text{min}^{-1}$ and $k_2' = 0.32\ M^{-1}\ \text{min}^{-1}$. The value of k_2' compares favorably with the value $0.23\ M^{-1}\ \text{min}^{-1}$ for the Np(IV)–U(IV) reaction at the same hydrogen ion concentration (57). With the values obtained for k_1' and k_2', the assumption $(k_2' - k_1')[\text{Np(IV)}]$ $> k_1[\text{Np(V)}]_0$ appears to be fair when $[\text{Np(IV)}] \approx 10^{-3}\ M$. Under these conditions $(k_2' - k_1')[\text{Np(IV)}]$ is five to ten times greater than $k_1'[\text{Np(V)}]_0$.

In the chloride media each k_2' must equal the k_2' of the mechanism given above for the perchlorate media times $[Cl^-]$. Remembering this distinction in k_2' and applying the mechanism to the reaction in chloride media, but without U(IV) in excess, the following integrated equation is obtained

$$\log \left(\frac{2[Np(IV)]}{2[U(IV)]_0 - [Np(IV)]} \right) = \frac{(2[U(IV)]_0 \, (k_2' - k_1') + k_1'[Np(V)]_0)t}{2.303}$$

$$+ \log \left\{ \frac{[Np(V)]_0}{[U(IV)]_0} \left(\frac{k_1'}{k_2' - k_1'} \right) \right\}. \quad (8.102)$$

The velocity constants k_1' and k_2' include hydrogen ion and, in the case of k_2', chloride ion dependence. It was this constant k_2' which was shown to have the chloride ion concentration dependence shown in Fig. 5.

In perchlorate medium $k_1'[H^+]^2$ was 3.9×10^{-4} M min^{-1} at 0.1 M H$^+$ and an ionic strength of 0.60. In the chloride medium $k_1'[H^+]^2$ was 8.9×10^{-5} M min^{-1} in the same H$^+$ concentration but at an ionic strength of 0.15. Since k_1' is independent of chloride ion concentration, it would be expected that k_1' would be the same in perchlorate and chloride ion media at the same hydrogen ion concentration and the same ionic strength. We can compare the two values of k_1' at the two ionic strengths given above using Eq. (1.97). Thus if k_1' in the perchlorate medium is designated as k_{1p}' and that in the chloride medium by k_{1c}', then from Eq. (1.97)

$$\log k_{1p}' - \log k_{1c}' = \frac{2z_A z_B A \sqrt{\mu_p}}{1 + \beta a_i \sqrt{\mu_p}} - \frac{2z_A z_B A \sqrt{\mu_c}}{1 + \beta a_i \sqrt{\mu_c}} \quad (8.103)$$

where $\mu_p = 0.60$, the ionic strength in the perchlorate medium, and $\mu_c = 0.15$, the ionic strength in the chloride medium. The other terms in the equation have their usual meanings. Inserting the valehce of 4 for U(IV), since its salts are generally completely dissociated and it is unhydrolyzed in acid solution; the valence of 1 for Np(V) which exists as NpO$_2^+$, and the requisite ionic strengths into Eq. (8.103), one calculates the ratio of k_{1p}'/k_{1c}' to be 4.6. The experimentally observed ratio of k_{1p}'/k_{1c}' is $39 \times 10^{-5}/8.9 \times 10^{-5}$ or 4.4. In the calculation of the ratio of the k'-values a_i was taken as the sum $1.96 + 0.93$ Å or 2.89 Å, where 1.96 Å is the neptunium-oxygen distance (59) and 0.93 Å is the radius of U(IV) ion (60). The calculated ratio of the values of k_1' in the two media agree surprisingly well with the experimental values of the ratio, especially when it is remembered that the two ionic strengths at which the comparison is made are far greater than ionic strengths at which the Debye-Hückel theory can be applied accurately.

The answer is perhaps that in the ratio the deviations are compensated for and cancel each other. Nevertheless, this is a striking example of the application of electrostatic theory to chemical phenomena in solution.

j. *The Solvent and Recent Theories of Electron Exchange Reactions*

All of the recent theories of electron exchange reactions include the effect of the solvent on the rates of these reactions. The contribution to the total free energy barrier includes the work to bring the reactants to within the reaction distance of each other. This work is calculated from electrostatics and is given by the expression $Z_A Z_B \varepsilon^2 / D_s R$. This expression includes the dielectric constant D_s of the solvent and indicates a solvent dependence of the rate. The total free energy barrier also includes the free energy of reorganization at the distance R. This free energy of reorganization comprises changes in solvation free energy arising from change in ion size due to changes in bond lengths and angles in the inner coordination shells. This may be a small effect but is evidently solvent dependent. Change in solvation free energy will depend on the solvent involved. Further the contribution to the total free energy barrier arising from fluctuations in the vibration orientation polarization at any point outside the inner coordination shell is certainly solvent dependent. This change of polarization of the medium on the formation of the activated complex will depend on the charges of the ions and complex involved, but will also depend on the medium or solvent, since the polarization of a substance is governed by its dielectric constant and on its molar volume, and will be different, unless by coincidence, for each solvent under a given set of conditions.

These recent developments in the theory have carefully considered the various aspects of electron exchange reaction. In general the assumptions made have been reasonable. The theories are complex, however, and contain many variables.

Numerous complexities arise in various electron exchange reactions as, for example, the often encountered change in orders and in rates with change in composition of the solvent when mixed solvents are used. In these cases specific effects of the solvent outside of simple solvation, polarization, and electrostatics, govern the rates, orders, and, therefore, the mechanisms of the reactions. These specific effects include solvolysis, cohesion, hydrogen bonding, cage phenomena, selective solvation, and no doubt many other specific solvent influences.

The complexities presented above and others so commonly encountered cannot be explained satisfactorily on only the bases of solvent polarization and solvent rearrangement. Only those electron exchange reactions which are free of specific effects of ions and solvent are amenable to satisfactory treatment by theories dealing with standard free energies of activation, solvent polarization, and solvent rearrangement.

Electrostatic Considerations

In spite of prevailing opinion that, in general, the electrostatic approach to the theoretical interpretations of chemical phenomena is inadequate due to the complexity of these manifestations, yet much can be done with electrical forces among different charge types of chemical particles to elucidate their behavior in solutions. This behavior includes not only chemical reaction rates, but other phenomena as well, such as electrical conductance and electrode potentials.

In *Chemical Abstracts* [59 (1963)] under the one heading "Dielectric Constant" the author found without too careful scrutiny 18 articles dealing with the application of electrostatics, mainly with respect to dielectric constant and polar moments, to chemical phenomena in solution. Many other applications to the solid state and surface chemistry were mentioned. Of the references to solution phenomena an important fraction were concerned with electrostatic influences on reaction rates. Only a few of these will be mentioned to illustrate the point.

Katiyar (61) found that the kinetics of the reaction of malachite green with acids were in quantitative agreement with the Brønsted-Christiansen-Scatchard equation with respect to the influence of the dielectric constant upon the reaction rate.

The hydrolyses of phthalic and terephthalic diacid chlorides were investigated by Entelis, *et al.* (62). For both hydrolyses in water-dioxane mixtures the rate of the reaction was first order for each reagent and second order overall. The rate of hydrolysis in both instances increased with increase of polarity of the medium according to the Kirkwood relationship. Dipole moments of the phthalic and terephthalic diacid chloride complexes were calculated to be, respectively, 6.85 and 6.95 Debye units. These values are certainly reasonable. The value of the isodielectric constant energy of activation was found to increase with increasing dielectric constant of the media.

The kinetics and mechanism of the Menshutkin reaction as affected by the chemical structure of the solvent was studied by Stepukhovic *et al.* (63). The reaction studied was

$$PhNEt_2 + EtBr \rightleftarrows PhNEt_3Br \tag{8.104}$$

at 70° to 98°C. It was found in water-dioxane media that the enthalpy of activation increased linearly with dielectric constant, whereas the heat of solvation decreased linearly. These investigators used the above observations to support the concept that the reaction mechanism is related to the formation of a solvated activated complex.

Shanker and Joshi (64) studied the oxidation of benzyl alcohol by vanadium (V) in 60 to 80% HOAc at 25°–45°C. The reaction was first order with respect to both benzyl alcohol and vanadium (V). A linear relation between the second-order rate constant and the reciprocal of the dielectric constant, in the range of dielectric constant values of 47.48 to 19.96, was found, suggesting an ion-dipole interaction.

Examples of other influences of electrostatics on chemical phenomena in the articles scanned were: the linear relationships of the free energy and enthalpy of association of certain amides with the reciprocal of the dielectric constant of the solvent media (65); the linear relationships of the pK and the electrical thermodynamic constants of acetyl acetone with the reciprocal of the dielectric constant of the solvent (66); and the prediction, from conductance measurements, of a dipole solvate between picrate ion and p-nitroaniline, stabilized by electrostatic attraction between ion and dipole (67).

The most striking example of the continued validity of the application of electrostatics to chemical phenomena in solution is the waxing, waning, and a later resurgence of the primary kinetic salt effect, a recent application of which has been made by Perlmutter-Hayman and Weissman (68) with respect to the bromopentamminecobalt (III) and hydroxide ion. It will be remembered from former discussion how Fig. 3 in Chapter I was accepted as convincing proof of the Brønsted-Bjerrum treatment of the primary salt effect using activity coefficients. Then the Olson and Simonson (8) theory threw doubt on the activity coefficient treatment of the primary salt effect. However as discussed above it was shown that, when quantitative allowance is made for the role played by ion pairs, the concept of activity coefficients must be retained without question. The electrostatic theory, like the phoenix, seems to be able to rise in youthful freshness from its own ashes. The first three chapters of this book presented electrostatic theory and gave pertinent examples in its support. Exhaustive listing of applications would have been superfluous, since the purpose was to make the reader aware of this approach to the explanation of these phenomena, give examples of where this method has been successfully applied, and prepare the reader to recognize such solvent influences and to know how to calculate them.

Chapter VI presents several empirical correlations of reaction rates with various properties of the solvent and the solute. These properties include acidity, basicity, substituent properties, inductometric effect, electrometric effect, ionizing power of the solvent, electrophilicity, and nucleophilicity. These empirical approaches are not to replace the theoretical approaches, electrostatic or otherwise, but are to supplement other approaches and to explain specific effects not accounted for in the theoretical approaches.

While many of the phenomena of solvent effects have been verified by laboratory experiments, and have been subjected to theoretical explanations, much remains to be explained satisfactorily. Differences of opinion exist as to the correct explanations of many phenomena. In many cases various influences are at work simultaneously, and how to separate these, even in a qualitative manner, is questionable. As the author has often remarked it is amazing that some of the theories work as often as they do, and that they do work so often may really be due to a concurrent effect of many influences which tend to mollify or modify each other so that one influence seems to dominate. Much needs to be learned about solutes and solvents in the solution state. In particular, the microscopic regions around solute particles need to be elucidated. For example, what is the dielectric constant in the immediate neighborhood of an ion in water or in a water-dioxane solvent? How does this microscopic dielectric constant vary from ion to ion, from solvent to solvent, and with composition of mixed solvent? How does the extent and nature of solvation of a solute particle vary with solvent, composition of mixed solvent, and with temperature? How does the electronic configuration and the reactivity of solutes change with different solvents and composition of mixed solvents due to solvation, selective solvation, solvolysis, and other solvent influences? The answers to these and other questions remain vague, and the answers proposed are often conflicting. The solution chemist is still only able to be fed on milk; he must mature markedly before he can safely indulge in a diet of meat. Tools, such as NMR (69) and ESR (70), are available which will enable him to answer some of the questions which are pertinent to an understanding of solution chemistry and therefore of kinetics of reactions in solution. Laidler (71) feels that ion-dipole activated complexes have complicated distribution of charges, so that specific hydration plays an important role. He thinks that in solvent rate theories dielectric saturation should be taken into account, the solvent still being treated as continuous. Also he proposes that careful consideration be given to activated complexes together with their surrounding solvent shells from the standpoint of a variety of models corresponding to various structures of the solvation shell. In water solvent the molecules farther away could be satisfactorily dealt with using the continuum theory and considering the dielectric constant to be constant though it would be necessary to treat the disorder zone explicitly. Let us hope that, by the use of the above-mentioned and other tools and methods which will become available, by the exercise of ingenuity, by never-flagging interest, and by unstinting effort, the solution kineticist will progressively unravel the mysteries of the influence of solvents on rates and mechanisms of chemical reactions.

REFERENCES

1. J. S. Hoppé and J. E. Prue, *J. Chem. Soc.* (1957), 1775.
2. A. Indelli and E. S. Amis, *J. Am. Chem. Soc.* 82, 3233 (1960).
3. C. W. Davies, H. W. Jones, and C. B. Monk, *Trans. Faraday Soc.* 48, 921 (1952).
4. J. Faucherre, *Compt. rend.* 227, 1367 (1958).
5. J. D. Hefley and E. S. Amis, *J. Phys. Chem.* 64, 870 (1960).
6. A. Skrabal and S. R. Webertsch, *Monatsch.* 36, 211 (1915).
7. E. A. Moelwyn-Hughes, private communication.
8. A. R. Olson and J. R. Simonson, *J. Chem. Phys.* 17, 1167 (1949).
9. C. W. Davies, *Trans. Faraday Soc.* 23, 351 (1927); "Ionic Association." Buttersworth, London and Washington, D. C., 1962.
10. B. Holmberg, *Z. physik. Chem.* 79, 147 (1912).
11. H. C. Robertson and S. F. Acree, *J. Am. Chem. Soc.* 37, 1902 (1915).
12. A. Indelli and E. S. Amis, *J. Am. Chem. Soc.* 82, 332 (1960).
13. V. Carassiti and C. Dejak, *Atti Acad. Sci., Ist. Bologna, Class. Sci. Fis. Rend.* 11, 2 (1955).
14. W. J. Howells, *J. Chem. Soc.* 1946, 203.
15. P. G. M. Brown and J. E. Prue, *Proc. Roy. Soc.* (*London*) A 232, 320 (1955).
16. E. A. Guggenheim, *Discussions Faraday Soc.* 24, 53 (1957).
17. K. H. Stern and E. S. Amis, *Chem. Rev.* 59, 1 (1959).
18. J. O. Wear and E. S. Amis, *J. Inorg. & Nucl. Chem.* 24, 903 (1962).
19. E. S. Amis, *J. Phys. Chem.* 60, 428 (1956).
20. L. Onsager, *Physik Z.* 27, 388 (1926); 28, 277 (1927); *Trans. Faraday Soc.* 23, 341 (1927).
21. L. G. Pedersen and E. S. Amis, *Z. physik. Chem. NF.* 36, 199 (1963).
22. H. S. Harned and B. B. Owen, "The Physical Chemistry of Electrolytic Solutions." Reinhold, New York, 1943; R. M. Fuoss, *J. Am. Chem. Soc.* 79, 330 (1957); C. W. Davies, "Ion Association." Butterworth, London and Washington, D. C., 1962.
23. E. A. Moelwyn-Hughes, *Z. Naturforsch.* 18a, 202 (1963).

24. N. Bjerrum, *Ergeb. exakt. Naturw.* **6**, 125 (1926).

25. J. T. Denison and J. E. Ramsey, *J. Am. Chem. Soc.* **77**, 2615 (1955).

26. R. M. Fuoss and C. A. Kraus, *J. Am. Chem. Soc.* **79**, 3304 (1957).

27. Y. H. Inami, H. K. Bodenseh, and J. B. Ramsey, *J. Am. Chem. Soc.* **83**, 4745 (1961).

28. F. Accascina, A. D'Aprano, and R. M. Fuoss, *J. Am. Chem. Soc.* **81**, 1058 (1959).

29. F. Accascina, S. Petrucci, and R. M. Fuoss, *J. Am. Chem. Soc.* **81**, 1301 (1959).

30. H. K. Bodenseh and J. B. Ramsey, in press.

31. R. M. Fuoss and T. Shedlavsky, *J. Am. Chem. Soc.* **71**, 1496 (1949).

32. D. Roach and E. S. Amis, *Z. physik. Chem. NF* **35**, 274 (1962).

33. H. S. Harned, *J. Am. Chem. Soc.* **51**, 416 (1929).

34. H. S. Harned and R. W. Ehlers, *J. Am. Chem. Soc.* **54**, 1350 (1932).

35. H. S. Harned and M. E. Fitzgerald, *J. Am. Chem. Soc.* **58**, 2624 (1936).

36. J. D. Hefley and E. S. Amis, *J. Electrochem. Soc.* **112**, 336 (1965).

37. E. S. Amis, *J. Electroanal. Chem.* **8**, 413 (1964).

38. G. Corsaro and H. L. Stephens, *J. Electrochem. Soc.* **104**, 512 (1957).

39. H. S. Harned and H. C. Thomas, *J. Am. Chem. Soc.* **57**, 1666 (1935).

40. H. S. Harned, F. Walker, and C. Calmon, *J. Am. Chem. Soc.* **61**, 44 (1939).

41. N. Goldenberg and E. S. Amis, *Z. physik. Chem. NF* **30**, 65 (1961).

42. S. Glasstone, K. J. Laidler, and H. Eyring, "The Theory of Rate Process," p. 432. McGraw-Hill, New York, 1941.

43. H. S. Harned and B. B. Owen, "The Physical Chemistry of Electrolyte Solutions," 3rd ed., Chaps. 3 and 11. Reinhold, New York, 1958.

44. N. Sutin, *Ann. Rev. Nuclear Sci.* **12**, 285 (1962).

45. R. A. Marcus, *Ann. Rev. Phys. Chem.* **15**, 155 (1964).

46. R. A. Marcus, *Discussions Faraday Soc.* No. 29, 21 (1960); *J. Chem. Phys.* **26**, 872 (1957).

47. J. Halpern, *Quart. Revs. (London)* **15**, 207 (1961).

48. N. S. Hush, *Trans. Faraday Soc.* **57**, 557 (1961).

49. V. G. Levich and R. R. Dogonadze, *Doklady Akad. Nauk S.S.S.R.* **133**, 158 (1960); *Proc. Acad. Sci. U.S.S.R., Sect. Phys. Chem. (English Translation)* **133**, 591 (1960); *Collection Czechoslov. Chem. Communs.* **26**, 193 (1961); Transl., Boshko, O., Univ. of Ottawa, Ontario.

50. R. A. Marcus, *J. Phys. Chem.* **67**, 853 and 2889 (1963).

51. R. R. Dogonadze, *Doklady Akad. Nauk S.S.S.R.* **144**, 1077 (1962); **145**, 848 (1962); *Proc. Acad. Sci. U.S.S.R., Sect. Phys. Chem. (English Translation)* **144**, 163 (1962); **145**, 563 (1962).

52. N. Sutin and M. Wolfsberg, unpublished calculations.

53. K. J. Laidler, *Can. J. Chem.* **37**, 138 (1959).

54. E. Sacher and K. J. Laidler, *Trans. Faraday Soc.* **59**, 396 (1963).

55. K. J. Laidler and E. Sacher, in "Modern Aspects of Electrochemistry" (J. O'M. Bochris, ed.), Vol. 3, pp. 1–42. Butterworth, London and Washington, D. C. 1964.

56. J. O. Wear, N. K. Shastri, and E. S. Amis, *J. Inorg. & Nuclear Chem.* in press.

57. J. O. Wear, *J. Phys. Chem.* in press.

58. J. O. Wear, U. S. At. Energy Comm. *SC*-4972 (*RR*).

59. L. H. Jones and R. A. Penneman, *J. Chem. Phys.* **21**, 542 (1953).

60. W. H. Zachariasen, "The Actinede Elements," Natl. Nuclear Energy Series, Div. IV, Vol. 14A, Chap. 18, 1954.

61. S. S. Katiyar, *Z. physik. Chem. (Frankfurt)* **34**, 346 (1962).

62. S. G. Entelis, R. P. Tiger, E. Ya. Nevel'skii, and I. V. Epel'baum, *Izv. Akad. Nauk SSSR Otd. Khim. Nauk* **1963**, 245, 429.

63. A. D. Stepukhovic, N. S. Lapshova, and T. D. Epimova, *Zhur. Fiz. Khim* **35**, 2532 (1961).

64. R. Shanker and S. N. Joshi, *Indian J. Chem.* **1**, (7), 289 (1963).

65. J. S. Frazen and R. E. Stephens, *Biochemistry* **2** (6), 1321 (1963).

66. P. S. Gentile, M. Cefola, and A. V. Celiano, *J. Phys. Chem.* **67** (7), 1447 (1963).

67. A. D'Aprano and R. M. Fuoss, *J. Phys. Chem.* **67**, 187 (1963).

68. B. Perlmutter-Hayman and Y. Weissman, *J. Phys. Chem.* **68**, 3307 (1964).

69. F. Huska, E. Bock, and T. Shaefer, *Symposium on Solvation Phenomena* p. 16. The Chemical Institute of Canada, Calgary Section, July 1963.

70. M. C. R. Symons and M. J. Blandamer, *Symposium on Solvation Phenomena* p. 7. The Chemical Institute of Canada, Calgary Section, July 1963.

71. K. J. Laidler, *Symposium on Solvation Phenomena* p. 16. The Chemical Institute of Canada, Calgary Section, July 1963.

AUTHOR INDEX

Numbers in parentheses are reference numbers and indicate that an author's work is referred to although his name is not cited in the text. Numbers in italic show the pages on which the complete references are listed.

SUBJECT INDEX

A

α-Amylase-catalyzed hydrolysis of amylose, 206

a, values, 173, 174

crude determination of, 173

Absolute rate expression, 18

Acceptor-donor interactions, 178

Acetolysis of benzhydryl chloride and bromide, 166

p-toluenesulfonates of stereoisomers, 229

rearrangement of α, α-dimethylallyl chloride, 166

Acetyl peroxide, 203

thermal decomposition of, 198

Acid and basic hydrolysis of esters, 234

models for, 49-51

values in angstroms of b_x and function in, 49-51

Acid catalyzed inversion of sucrose in water-ethanol solvents, 40

Acid dissociation constant, equation for, 148

Acid hydrolysis

equation for, 50, 159

mechanism for, 63

of esters, 43, 62

of ethyl orthoformate, 60

Action of immersed dipole on distant charges, 35

Activated complex, 286

separation distance in, 286

Activation

energy of, 123

and electrostatic contribution, equation for, 10

and ionic atmosphere contribution, 12

apparent, 9, 68

at D_2 related to energy of activation at D_1, equation for, 65

calculation of difference between E at constant dielectric constant and at any dielectric constant, equation for, 11

components of, 8, 16, 17, 52

effect of pressure on, 139

for iodide ion-methyl fluoride and iodide ion-methyl bromide reactions, 54

equation for, 12

Nathan and Watson empirical equation for, 52

total, 17

true energy of, 10

enthalpy of, 73, 124

entropy of, 73, 124

and Arrhenius frequency factor, 18

and ionic strength, 18

equation for, 18

free energy of, 56, 78, 79

equation for, 17, 18

factors contributing to, 74

volume of, 134

Activity coefficient(s), 2, 3, 5, 6, 31, 48, 74, 259-261, 273-275, 281

equation for, 2, 3, 31, 32, 38, 126

electromotive force and, 275, 276

for dilute solutions, equation for, 127

in dilute gases, 5

joint role with ion pairing, 261

reaction rate and, 261

Kirkwood and Westheimer's equation for, 48

of arbitrary distribution of charges, 48

of a solute, equation for, 126

of dipolar molecules, equation for, 32

split with reference to standard states, 6

312

B